Sensory Perception and Transduction in Aneural Organisms

NATO ASI Series
Advanced Science Institutes Series

A series presenting the results of activities sponsored by the NATO Science Committee, which aims at the dissemination of advanced scientific and technological knowledge, with a view to strengthening links between scientific communities.

The series is published by an international board of publishers in conjunction with the NATO Scientific Affairs Division

A	**Life Sciences**	Plenum Publishing Corporation
B	**Physics**	New York and London
C	**Mathematical and Physical Sciences**	D. Reidel Publishing Company Dordrecht, Boston, and Lancaster
D	**Behavioral and Social Sciences**	Martinus Nijhoff Publishers
E	**Engineering and Materials Sciences**	The Hague, Boston, and Lancaster
F	**Computer and Systems Sciences**	Springer-Verlag
G	**Ecological Sciences**	Berlin, Heidelberg, New York, and Tokyo

Recent Volumes in this Series

Series A: Life Sciences

Sensory Perception and Transduction in Aneural Organisms

Edited by

Giuliano Colombetti
Francesco Lenci

National Research Council
Pisa, Italy

and

Pill-Soon Song

Texas Tech University
Lubbock, Texas

Plenum Press
New York and London
Published in cooperation with NATO Scientific Affairs Division

Proceedings of a NATO Advanced Study Institute on
Sensory Perception and Transduction in Aneural Organisms,
held September 3–14, 1984,
in Volterra, Pisa, Italy

Library of Congress Cataloging in Publication Data

NATO Advanced Study Institute on Sensory Perception and Transduction in
    Aneural Organisms (1984: Volterra, Italy)
    Sensory perception and transduction in aneural organisms.

    (NATO ASI series. Series A, Life sciences; vol. 89)
    "Proceedings of a NATO Advanced Study Institute on Sensory Perception and
Transduction in Aneural Organisms, held September 3–14, 1984, in Volterra, Pisa,
Italy"—T.p. verso.
    "Published in cooperation with NATO Scientific Affairs Division."
    Includes bibliographies and index.
    1. Bacteria—Physiology—Congresses. 2. Protozoa—Physiology— Congress-
es. 3. Light—Physiological effect—Congresses. 4. Photoreceptors—Congresses. 5. Senses and sensation—Congresses. I. Colombetti, Giuliano. II. Lenci,
Francesco. III. Song, Pill-Soon. IV. North Atlantic Treaty Organization. Scientific
Affairs Division. V. Title. VI. Title: Aneural organisms. VII. Series: NATO ASI
series. Series A, Life sciences; v. 89. [DNLM: 1. Bacteria—physiology—congresses. 2. Photoreceptors—congresses. 3. Protozoa—physiology—congresses.
4. Sensation—congresses. QW 52 N279s 1984]
QR97.L5N37   1984   574.1'8   85-6466

ISBN-13: 978-1-4612-9511-2   e-ISBN-13: 978-1-4613-2497-3
DOI: 10.1007/978-1-4613-2497-3

PREFACE

This book is based on the lectures given at the NATO Advanced Study Institute on "Sensory Perception and Transduction in Aneural Organisms" held in Volterra (Pisa, Italy) from the third to the fourteenth of September, 1984. The Advanced Study Institute was planned as a high level course dealing with several aspects and problems of sensory perception and transduction of diverse environmental stimuli in aneural organisms. Scientists from different fields and cultural backgrounds were present at the meeting, both as lecturers and as students. The lectures and the discussions that followed represented a well integrated interdisciplinary approach to the questions considered. At the end of the Advanced Study Institute course, it was quite clear that, notwithstanding the apparent heterogeneity of the topics dealt with, unifying concepts and ideas already existed, among the most important being the role of membranes and their physicochemical properties. All this should be reflected in the content of this book.

We gratefully acknowledge the financial sponsorship of the Scientific Affairs Division of NATO (Brussels), that made both the Advanced Study Institute on "Sensory Perception and Transduction in Aneural Organisms" and this book possible. Finally, we are also indebted to Ms. Pat Parham Morgan who expertly retyped all the chapters of the book and Ms. Leslie Schmidt of Plenum Publishing Co. provided us valuable advice and suggestions on the preparation of this book.

G. Colombetti
F. Lenci
P. S. Song

CONTENTS

GENERAL SURVEY OF SENSORY TRANSDUCTION IN ANEURAL ORGANISMS

Wolfgang Haupt

Institut für Botanik und
Pharmazeutische Biologie
der Universität
Erlangen-Nürnberg
W. Germany

INTRODUCTION

A living organism is continuously exposed to physical and chemical factors of its environment. This necessitates the organism to perform responses so as to make the best use of these factors or as to avoid injurious effects by them. To be effective, such responses must be regulated by the external factors, and hence these factors have to be recognized or perceived.

In the simplest way, we can describe this behavior as an input, which is processed in a black box until an output results (Fig. 1, top). In more modern terms, we are dealing with (i) a stimulus, which is perceived by the organism, (ii) transduction, which is started after perception of the stimulus, and which results (iii) in the visible or measurable response (e.g. movement, Fig. 1, bottom).

In nature, the most important stimuli, in general, are light, chemical environment, gravity, mechanical forces, and temperature. Usually such a factor is called a stimulus only if its information is more important than its energy or its mass; a stimulus does not contribute to the requirement of an organism for energy or substances. As an example, for a photoautotrophic organism the energy of a short light stimulus does not measurably contribute to photosynthesis, but it can be transduced so as to result in a photomorphogenic response, i.e. a change in growth habit or in a movement response. Similarly, in animals and men chemical factors as detected by smell can change their behavior without contributing to nutritional requirements, and comparable behavior is known even in bacteria (cf. Macnab, this volume).

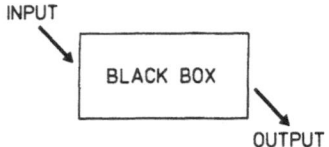

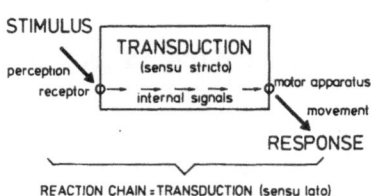

Fig. 1.   Formal presentation of a reaction chain in a simple and in
          a more sophisticated version.

     There are many different kinds of responses to external stimuli,
ranging, e.g. from enzyme synthesis to control of motility, from
rapid turgor variations to long-lasting regulation of growth processes. For analyzing the transduction, i.e. the black box between
stimulus and response, in principle each kind of response can be used
as the output. However, most convenient are those responses which
appear soon after the application of the stimulus. Hence, in this
book a high percentage of responses is selected out of the field of
movements.

     Transduction, finally, or the Black Box, denotes the sequence
of events which causally connect stimulus perception with response.
This sequence can consist of a few steps only, as I will show in one
example, or there can be a sequence of many steps, which makes dis-
entangling a difficult task. Many transduction chains to be presented
in this book share such a complexity.

TERMINOLOGICAL PROBLEMS

     In defining transduction properly, there is a terminological
difficulty in separating it clearly from perception and from response.
For the latter, gravitropism of higher plants can be taken as an
example. A horizontally placed coleoptile bends upward; this negative
gravitropism is the visible response. The immediate cause for this
bending is differential growth on opposite flanks. Thus, local regulation of growth may be taken as the last step of transduction.
However, consider a plant physiologist who is primarily interested
in growth and its regulation. For him, changes in growth are the
final response. Those growth regulations may be traced back to

differences in the phytohormone auxin, the growth-regulating effect
of which may be a change in proton secretion out of the cells. Thus,
gravity-induced differences in auxin concentration or auxin transport
and the resulting control of proton secretion are clearly late steps
in the transduction chain. But still different is the view of a
scientist who is interested only in the exciting phenomenon of auxin
transport and its regulation. He also may stimulate the coleoptile
gravitropically, but he certainly will consider the first sign of
transverse auxin transport as the response to be analyzed. Thus,
there is an open boundary between transduction and response, and in
detail terminology depends on the subjective view.

Similar difficulties can arise at the perception side, if one
compares, e.g. mechanosensing with gravisensing. In the latter case,
displacement of a statolith is followed, in higher plants, by its
pressure on a membrane; the resulting membrane deformation appears
to be the first step of transduction. In mechanosensing, however,
the stimulus acts on the membrane, without any step in between;
thus, the resulting membrane deformation is the perception process.
Or for light sensin, is perception restricted to generation of an
excited state of the pigment, which lasts only picoseconds, or does
it include subsequent dark relaxations in the range of nanoseconds,
micro-, milliseconds or even seconds?

These difficulties of flowing boundaries can be circumvented,
if the term "transduction" is used for the <u>whole</u> chain of events
("reaction chain"), <u>including</u> perception and response: "transduction
<u>sensu late</u>" (Fig. 1). In this meaning, the term has been used in
the title of a series of Gordon Conferences "Sensory Transduction in
Microorganisms," where always perception and response were included.
Similarly, this book devotes a good deal to responses without having
them in the title. On the other hand, in several of the chapters,
the term is used as "transduction <u>sensu stricto</u>," being restricted
to the Black Box, notwithstanding the difficulties pointed out above.

There is another potential terminological difficulty, which has
to be kept in mind. It concerns the term "signal." If a cell membrane undergoes deformation, e.g. by mechanic stimulation, an action
potential may be generated, and this is called an electrical signal,
which may act on neighboring cells or which may trigger a subsequent
step in the transduction chain. Similarly, chemical signals can be
produced at any stage in a transduction chain, e.g. cyclic AMP. On
the other hand, frequently the term "signal" is used synonymously
with stimulus, as can be seen in the commonly-used terms "signal
perception," "perception of light signals," etc. In the best case,
such a signal is specified as "external signal," to distinguish it
from the internal signals in the transduction chain (cf. Fig. 1).
Again, care has to be taken not to intermingle the different meanings
of the term.

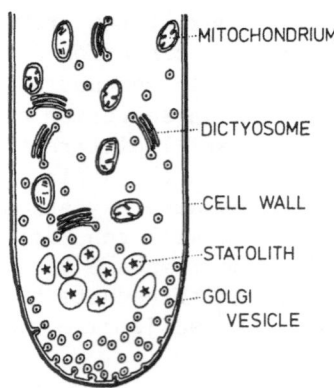

Fig. 2.   Apical portion of a rhizoid cell of <u>Chara</u>. For clarity,
          the endoplasmic reticulum has been omitted. The asterisks
          outside the rhizoids denote the approximate limits of the
          apical growth zone. Modified after Sievers and Volkmann
          (1979).

     After these theoretical considerations, two examples will be
given to illustrate the transduction concept. For both examples,
gravitropic response is selected, because gravitational stimuli will
otherwise be treated, in this volume, only marginally.

GRAVITROPISM IN <u>CHARA</u>

     The first example deals with positive gravitropism of the
rhizoid cell of <u>Chara</u> (cf. Sievers and Volkmann, 1979). If such a
rhizoid is placed horizontally, the continuously growing cell bends
downward. The first indication of bending can be observed after about
10 min, and after 2 to 3 h the vertical orientation of growth is
reached again. The stimulus is the asymmetric action of gravity, and
the response is a curvature, due to asymmetric growth. The whole reaction chain from perception to response occurs in one single cell.
To understand the important steps of this reaction chain, we have to
know the rhizoid cell in more detail with its characteristic properties.

     The rhizoid cell is characterized by its strong polarity, both
structural and physiological. Structurally, the polarity is recognized by the distribution of cell organelles (Fig. 2). The apical
region is nearly free of mitochondria and dictyosomes, which are
found in huge amounts more basally. Near the tip, only golgi vesicles
are found. Thus, these vesicles obviously move from the dictyosomes
to the tip, which is a manifestation of physiological polarity.
Accordingly, growth is restricted to the apical half-dome, where the
content of the golgi vesicles undergoes exocytosis, thus sustaining

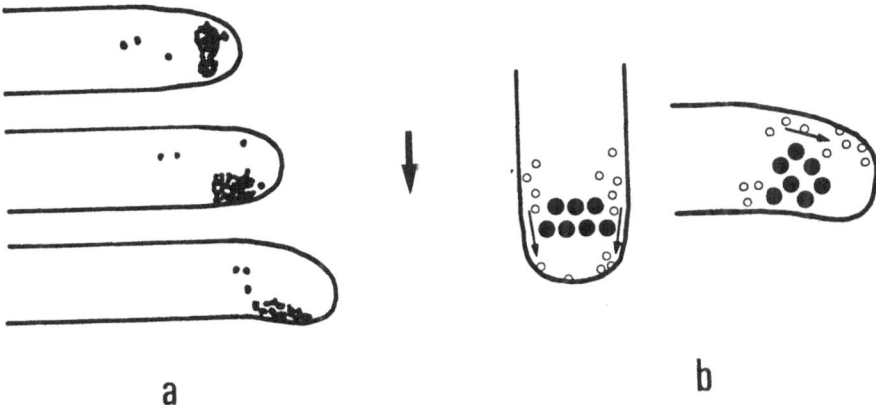

a   b

Fig. 3.   Rhizoid of <u>Chara</u> under the influence of gravity (arrow in
          the center). (a) Rhizoid 0, 6, and 15 min, respectively,
          after being placed horizontally. The black dots denote
          the statoliths. (b) Schematic drawing of the tip in vertical
          and horizontal position, showing the statoliths (black) and
          the acropetally moving golgi vesicles (white). Modified
          after Sievers and Volkmann (1979) and after Libbert (1979).

cell wall growth. Finally, between the apical region and the organ-
elle-containing region, there is a layer of membrane-bound vesicles,
which contain crystals of barium sulfate.

     The structures mentioned so far appear to be sufficient to ex-
plain the gravitropic reaction chain (Sievers and Volkmann, 1979).
Perception of the stimulus can immediately be observed as a displacement of the $BaSO_4$-containing vesicles; they sediment in the gravitational field and can be considered as "statoliths" (Fig. 3a). They
accumulate near the physically lower wall, thus blocking there the
migration of golgi vesicles. As a result, more vesicles arrive at
the physically upper wall in the growth zone, and asymmetric growth
ensues (Fig. 3b).

     It is true, there is still some doubt whether this model really
can explain all details of gravitropism in <u>Chara</u>. But even if it
were an oversimplification, it is a good example to show the principle
of a very short and simple transduction chain. Yet, this model is not
completely without complications. Although all processes, from perception to response, occur within one cell, close to its tip, perception and response are not localized at exactly the same site.
Perception is localized in the subapical layer of statoliths, but
response takes place in the very tip. Thus, some kind of transmission
of information is included in the transduction chain, also called
<u>signal transmission</u>.

In Chara rhizoids, the golgi vesicles are the "transmitters," the movement of which is modulated by the gravity-perceiving stato- liths. This obviously is a very simple mechanism, and usually more sophisticated mechanisms of signal transmission have evolved, as will be shown in the next example, which deals with a multicellular organ.

GRAVITROPISM IN ROOTS

Superficially, the response of roots to gravity is the same as that of the Chara rhizoid: when oriented horizontally, they bend downward. However, the whole reaction chain is quite different and more complicated. Bending is due to differential cell growth on opposite sides of the multicellular organ as compared to differential cell-wall growth in one cell; bending occurs a few millimeters behind the tip rather than at the tip proper. Moreover, stimulus perception is localized in the root cap, i.e. distally from the apical meristem. Thus, transduction has to include a signal transfer in longitudinal direction over many cells.

For gravitropism in roots, our present knowledge about stimulus perception and early steps of transduction is based mainly on recent experiments with cress roots (Lepidium staivum; cf. Volkmann and Sievers, 1979; Behrens et al., 1984), whereas for analyzing the later steps different species have been used preferentially (cf. Wilkins, 1979). Nevertheless, we can try to establish a preliminary and simplified model of the whole reaction chain.

In the center of the root cap, there is a group of cells con- taining particularly large amyloplasts, i.e. starch grains. They are considered as statoliths, and the whole tissue is called staten- chyma, consisting of statocytes (Fig. 4a). The statoliths sediment to the lower part of the statocyte, where they come to lay on a cushion of endoplasmic reticulum, and to exert a pressure on it (Fig. 4b). As soon as the root's orientation is not vertical, this pressure becomes asymmetric on opposite sides (Fig. 4c).

The pressure of the statoliths on the endoplasmic reticulum and its transversal difference may be considered as the first step of transduction sensu stricto, and this difference very probably results in differential changes of membrane potentials (Fig. 5): the stato- cytes at the lower flank depolarize within a few seconds, but those at the upper flank hyperpolarize (Behrens et al., 1984). This elec- tric asymmetry can be considered as the second transduction step.

At the output end of the Black Box, there is good evidence for an asymmetric phytohormone distribution in the growth zone, with a growth inhibitor accumulating at the lower flank. With some addi- tional observations and with some speculation, we can roughly bridge the unknown part of the Black Box (cf. Wilkins, 1979). It is known

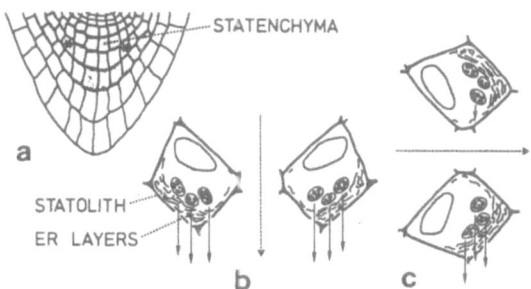

Fig. 4.   Root cap of Lepidium with the statocytes. (a) Longitudinal
           section of the whole cap, with the statenchyma shaded.
           (b) A pair of symmetrically situated statocytes, i.e. cells
           of the statenchyma, as indicated by the asterisks in (a).
           The root axis is indicated by the dashed arrow. In vertical
           orientation, the pressure of the statoliths due to gravity
           (arrows) on the endoplasmic reticulum (ER) is identical in
           both cells. (c) As in (b), but root in horizontal orienta-
           tion, the pressure differs in both cells (number and length
           of the arrows). Modified after Volkmann and Sievers (1979).

that an inhibitor is produced in the root cap; that it moves basipetally to the growth zone; and that it moves in higher concentration at
the lower flank of a horizontally placed root. Moreover, abscisic
acid (ABA) can be detected in root tips, this substance inhibits root
growth and induces curvature if applied unilaterally. Thus, it is
likely that in the root cap ABA is transversely transported (probably
as a result of the transverse electric potential), and thereafter
the basipetally directed ABA currents differ on the upper and lower
flank (Fig. 6). ABA would be, then, one of the transducing signals,
and its currents would establish signal transmission.

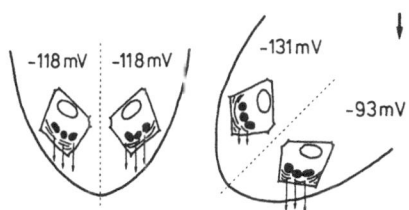

Fig. 5.   Cell potential in a selected pair of statocytes (for details
           cf. Fig. 4) in the cap of a Lepidium root, oriented vertical
           or inclined. The direction of gravity (plumb line) is in-
           dicated in the upper right corner. Modified after Behrens
           et al. (1984).

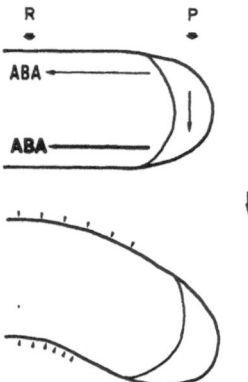

Fig. 6. Hypothetical control of asymmetric root growth by abscisic
         acid (ABA). In the perception zone (P; root cap) a gravity-
         induced transversal ABA transport is suggested, which re-
         sults in unequal basipetal ABA currents. In the response-
         zone (R; zone of the cell growth) the asymmetric ABA con-
         centrations can be recognized. The resulting differential
         growth is indicated by the pointed dashes at the lower
         root, which therefore bends downward. The direction of
         gravity (plumb line) is shown by the arrow at the right.

         Again, it should be stressed that in detail this model may still
be too simple; it certainly has to be supplemented here and there,
and it may need some correction. Nevertheless, it can be taken as a
good example for a more complicated transduction chain, with an obvious signal transmission between cells being involved. Additional
examples, with stimuli other than gravity, will be presented in
subsequent chapters.

LIGHT AS A STIMULUS

         For light as a stimulus, many aspects of perception and trans-
duction will be discussed in detail in this book. I will, therefore,
contribute only a few general considerations.

         In typical light-stimulated responses, a short light stimulus
is sufficient to start the transduction chain until the response is
manifested, usually long after the light stimulus has disappeared;
such light effects are called induction. A good example is phototropism in plants (cf. Dennison, 1979). If a unilateral light pulse
of seconds or even shorter is applied, bending towards the light is
observed after half an hour or later in shoots or coleoptiles of
higher plants. Thus, transduction also implies "remembering" of a
past stimulus.

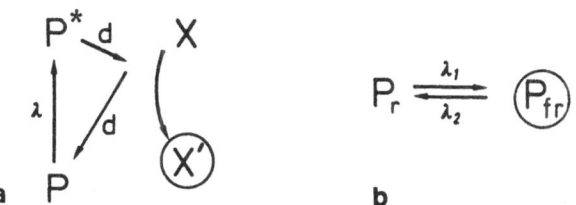

Fig. 7.   Early effects of light absorption, and "memory." (a) A
pigment molecule (P), by absorption of a photon (λ), is
driven to an excited state (P^*), from where it returns in
light-independent processes (d) to the ground state. The
coupled reacition generates a signal X', which may be stable
for some time, thus being the "memory" for the light stimulus
(encircled). (b) A photochromic pigment, e.g. phytochrome
(P_r), by absorption of a photon (λ_1; red light), is converted to its second form (P_{fr}), which remains stable for a
long time, unless it is photoreverted by another photon
of a different wavelength (λ_2; far-red light). Thus, the
memory resides in the pigment itself (encircled).

There are two fundamental alternatives how the stimulus can be
remembered. The first alternative is found in all well-investigated
blue-light responses of plants, including most cases of phototropism.
Here, the photoreceptor pigment, by absorption of a photon, is driven
to an excited state, from where it returns to its ground state, sometimes through a complex way. In this way, part of the pigment's
energy is transferred to another molecule, which starts the transduction (sensu stricto). Thus, after a very small fraction of a
second, the information about the stimulus does no longer reside in
the pigment molecule, and the latter is ready for new perception,
i.e. to absorb another photon (Fig. 7a).[**]

Completely different is the path of information in those responses in plants, which are induced by red light via the well-known
phytochrome system. Phytochrome can exist in two metastable forms,
which absorb red light (maximum 660 nm) and far-red light (maximum
730 nm), respectively, and which are called, therefore, P_r and P_{fr}.
They are phototransformed to each other according to

$$P_r \xrightarrow{\text{red}} P_{fr}$$
$$\xleftarrow{\text{far-red}}$$

[**] This mechanism probably is not restricted to that wide-spread blue-
light receptor in plants, which is assumed to be a flavin, but it
is concluded to apply also to non-flavin blue-light receptor
pigments.

In this system, P_r can be considered as the photoreceptor pigment, detecting the light stimulus, and P_{fr} is the so-called "active
form," i.e. P_{fr} is able to start the transduction chain, thus inducing, controlling or regulating the response, e.g. growth or morphogenesis in the plant. However, once P_{fr} has been formed, phytochrome remains in this form, sometimes many hours. Thus, the "memory"
resides for a long time in the pigment proper (Fig. 7b). In this
case, P_{fr} can continuously keep the transduction chain going, or it
can wait until the cell acquires its competence to respond to the
signal. For the latter case, an example will be given.

LIGHT-INDUCED FERN-SPORE GERMINATION

It is well known that the spores of many fern species have an
absolute light requirement for germination, with phytochrome as the
photoreceptor pigment (cf. Furuya, 1983). The effect of even 12 h
of red light can be reverted by subsequent far-red light (Mohr,
1956).

Spores sown on agar containing mineral nutrients according to
Etzold (1965) are particularly sensitive to light and can be induced
by a short pulse of red light already. They acquire light sensitivity
in the first min after sowing, as shown by Haupt (1984). If freshly
sown spores are kept in light for 5 min before being placed in darkness, some germination is observed after 5 days; and with increasing
exposure to light, a higher percentage of germination is obtained
(Fig. 8). This effect, however, is completely reversible by far-red
as late as 18 h after sowing. Thus, the first product of stimulus

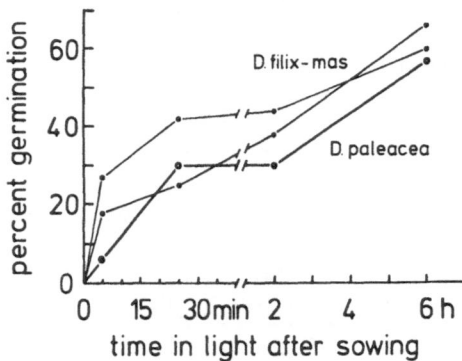

Fig. 8. Germination of light-requiring fern spores (Dryopteris),
         which were exposed to diffuse daylight in the laboratory,
         starting with sowing, for the time given at the abscissa.
         At zero point, the spores received less than half a min
         light. For D. filix-mas, two samples with different origin
         were used. After Haupt (1984, and unpublished data).

perception, viz. P_{fr}, is "stored," and obviously no further processes are started until 18 h later. Only if far-red is given still later, e.g. after 24 h, reversion is no longer possible. Obviously, P_{fr} as an internal signal cannot act before 18 h after sowing, the cell is not yet competent for this signal (the transduction chain is not ready to accept the signal). After that time, competence appears relatively fast, the transduction chain is started, and soon the reaction chain has escaped from far-red reversibility.

In photomorphogenesis of higher plants, many examples of this kind are well known, and it has been shown that in the same seedling different P_{fr}-dependent transduction chains have fundamentally different time courses for the appearance of P_{fr} competence (cf. Mohr, 1983).

In conclusion, in phytochrome, the result of light perception, P_{fr}, is relatively stable for several hours. This stability ensures that a response can be controlled by a light pulse even if the latter is given long before competence for the signal has been reached. Such a long-living "memory" of the stimulus, however, also has a disadvantage. A new stimulus cannot be perceived as long as the memory has not yet been cleared. This problem will be discussed in the next section.

CLEARANCE OF THE PERCEPTION SYSTEM FOR A NEW STIMULUS

The problem will be approached for light as a stimulus. We consider first those systems which make use of a blue-light-photoreceptor pigment. For short pulses, there is no problem with perception, because immediately after "light-off" the pigment molecule returns to its ground state and is ready to perceive a new stimulus (cf. Fig. 7a). If there is a memory which has to be cleared, it resides at a later point in the transduction chain. This has been found, e.g. in the first phototropic curvature of oat coleoptiles (Avena sativa) or in polarity induction of horsetail spores (Equisetum spp), where a second light stimulus becomes fully effective only, if separated from the first one by at least 20 min (cf. Meyer, 1969; Haupt, 1958). Thus, we are not dealing with the perception system as the site of the memory.

Perception can be involved, however, if continuous light is applied. In this case, the pigment molecules are continuously cycling between ground state and excited state. With increasing fluence rate (intensity), an increasing percentage of the pigment is in the excited state, the dark relaxation being continuously balanced by photoexcitation (Fig. 9).

There are two possibilities how this continuous stimulation feeds into the transduction chain.

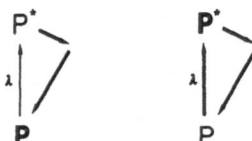

Fig. 9.   Cycling of a pigment between ground state (P) and excited
          state (P*) in continuous light of low (left) and high in-
          tensity (right). The fraction of P or P* undergoing trans-
          formation per unit time is indicated by the thickness of the
          arrows.

    (i) The stomata opening of higher plants is taken as the first
example (cf. Raschke, 1979). Under given conditions the width of the
stomata depends on the stationary fluence rate of light, whatever
the duration of illumination (Fig. 10a). Obviously, with continuous
stimulation, a continuous transduction proceeds, and its intensity
depends on the light intensity.

    (ii) As a second example, we consider the sporangiophore of
Phycomyces (cf. Delbrück and Reichardt, 1956), which at its IV b
stage has a fairly constant growth rate. If a dark-growing sporangiophore is transferred to continuous and constant illumination, it responds by growing faster. Remarkably, however, this "light-growth
response" is only transient, and after about 15 min the organ returns
to its original growth rate, in spite of the stimulus continuing to
act (Fig. 10b). The transduction chain has adapted to the stimulus,
i.e. the stimulus now is considered as being "no stimulus." In consequence, an increase in fluence rate again induces a light-growth
response, which, of course, is transient as before. And finally,
reducing the fluence rate results in a transient growth retardation;
this change in fluence rate obviously is transduced as a "negative
stimulus."

    Theoretically, adaptation may concern the perception system
proper, e.g. bleaching of the photoreceptor pigment; or adaptation
may occur at any stage in the transduction chain. This has to be
analyzed in each particular case, but for light as a stimulus, there
are not many examples where we have full understanding. However, in
chemosensing, much progress has been made in elucidating the mechanisms of adaptation (cf. Macnab, this volume).

    Let us turn back to the original problem of this section:
clearance of the perception system for a new stimulus. In contrast
to blue-light-absorbing pigments, in phytochrome the photoreceptor
is not reestablished immediately after light-off. Instead, photoconversion depletes the pool of P_r for a longer time, which makes the
cell unable to receive another stimulus. At least two possibilities
exist to balance this effect after some time (cf. Hartmann and
Haupt, 1983; Smith, 1983):

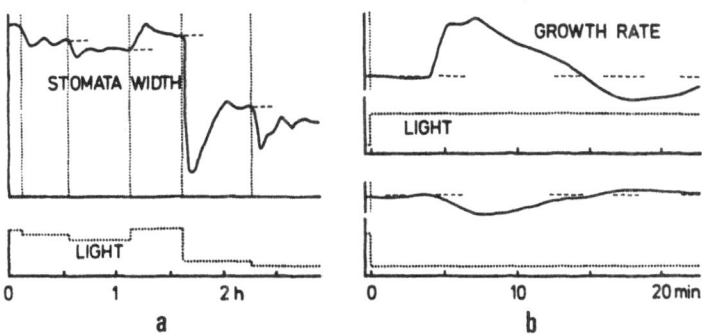

Fig. 10. Responses to continuous light by systems without (a) and
          with (b) adaption. (a) Light-controlled stomata opening
          in Zea mays, as measured by the porometer flux (upper
          graph, solid curve). At each vertical dotted line, the
          light intensity (heavy dots, lower graph) has been changed.
          The steady-state opening in the preceding period is ex-
          tended by the dashed line. (b) Light-growth response in
          the sporangiophore of Phycomyces upon a step-up and a
          step-down change of light intensity (upper and lower pair
          of graphs, respectively). Otherwise as in (a). Modified
          after Raschke (1966) and Foster and Lipson (1973).

     (i) P_r is currently biosynthesized and P_{fr} undergoes a slow de-
struction in darkness.

     (ii) P_{fr} can slowly revert to P_r in darkness with a halflife of
about 45 min; this has been observed especially in dicotyledons.

     Besides, there is some indication that the active phytochrome
P_{fr} can regulate the rate of de-novo biosynthesis of P_r; however,
an inhibition rather than an acceleration has been reported (Gottmann
and Schäfer, 1982), which, of course, does not solve the above problem.

     In the two former cases (i.e., destruction and reversion), the
P_r/P_{fr} ratio is increased; a new red-light stimulus, therefore, with
the tendency to reestablish the $P_r \rightleftharpoons P_{fr}$ photoequilibrium, has to
produce additional P_{fr}, which may start the transduction chain again.
Considering the slow kinetics of all the three processes mentioned
above, it is evident that in phytochrome-mediated systems "clearance
of the memory" is very slow as compared to the blue-light effects.

     In conclusion, if a manufacturer were to construct a perception-
transduction system with either phytochrome or a blue-light photoreceptor, he would have to decide what is more important for the particular purpose: a good memory to a past stimulus, or the possibility
to detect new stimuli rather soon after a preceding one.

There is, however, another fundamental difference between phytochrome and the blue-light systems, which has to be considered. This deals with "color vision." To fully understand it, we first have to learn which parameters of light can act as a stimulus.

PARAMETERS OF LIGHT BEING DETECTABLE BY A PERCEPTION SYSTEM

There is no doubt that the fluence rate or intensity of light can be perceived. In fact, this is the most trivial parameter to be detected, and we have referred to it previously. Thus, light intensity can be considered as the stimulus.

Besides its quantity, light is characterized by its quality, i.e. by its spectral properties. We are familiar with color vision in man, which, however, requires more than one type of photoreceptor pigments simultaneously. The questions arises whether, e.g. plants and microorganisms, can distinguish between and hence recognize colors. We will postpone this question till the end of this section.

At least in the laboratory, light may be characterized by its polarization also. There are photoreceptor systems, in animals, in plants and in single cells, which can detect the orientation of the electrical vector of linearly polarized light (cf. Etzold, 1965). If, then, the response has a relation to light polarization, it is reasonable to consider this parameter as the stimulus.

Another interesting parameter is the temporal pattern of light intensity. Coming back to the light-growth response of Phycomyces, w have learned that it is not the absolute fluence rate (intensity) of the light which controls the growth rate, but a sudden change in intensity. It is tempting to consider, in those cases, this temporal pattern as the stimulus, i.e. a step-up or step-down stimulus; however, there is still some debate whether this is an acceptable terminology; in the worst case, it is a good description of the situation. Besides the light-growth response in Phycomyces, the photophobic response in motile organisms is a good example for perception of step-up or step-down light stimuli, and in chemosensory behavior of bacteria, perception of concentration jumps is well developed. I can refer to several chapters in this volume (e.g. Diehn, Feinleib, Hildebrand, Macnab and Spudich).

Finally, an important parameter of light is its direction. And, indeed, there are so many examples of responses in relation to the light direction that more than one whole lecture could be devoted only to the basic facts. Phototropic curvature of growing plant organs or phototaxis of motile organisms may just be mentioned. Obviously, the cell or the organism has the possibility to perceive the light direction, and indeed, there are several possibilities how light direction is transformed to a spatial gradient of light absorption in a cell or in an organ (cf. Haupt, 1965). It is reasonable, therefore, to consider light direction as the stimulus in these cases.

In all these examples, the stimulus would insufficiently or improperly be described, if we just would term it "light stimulus" without adding the particular parameter being perceived for the transduction chain in questions.

We have to return to the still open question: Can aneural organisms distinguish colors - or more scientifically - Can they perceive the spectral region of a light stimulus? Superficially, the questions appears trivial. A yellow-pigment photoreceptor can perceive a blue-light stimulus better than a green one, and yellow to red light cannot be detected at all. However, this is still a pure matter of intensity, as the system cannot distinguish between high-intensity green and low-intensity blue. Nevertheless, there is indeed a possibility for "color vision," if the photoreceptor pigment is photochromic, i.e. two forms of the pigment are interconverted by light according to:

$$P_1 \xrightleftharpoons[\lambda_2]{\lambda_1} P_2$$

as has been shown for phytochrome and as will be supplemented in the next paragraph (cf. Hartmann, 1983; Hartmann and Haupt, 1983).

The absorption spectra of the two forms of phytochrome, P_r and P_{fr}, overlap (Fig. 11). Thus, light is always absorbed by both forms simultaneously, and in saturation, a photostationary state is established, usually described as the fraction of P_{fr} to the total phytochrome: $P_{fr}/(P_r + P_{fr})$ or P_{fr}/P_{tot}. This photostationary state depends on the wavelength or on the ratio of red to far-red light; the higher the relative content of red, the higher P_{fr}/P_{tot}. As long as P_{tot} is constant, a change in P_{fr}/P_{tot} also means a change in P_{fr} concentration. As we have learned, P_{fr} is the first internal signal in transduction, and its concentration determines the response. Thus, a phytochrome-containing plant can recognize the ratio of red to far-red light, and indeed, it responds to it. Under a canopy of green leaves, e.g. in a forest, most of the red light is lost by absorption in chlorophyll, but far-red is hardly reduced. In this "green shade," which for phytochrome is equivalent to nearly pure far-red, the plant behaves completely different from growing in "neutral shade." It etiolates, and light-requiring seeds or spores do not germinate (Hartmann and Haupt, 1983).

Thus, aneural organisms can collect information about their spectral environment, if they have a photochromic photoreceptor pigment available. Insofar, there is a fundamental difference between

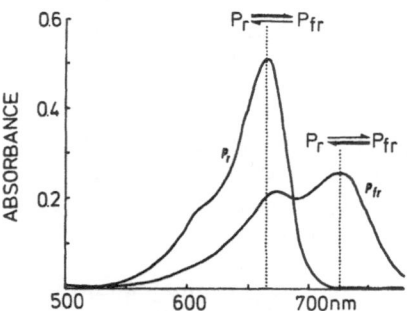

Fig. 11. Absorbance spectra of phytochrome in its two metastable
          forms, P_r and P_{fr}. The photoequilibrium, established by
          monochromatic light, is qualitatively shown at two selected
          wavelengths (dotted lines). Curves after Smith (1983).

the blue-light photoreceptor pigments of plants and phytochrome; only
the latter is able to perceive a "light-quality stimulus" in addition
to the other types of light stimuli.

     Recently, there is indication that, besides phytochrome, other
photochromic pigments exist in plants (including the cyanobacteria
or blue-green algae; cf. Björn and Björn, 1982), and it may be predicted that in those cases also "color vision" by a single pigment
system will be discovered.

     In the purple-colored Halobacterium, finally, a photochromic
pigment appears to be responsible for different responses in different spectral ranges, as will be shown by Spudich (this volume).

FINAL REMARKS

     It was the aim of this survey to present an outline about the
principles of the sequence from perception via transduction to response, using examples which were not too complicated. This has
been extended to introduce various aspects which make the systems
increasingly more complicated and which are also responsible for
terminological problems. But even these more complicated considerations were more or less arbitrary examples, and in nature, much more
diversity exists. That diversity will become obvious during this
meeting, and you will become aware that some of the examples do not
fit completely into the outlines given before. I urgently hope you
become sensitized to realize those complications and to detect the
limitations. Only then are models valuable and useful.

REFERENCES

Behrens, H. M., Gradmann, D., and Sievers, A., 1984, Membrane-potential responses following gravistimulation in roots of Lepidium sativum L., Planta, in press.

Björn, L. O., and Björn, G. S., 1980, Photochromic pigments and photoregulation in blue-green algae, Photochem. Photobiol., 32:849-852.

Delbrück, M., and Reichardt, W., 1956, System analysis for the light growth reactions of Phycomyces, in: "Cellular Mechanisms in Differentiation and Growth," D. Rudnick, ed., Princeton University Press, Princeton.

Dennison, D. S., 1979, Phototropism, in: "Encycl. Plant Physiol., N. S., Vol. 7," W. Haupt and M. E. Feinleib, eds., Springer, Berlin-Heidelberg-New York, pp. 506-566.

Etzold, H., 1965, Der Polarotropismus und Phototropismus der Chloronemen von Dryopteris filix-mas (L.) Schott, Planta, 64:254-280.

Foster, K. W., and Lipson, E. D., 1973, The light growth response of Phycomyces, J. Gen. Physiol., 62:590-617.

Furuya, M., 1983, Photomorphogenesis in ferns, in: "Encycl. Plant Physiol. Vol. 16," W. Shropshire, Jr, and H. Mohr, eds., Springer, Berlin-Heidelberg-New York-Tokyo, pp. 569-602.

Gottmann, K., and Schäfer, E., 1982, In vitro synthesis of phytochrome apoprotein directed by mRNA from light and dark-grown Avena seedlings, Photochem. Photobiol., 35:521-525.

Hartmann, K. M., 1983, Action spectroscopy, in: "Biophysics," W. Hoppe, W. Lohmann, H. Markl, and H. Ziegler, eds., Springer, Berlin-Heidelberg-New York-Tokyo.

Hartmann, K. M., and Haupt, W., 1983, Photomorphogenesis, in: "Biophysics," W. Hoppe, W. Lohmann, H. Markl, and H. Ziegler, eds., Springer, Berlin-Heidelberg-New York-Tokyo.

Haupt, W., 1958, Über den Primä vorgang bei der polarisierenden Wirkung des Lichtes auf keimende Equisetum-Sporen, Planta, 51:74-83.

Haupt, W., 1965, Perception of environmental stimuli orienting growth and movement in lower plants, Annu. Rev. Plant Physiol., 15: 267-290.

Haupt, W., 1984, Effect of nutrients and light pretreatment on phytochrome-mediated fern-spore germination, Planta, in press.

Libbert, E., 1979, "Lehrbuch der Pflanzenphysiologie," 3rd edition, VEB G. Fischer, Jena.

Meyer, A. M., 1969, Versuche zur Trennung von 1. positiver und negativer Krümmung der Avena koleoptile, Z. Pflanzenphysiol., 60: 135-146.

Mohr, H., 1956, Die Beeinflussung der Keimung von Farnsporen durch Licht und andere Faktoren, Planta, 46:534-551.

Mohr, H., 1983, Pattern specification and realization in Photomorphogenesis, in: "Encycl. Plant Physiol. N. S., Vol. 16," W. Shropshire, Jr., and H. Mohr, eds., Springer, Berlin-Heidelberg-New York-Tokyo, pp. 336-357.

Raschke, K., 1966, Die Reaktionen des CO_2-Regelsystems in den
    Schliesszellen von Zea mays auf weiBes Licht, Planta, 68:111-
    140.
Raschke, K., 1979, Movements of Stomata, in: "Encycl. Plant Physiol
    N. S., Vol. 7," W. Haupt and M. E. Feinleib, eds., Springer,
    Berlin-Heidelberg-New York, pp. 383-441.
Sievers, A., and Volkmann, D., 1979, Gravitropism in single cells,
    in: "Encycl. Plant Physiol. N. S., Vol. 7," W. Haupt and
    M. E. Feinleib, eds., Springer, Berlin, pp. 567-572.
Smith, W. O., 1983, Phytochrome as a molecule, in: "Encycl. Plant
    Physiol. N. S., Vol. 16," W. Shropshire, Jr., and H. Mohr,
    eds., Springer, Berlin-Heidelberg-New York-Tokyo, pp. 96-118.
Volkmann, D., and Sievers, A., 1979, Graviperception in multicellular
    organs, in: "Encycl. Plant Physiol. N. S., Vol. 7," W. Haupt
    and M. E. Feinleib, eds., Springer, Berlin, pp. 573-600.
Wilkins, M. B., 1979, Growth-control mechanisms in gravitropism, in:
    "Encycl. Plant Physiol. N. S., Vol. 7," W. Haupt and M. E.
    Feinleib, eds., Springer, Berlin-Heidelberg-New York, pp. 601-
    626.

PHYSICS OF BACTERIAL CHEMOTAXIS

Howard C. Berg

Division of Biology 216-76
California Institute of Technology
Pasadena, CA 91125

ABSTRACT

Physical constraints limit the way in which an organism as
small as Escherichia coli can interact with its surroundings. Cells
are propelled by the movement of thin helical flagella, because
motion is dominated by viscous rather than inertial forces. Cells
are unable to swim in straight lines because of perturbations due to
rotational Brownian movement. Cells are unable to improve their lot
locally by swimming or stirring, because transport of small molecules
is effected by diffusion rather than bulk flow. Cells must sense
gradients temporally rather than spatially, because comparison between concentrations in front and behind are overwhelmed by diffusive
currents due to the cells' motion. Finally, the precision with which
cells can make temporal comparisons are limited by statistical
fluctuations. A survey of these constraints is given, followed by
a description of how E. coli has optimized its chemotaxis machinery
to meet them. This optimization is revealed by measurements of the
chemotactic impulse response.

PHYSICAL CONSTRAINS

As you follow this lecture, imagine yourself as a tiny living
thing no more than 10^{-4} cm in diameter by about 2×10^{-4} cm long,
the size of the intestinal bacterium Escherichia coli. You are
surrounded by water, a fine-grained substance of seemingly inex-
haustible extent, whose component particles display continuous
riotous activity. Although small on your scale, roughly 10^{-4} of
your diameter, their incessant collisions drive you this way and
that at speeds approaching 10^3 cell diameters per second, effecting

19

changes in course some 10^6 times per second. The accelerations involved are more than a hundred times that of gravity! As a result
of this chaotic motion, each second you find yourself displaced
about 1 diameter in a randomly-chosen direction and tumbled some 60
degrees around a randomly-chosen axis. The only constant feature
of the landscape is the pull of gravity that produces a steady
downward drift at a rate of about half a cell diameter per second.
Add to this stark picture hunger (or some primitive equivalent),
the desire to ingest certain molecules such as amino acids and
sugars by which you are continously bombarded. Given that Nature
has come to your rescue and provided a sense of taste, what measurements should you make to determine whether life is getting
better or worse? What propulsive means should you devise, and how
should you modulate those means to seek out regions in the environment that are more favorable? These are the problems that E. coli
and other bacteria have solved over many millennia.

Flagellar Propulsion

A bacterium soon finds that its progress is resisted, not by
inertial forces, but rather by viscous forces. Its forward momentum
is rapidly transferred to the surrounding medium that undergoes
viscous shear. The bacterium drags nearby molecules along with it,
and it must move some distance before shedding this local environment (cf. Purcell, 1977). A hydrodynamicist, taking a macroscopic
point of view, would say that we are dealing with fluids at low
Reynolds number. The Reynolds number is a dimensionless parameter
that indicates the relative size of inertial forces (forces required
to accelerate masses) and viscous forces (forces due to viscous
shear). The Reynolds number is

$$R = vL\rho/\eta \qquad\qquad (1)$$

where v is the velocity of the fluid (the velocity of a particle
moving through the fluid), L is the linear scale of the motion (the
size of the particle), ρ is the specific gravity of the fluid, and
η is its viscosity. For a man in a swimming pool, $v \simeq 10^2$ cm/sec,
$L \simeq 10^2$ cm, $\rho \simeq 1$ g/cm^3, and $\eta \simeq 10^{-2}$ g/cm sec, so that $R \simeq 10^6$.
For a bacterium, $v \simeq 10^{-3}$ cm/sec, $L \simeq 10^{-4}$ cm, $\rho \simeq 1$ g/cm^3, and
$\eta \simeq 10^{-2}$ g/cm sec, so that $R \simeq 10^{-5}$. The Reynolds number for the
man is very large, while that for the bacterium is very small. The
man propels himself by accelerating water, the bacterium by taking
advantage of viscous shear. The two live in very different hydrodynamic worlds.

The most dramatic way that I know to drive this point home is
to compute the distance that a bacterium the size of E. coli coasts
upon turning off its flagellar motors. The result of this calculation, given elsewhere (pp. 76-77 of Berg, 1983), is about 0.04 Å,

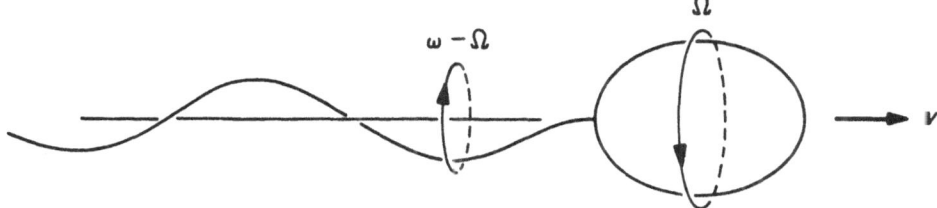

Fig. 1. A bacterium swimming to the right at velocity v. A rotary
         motor at the base of the flagellum turns the helical fila-
         ment relative to the body of the cell at angular velocity
         ω. When the cell is viewed from behind, the body turns
         clockwise at angular velocity Ω, and the filament turns
         counterclockwise at angular velocity ω - Ω; ω is larger
         than Ω. The filament is left-handed; when the helix is
         viewed end-on, a particle moving along it away from the
         observer turns counterclockwise. From p. 78 of Berg (1983).

a distance small compared to the diameter of a hydrogen atom! Of
course, the bacterium continues to move randomly, buffeted by the
surrounding water, but the unidirectional drift due to the propulsion
rapidly damps out.

    A bacterium swims by continuously rotating a long thin asymmetric
structure, a shallow helical filament (or set of helical filaments),
as shown in Fig. 1. It does not utilize a rigid reciprocating motion,
because this turns out to be futile at low Reynolds number (cf.
Purcell, 1977). The thrust generated by the filament(s) is balanced
by viscous drag due to translation of the cell body, while torque
generated by the filament(s) is balanced by viscous drag due to
counter-rotation of the cell body. The net force acting on the cell
is zero. The fundamental reason for the success of this method lies
in the fact that viscous drag on a thin rod moving slowly at a given
speed through a viscous medium is larger when the rod moves sideways
than when it moves end-on, by about a factor of 2. The flagellar
filament (or bundle of filaments) revolves many times before the
cell body advances one length; therefore, each rod-like element of
the filament is dragged slantwise through the medium. The longitudinal component of the drag on each element contributes to the thrust.
The circumferential component of the drag on each element contributes
to the torque. A detailed analysis is given on p. 79 of Berg (1983).
For a treatment based on considerations of symmetry, see Purcell
(1977). For a general review of the fluid mechanics of propulsion
by cilia and flagella, see Brennen and Winet (1977).

    From a hydrodynamic point of view, it doesn't really matter
whether the bacterium rotates its flagellar filaments rigidly by
driving them at their base or bends them in a helical fashion; in

either case, the torque and thrust are nearly the same (Berg and
Anderson, 1973; Berg, 1975). In fact, bacteria rotate their flagellar filaments (Berg and Anderson, 1973; Silverman and Simon,
1974). The filaments are only about 100 Å in diameter. They do
not contain the machinery required for the local conversion of chemical to mechanical energy. Species that have several filaments form
flagellar bundles, in which the filaments rotate side by side. For
discussions of the mechanics of this situation, see Berg and Anderson (1973), Anderson (1975), and Macnab (1977). This situation is
complicated further by the fact that the motors that drive each
filament are autonomous two-state systems that spin either clockwise
or counterclockwise, changing states at random (Macnab and Han, 1983;
Ishihara et al., 1983). How a cell equipped with such motors manages
to form a flagellar bundle is an unresolved questions; see the
appendix of Ishihara et al. (1983).

Rotational Brownian Movement

    Collisions between a cell and the surrounding water molecules
lead to changes in angular orientation as well as in displacement.
In time t, the mean-square angular deviation of the cell about a
given axis is

$$<\theta^2> = 2D_r t \qquad\qquad (2)$$

where D_r is the rotational diffusion coefficient in units radians^2/
sec (cf. pp. 81-85 of Berg, 1983). For a sphere of radius a,

$$D_r = kT/8\pi\eta a^3$$

where k is Boltzmann's constant and T is the absolute temperature--
the value of kT at 300°K (27°C) is 4.14 x 10^{-14} g cm^2/sec^2. If we
approximate the bacterial cell body as a sphere of radius 10^{-4} cm,
we find $D_r \simeq 0.2$ radians^2/sec. Since a bacterium swimming along the
z-axis can go off course by rotating about either the x- or y-axis,
we expect in time t sec a root-mean-square angular deviation of
about $(4D_r t)^{1/2} \simeq (0.8\ t)^{1/2}$ radians. This implies that a cell will
wander off course by as much as 90 degrees in 3 sec. The flagellar
filaments, which can be several body lengths long, will reduce this
deviation somewhat, but experiments with the tracking microscope
indicate that the cells do, in fact, meander as they swim by nearly
the predicted amount (Berg and Brown, 1972). Thus, in a period of
time of about 10 sec, a bacterium forgets the direction in which it
was swimming. If it has not measured changes in its environment and
acted on that information within this time, it is too late. Rotational Brownian movement sets a fundamental limit on acceptable
signal-processing times in bacterial chemotaxis.

Fig. 2.   A bacterial cell shown as a short rod on the left or a
          small fish on the right, attempting to improve its intake
          of small molecules (shown as dots) by reaching out with
          a flagellum and stirring the medium (left) or moving
          through the medium with an open mouth (right). The mole-
          cules diffusive out of the bacterium's grasp (arrow, left)
          or move away from its mouth via laminar flow (arrow,
          right).

Diffusive Intake

The molecules that a bacterium tastes or ingests reach its
surface by diffusion. In time t, the mean-square displacement of
such a molecule along a given axis is

$$<x^2> = 2Dt$$

where D is the diffusion coefficient in units cm^2/sec (cf. pp. 5-16
of Berg, 1983). Bacterial chemoattractants or repellents are small
molecules such as amino acids or sugars. Their diffusion coefficients fall in the range 5×10^{-6} to 10^{-5} cm^2/sec. Naively, one
might think that a cell could ingest more of these molecules by
stirring the medium with its flagellar filaments or by swimming
open-mouthed, as it were, as illustrated in Fig. 2. However, for
a cell as small as E. coli, this proves to be futile (Berg and
Purcell, 1977). Small molecules diffuse out of a flagellum's grasp,
so to speak, much sooner than the motion of the flagellum can bring
them to the surface of the cell. As the cell moves through the
medium, molecules in front move around to the sides entrained in
the laminar flow, so they still must reach the surface of the cell
by diffusion. The work that the cell would have to do to increase
its intake by stirring or swimming is much larger than the energy
available for propulsion.

There is one important caveat: while a completely-absorbing
cell moving at a constant velocity through a viscous medium cannot
easily improve its net intake, the motion raises the intake over
the front of the cell and lowers it over the back (Berg and

Purcell, 1977). As a result, a rapidly moving cell cannot make
spatial comparisons of the concentrations of chemicals in its environment, i.e. compare concentrations in front to those in back,
because it always thinks that the concentration in front is larger.
This is true regardless of the direction in which it chooses to
swim. Therefore, the best such a cell can do is measure the arrival
of molecules of interest over its entire surface and ask, as it
moves about, whether that flux increases or decreases.

Statistical fluctuations limit the precision with which a
bacterium can make such measurements. The cell deals with this
problem by integrating the input over an extended period of time.
To get the gist of this argument, consider the bacterium as a
sphere of radius a that can count every molecule of interest in its
volume $4\pi a^3/3$. The result of one count is $n = 4\pi a^3 C/3$, where C is
the average concentration of molecules (particles/cm^3) in the vicinity of the cell. The standard deviation in this count is $n^{1/2}$
(cf. pp. 88-90 of Berg, 1983). For example, if $a = 10^{-4}$ cm and
$C = 10^{-8}$ M (6×10^{12} particles/cm^3), $n = 25$, $n^{1/2} = 5$, and the precision of the measurement is about 20%. The cell can do better by
repeating the count, but only by waiting long enough that the first
set of molecules have diffused away and another set of molecules have
diffused in. This ensures that the second count is statistically
independent of the first, or that the cell is able to count a total
of twice as many molecules as before. The time that it must wait
to do this is of order $t = (2a)^2/6D$, i.e. the time required for
diffusion in three dimensions a distance roughly equal to the diameter of the cell. For a molecule of diffusion coefficient 7×10^{-6}
cm^2/sec, this time is about 10^{-3} sec. Therefore, by waiting 0.1
sec, the cell can repeat the count 100 times and increase the precision of the measurement by a factor of 10. The relative error
decreases inversely as the square-root of the total count, i.e.
as $(2\pi DaCt)^{-1/2}$. Since the bacterium really wants to know whether
the concentration is increasing or decreasing, it must make two
such measurements and consider the difference. If the difference
is substantially smaller than the standard deviation in either measurement, it will not be able to decide whether life is getting
better or worse. Thus, the bacterium is well advised to use as
much time as possible, within the time constraint imposed by rotational Brownian movement. For a more rigorous treatment of this
problem, see Berg and Purcell (1977).

AN OPTIMUM SOLUTION

Biased Random Walk

Escherichia coli is propelled by about six flagellar filaments
arising at random points on the surface of the cell. Each filament

is powered by a rotary motor at its base (Berg and Anderson, 1973;
Silverman and Simon, 1974). When the motors turn counterclockwise
(CCW), the filaments work together in a bundle that drives the cell
steadily forward--the cell runs; when the motors turn clockwise
(CW), the bundle flies apart, and the motion is highly erratic--
the cell tumbles (Larsen et al., 1974; Macnab and Ornston, 1977).
Runs and tumbles appear in an alternating sequence, each run con-
stituting a step in a three-dimensional random walk (Berg and
Brown, 1972). When the cell swims in a spatial gradient of a chemical attractant, runs up the gradient are extended; this imposes a
bias on the random walk that carries the cell in a favorable direction (Berg and Brown, 1972; Macnab and Koshland, 1972). Changes
in concentration of attractants are sensed by specific receptors
(Adler, 1969). CCW rotation is favored as more attractant is
bound (Berg and Tedesco, 1975; Spudich and Koshland, 1975). The
bias of a flagellar motor (the fraction of time spent spinning
CCW) increases in proportion to the rate of change of receptor
occupancy (Block et al., 1983). The signal that controls the direction of flagellar rotation has not been identified, but it is
known that it affects the bias of each motor, not the particular
times at which transitions between CW and CCW states occur (Macnab
and Han, 1983; Ishihara et al., 1983).

In executing the random walk, a cell swims steadily 1 to 2 x
10^{-3} cm/sec along gently curved paths. As noted above, the curvature is about what one would expect from rotational diffusion (Berg
and Brown, 1972). Runs are interrupted by tumbles: changes in
course are picked at random from a distribution peaked slightly in
the forward direction (Berg and Brown, 1972). Run and tumble intervals are distributed exponentially, with means of about 1 sec
for runs and 0.1 sec for tumbles. In the absence of a chemotactic
stimulus, the probability per unit time that a run (or tumble) will
end is constant--for a discussion of this kind of statistics, see
pp. 86-92 of Berg (1983). The mean run interval is long enough to
allow a cell to sense changes in concentration in the presence of
fluctuations, but not so long that the curvature induced by rotational diffusion causes it to deviate significantly from its
original course. When runs are extended, most still fall within
this time limit (ca. 10 sec).

As expected, a cell measures concentrations as a function of
time (Macnab and Koshland, 1972; Brown and Berg, 1974). It responds not to the ambient concentration of the chemical but to
temporal changes in that concentration. Analysis of fluctuations
in concentration encountered by a cell the size of E. coli shows
that a spatial gradient can be detected near threshold in a time
span of about 1 sec, provided that receptors of a given kind
(roughly a thousand) are distributed widely over the surface of
the cell (Berg and Purcell, 1977).

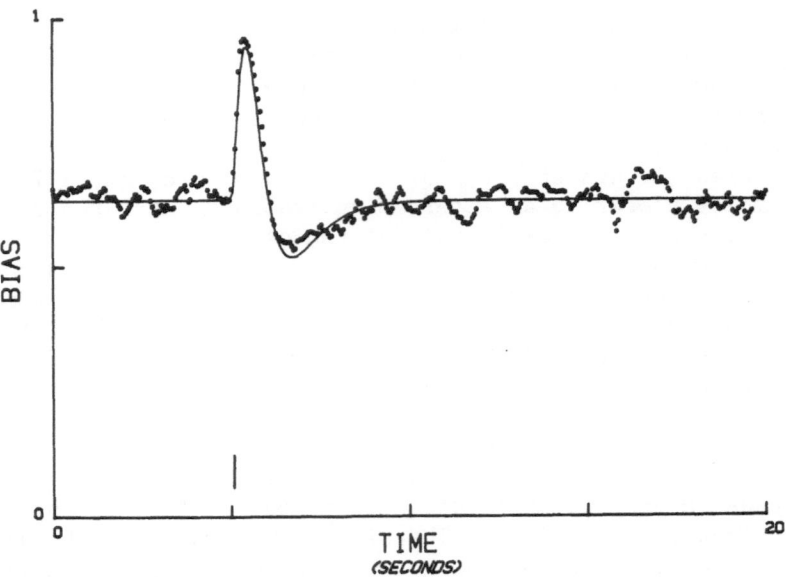

Fig. 3.   The impulse response of wild-type cells of E. coli to the
          attractants asparate or α-methylasparate. The rotational
          bias (the probability of spinning CCW) is plotted as a
          function of time. Pulses of attractant were delivered
          iontophoretically at time 5.06 sec (vertical bar). The
          graph was constructed from 378 trials with 17 different
          cells. The mean bias within a 0.05 sec window was de-
          termined and plotted every 0.05 sec. The smooth curve is
          a fit to these data of the sum of six exponentials
          (Segall, 1984).

Impulse Response

     Studies of responses of tethered cells (cells attached to
glass by a single flagellar filament) subjected to impulsive stimuli show that a cell compares concentrations measured over the past
second or so with concentrations measured about 3 sec before that;
the difference in these two averages determines the probability
that the flagella spin CW or CCW (Block et al., 1982). The impulse
response for wild-type cells is shown in Fig. 3. The pulse of
attractant giving rise to this response is quite short, about 0.1
sec or shorter, while the response itself is relatively long,
about 4 sec. The response is biphasic: it has two lobes of equal
area. The positive lobe lasts about 1 sec, the negative lobe
another 3 sec.

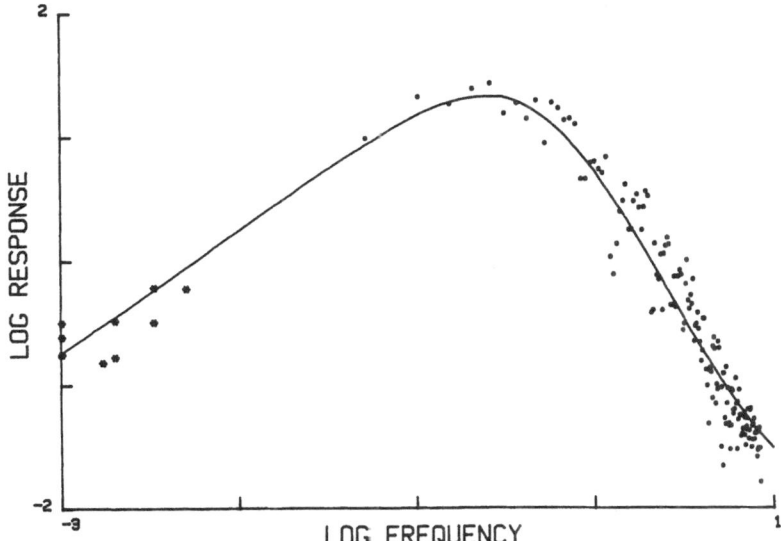

Fig. 4.   The Fourier transform of the impulse response for wild-type
          cells. A Bode plot--$\log_{10}$ (response amplitude) versus
          $\log_{10}$ (stimulus frequency)--of the absolute magnitude of
          the complex Fourier transform of the impulse response shown
          in Fig. 3. The dots are derived from the dots of Fig. 3,
          and the line from the line of Fig. 3. The asterisks are
          derived from data (Block et al., 1983) of cells exposed
          to sine-wave changes in receptor occupancy. See Segall
          (1984).

     Given an impulse response, one can predict the response of a
cell to any arbitrary stimulus, by decomposing that stimulus into a
sequence of impulses of different amplitude and time and summing the
corresponding impulse responses. There is a proviso, namely, that
the stimulus must be small enough that the cell operates in a linear
domain, i.e. that its responses to earlier impulses do not markedly
affect its responses to more recent ones (cf. Block et al., 1982).
The impulse response shown in Fig. 3 acts as a weighting function for
earlier stimuli. The cell looks at the concentration over the past
second with a positive weighting, determined by the first lobe, and
at the concentration for 3 sec before that with a negative weighting,
determined by the second lobe, and then shifts its present rotational
bias in proportion to the difference. Since the mean run interval
is 1 sec, it is fair to say that the cell, on average, determines
the mean concentration during the past run and compares it with the
mean concentration determined a few runs before that. It uses information obtained only over the past 4 sec, a time within the limit set
by rotational Brownian movement. It averages concentrations over
periods of at least 1 sec in order to minimize the uncertainty due
to random fluctuations in concentration. If a cell is exposed to a

Table 1. Contrasts in the Behavior of a Bacterium
          and a Leukocyte.

	Bacterium	Leukocyte
Size	~1 μm	~10 μm
Movement	swims constantly	crawls when activated
Speed	>10 μm/sec	<1 μm/sec
Motor	rotary	linear sliding (?)
Energy source	protonmotive force	ATP
Chemotactic	amino acids	peptides
toward	sugars	proteins
Receptor K_D	~10^{-6} M	~10^{-12} M
Sensing mode	temporal	spatial (?)
Direction change	all-or-none	graded
Genetics and	easy	hard
biochemistry		

step change in concentration, the bias rises for 1 sec and then falls
for another 3 sec, returning to its initial value. The cell is sensitive to changes in concentration, not to the ambient concentration.
However, changes that occur at high frequency are integrated out.

An equivalent description of the behavior of the cell can be
given in the frequency domain by reference to the Bode plot obtained
from the Fourier transform of the impulse response, as shown in
Fig. 4. The cell acts as a bandpass filter. The response falls off
slowly at low frequencies, where it is proportional to the first
derivative of concentration with respect to time. It falls off more
steeply at high frequencies, where noise due to fluctuations in concentration is most severe. The response peaks at a frequency of
about 0.25 Hz, a frequency characteristic of stimuli generated when
cells execute a random walk in a linear spatial gradient. In short,
the cell has tuned its sensitivity to the sensory task that it is
required to perform.

OTHER SOLUTIONS

What is optimum for E. coli is not necessarily optimum for
larger cells, particularly those with other means of locomotion.
Consider, for example, the dramatically different style of chemotaxis adopted by a leukocyte, Table 1 (for reviews, see Snyderman
and Goetzl, 1981; Schiffmann, 1982). Food for thought!

ACKNOWLEDGEMENTS

Work in this laboratory on bacterial motility and chemotaxis has been supported by grants from the US National Science Foundation, the US National Institutes of Health, the Research Corporation, and the Gustavus and Louise Pfeiffer Research Foundation. I thank Steve Block, Markus Meister and Jeff Segall for their comments on the manuscript.

REFERENCES

Adler, J., 1969, Chemoreceptors in bacteria, Science, 166:1588-1597.

Anderson, R. A., 1975, Formation of the bacterial flagellar bundle, in: "Swimming and Flying in Nature," T. Y.-T. Wu, C. J. Brokaw, and C. Brennen, eds., Plenum, New York, 1:45-56.

Berg, H. C., 1975, Bacterial movement, in: "Swimming and Flying in Nature," T. Y.-T. Wu, C. J. Brokaw, and C. Brennen, eds., Plenum, New York, 1:1-11.

Berg, H. C., 1983, "Random Walks in Biology," Princeton, Princeton.

Berg, H. C., and Anderson, R. A., 1973, Bacteria swim by rotating their flagellar filaments, Nature, 245:380-382.

Berg, H. C., and Brown, D. A., 1972, Chemotaxis of Escherichia coli analysed by three-dimensional tracking, Nature, 239:500-504.

Berg, H. C., and Purcell, E. M., 1977, Physics of chemoreception, Biophys. J., 20:193-219.

Berg, H. C., and Tedesco, P. M., 1975, Transient response to chemotactic stimuli in Escherichia coli, Proc. Natl. Acad. Sci. USA, 72:3235-3239.

Block, S. M., Segall, J. E., and Berg, H. C., 1982, Impulse responses in bacterial chemotaxis, Cell, 31:215-226.

Block, S. M., Segall, J. E., and Berg, H. C., 1983, Adaptation kinetics in bacterial chemotaxis, J. Bacteriol., 154:312-323.

Brennen, C., and Winet, H., 1977, Fluid mechanics of propulsion by cilia and flagella, Annu. Rev. Fluid Mech., 9:339-398.

Brown, D. A., and Berg, H. C., 1974, Temporal stimulation of chemotaxis in Escherichia coli, Proc. Natl. Acad. Sci. USA, 71: 1388-1392.

Ishihara, A., Segall, J. E., Block, S. M., and Berg, H. C., 1983, Coordination of flagella on filamentous cells of Escherichia coli, J. Bacteriol., 155:228-237.

Larsen, S. H., Reader, R. W., Kort, E. N., Tso, W.-W., and Adler, J., 1974, Change in direction of flagellar rotation is the basis of the chemotactic response in Escherichia coli, Nature, 249:74-77.

Macnab, R. M., 1977, Bacterial flagella rotating in bundles: A study in helical geometry, Proc. Natl. Acad. Sci. USA, 74: 221-225.

Macnab, R. M., and Han, D. P., 1983, Asynchronous switching of flagellar motors on a single bacterial cell, Cell, 32:109-117.

Macnab, R. M., and Ornston, M. K., 1977, Normal-to-curly flagellar
    transitions and their role in bacterial tumbling. Stabilization
    of an alternative quartenary structure by mechanical force, J.
    Mol. Biol., 112:1-30.
Purcell, E. M., 1977, Life at low Reynolds number, Am. J. Phys., 45:
    3-11.
Schiffmann, E., 1982, Leukocyte chemotaxis, Annu. Rev. Physiol., 44:
    553-568.
Segall, J. E., 1984, Chemotaxis of Escherichia coli studied using
    iontophoretic stimulation, Ph.D. Thesis, California Institute
    of Technology, Pasadena.
Silverman, M., and Simon, M., 1974, Flagellar rotation and the mech-
    anism of bacterial motility, Nature, 249:73-74.
Snyderman, R., and Goetzl, E. J., 1981, Molecular and cellular mech-
    anisms of leukocyte chemotaxis, Science, 213:830-837.
Spudich, J. L., and Koshland, Jr., D. E., 1975, Quantitation of the
    sensory response in bacterial chemotaxis, Proc. Natl. Acad. Sci.
    USA, 72:710-713.

BIOCHEMISTRY OF SENSORY TRANSDUCTION IN BACTERIA

Robert M. Macnab

Department of Molecular Biophysics and Biochemistry
Yale University
New Haven, CT 06517

INTRODUCTION

Bacterial taxis is a complex phenomenon, involving sensory and motor aspects, and a variety of experimental approaches and points of view (for other reviews of the topic, see Koshland, 1981; Boyd and Simon, 1982; Macnab, 1982; Parkinson and Hazelbauer, 1983; Macnab and Aizawa, 1984). In this paper, only Escherichia coli and Salmonella typhimurium, both peritrichously flagellated species, will be reviewed. I will emphasize the chain of information from the receptors to the motors, and what is known regarding the molecular events that are involved.

THE STIMULI

Before considering the tactic response, it is necessary to review the types of stimuli that can elicit it, and the nature of the primary detection of these stimuli. Bacteria respond to a variety of environmental parameters: light, oxygen, pH, temperature, and a range of small molecules, primarily organic compounds. Phototaxis is being addressed in other papers in this volume, and will not be discussed here.

Oxygen taxis is a consequence of the use of oxygen in electron transport, so that the terminal cytochrome o is not a "receptor" in the usual sense since mere binding cannot elicit a response. Rather, it is the influence of oxygen levels upon the cell's proton motive force that is responsible (Taylor, 1983). Oxygen taxis is thus a member of a class of "energy taxis" responses that include responses to other electron acceptors such as fumarate or nitrate, at least

31

some phototactic responses, and presumably other manipulation of the
cell's energy levels by the presence or absence of metabolites,
although this has not been clearly demonstrated. Thus, the original
Links-Clayton hypothesis for tactic response, disproven for many
chemotactic stimuli, is valid for these stimuli (cf. Macnab, 1982).

pH taxis (Kihara and Macnab, 1981; Repaske and Adler, 1981) and
thermotaxis (Maeda and Imae, 1979) do appear to employ conventional
stimulus reception in that they are a consequence of the pH-dependent
or temperature-dependent properties of a specific protein in the cell
(methyl-accepting chemotaxis protein I, MCP I, see below) rather than
of some general physiological disruption. This protein also functions as a receptor/transducer for chemical stimuli and will be discussed extensively later on. Among organic repellents, only one
class is well-understood; these are lipophilic weak acids, which act
by shunting protons into the cytoplasm and hence perturbing cytoplasmic pH. This in turn has a specific effect on a specific protein
which happens to be the same one, MCP I, as is employed in pH sensing
(Kihara and Macnab, 1981; Repaske and Adler, 1981; Slonczewski et al.
1982). Thus, pH sensing and weak-acid sensing are mechanistically
rather similar--we can regard the receptor in both instances as a
proton sensor.

How other repellents operate is unclear (Tso and Adler, 1974).
They too perturb the properties of one or more of the MCPs, but it
is questionable whether there are binding sites for any given organic
repellent, in view of the high concentrations (typically 10 mM or
higher) that are required (Oosawa and Imae, 1983). A reasonable
guess would be that they are affecting some cellular parameter, say
membrane fluidity, and that the sensing protein is responding to that
rather than to the chemical compound per se.

Finally, we come to the best-studied class of stimuli, the organic attractants. These are small molecules typically in the 100-
dalton range, most of them amino acids or sugars (Mesibov and Adler,
1972; Adler et al., 1973). These are all detected via binding to
specific receptors with a high affinity (K_d typically around 1 to
10 μm) and specificity. The binding site is external to the cell
and the information is transmitted to the interior without the need
for uptake of the stimulating molecule itself (Adler, 1969). This
remark needs to be qualified in the case of the sugars detected and
transported by the PTS system, as will be discussed later.

MACROMOLECULAR COMPONENTS

Having surveyed the stimuli, we next turn to the macromolecular
components of the sensory transduction system. Identification of
these has, for the most part, proceeded by the indirect route that
is usual in the study of prokaryotic systems--isolation and mapping

of mutants, followed by the search for the gene products (see, for example, Parkinson, 1981). Four broad classes were discovered: Mutants defective in response to a specific chemical (receptor mutants), mutants defective in response to all or a range of chemicals (sensory transduction, or che, mutants), mutants defective in motor rotation (paralyzed, or mot, mutants) and mutants defective in the assembly of the flagellar apparatus (non-flagellate, or fla, mutants). These were usually isolated by failure to swarm on semisolid agar plates, or by some other chemotactic assay. In more recent years, many of the genes have been cloned, have been or are being sequenced, and amplified expression of the gene products is being achieved (for example, Matsumura et al., 1977; DeFranco et al., 1979; Boyd et al., 1981; DeFranco and Koshland, 1981; Bartlett and Matsumura, 1984; Dean et al., 1984).

Receptors and Transducers

Among receptor mutants, the first class to be well-understood were found to be defective in genes for periplasmic sugar-binding proteins, e.g. the ribose or galactose binding proteins (Ordal and Adler, 1974a, b; Strange and Koshland, 1976; Zukin et al., 1979). These were rather easily isolated from the periplasmic space and purified. They had high-affinity binding sites for their substrate and underwent conformation changes upon binding. The target of the occupied receptor will be discussed shortly. These proteins were also components of corresponding transport systems.

Subsequently, a great deal of effort went into searching through the uptake systems for serine and aspartate (the two principal amino acid attractants) for the component that would prove to be the chemoreceptor (Aksamit et al., 1975). The efforts were unsuccessful; when the receptors in these cases were finally identified, they turned out to be dedicated components of the chemotaxis system, with no function in transport at all (Reader et al., 1979; Clarke and Koshland, 1979; Wang and Koshland, 1980). They were integral membrane proteins that underwent specific and reversible methylation (see below) and hence were called methyl-accepting chemotaxis proteins, or MCPs (Kort et al., 1975; Springer et al., 1977). Four such proteins are presently recognized: MCP I, which mediates tactic responses to serine and some repellents and is encoded by the tsr gene; MCP II, which mediates tactic responses to aspartate and some other repellents, and is encoded by the tar gene; MCP III, which is the target site for the ribose and galactose receptors, and is thus a secondary component in that sensory transduction chain, and is encoded by the trg gene (Kondoh et al., 1979); and finally a rather mysterious protein, MCP IV, for which no significant role in any tactic response has been established (and which was only discovered by investigations at the molecular genetic level, since there is no known mutant phenotype), which is encoded by the tap gene, immediately adjacent to the tar gene on the chromosome (Boyd et al., 1981; Slocum and Parkinson, 1983; Wang et al., 1982). There

is extensive homology among all four genes and gene products
(Krikos et al., 1983). The significance of this will be discussed
in terms of mechanisms of sensory transduction later on.

   Whereas receptor mutants were, with a few exceptions, competent
in their responses to compounds not directed against that particular
receptor, a large number of mutants were found to be generally defective (Armstrong and Adler, 1969; Aswad and Koshland, 1975b;
Parkinson, 1976; DeFranco et al., 1979). It was noticed that they
tended to differ from wild-type cells in their motility pattern,
either by swimming more smoothly with few tumbles, or by tumbling
more frequently. It had been established that modulation of
tumbling frequency was the key to the tactic response of flagellated
bacteria (Berg and Brown, 1972; Macnab and Koshland, 1972) and so it
was not surprising that an abnormal unstimulated frequency would
prevent a normal response to stimuli. A more precise statement regarding their phenotype could be made as a result of the discovery
that the motor was a rotary device (Berg and Anderson, 1973; Berg,
1974; Silverman and Simon, 1974), and that its rotation could be
monitored by tethering a cell by one of its flagella. The unstimulated probability of counterclockwise (CCW) rotation was either
higher or lower than in wild-type cells (Larsen et al., 1974). In
almost all cases, a given gene only gives rise to one of those two
possible mutant motility phenotypes, suggesting that the gene
product is functioning in one of two parallel pathways, one for
placing the motor in the CCW state, and the other for placing it in
the clockwise (CW) state.

Other Chemotaxis Genes and Proteins

   To date, nine complementation groups have been found to be
associated with general chemotaxis deficiencies. All appear to
derive from genes that are dedicated to the motility/chemotaxis
system; no other function has been found for these genes, and they
for the most part map in clusters that include obviously motility-
related genes such as fla and mot. They are cheA, cheB, cheC, cheD,
cheR, cheV, cheW, cheY, and cheZ (DeFranco et al., 1979). However,
cheD is in fact the tsr gene already alluded to; certain mutant
alleles apparently lock the MCP I protein in an extreme state that
prevents it from functioning as a transducer, but also interferes
with other components in some way that is still not well-understood
(Parkinson, 1980). Use of the symbol cheD is now restricted to
these alleles, and tsr is the preferred symbol for the gene. [An
analogous symbol, cheM, was applied to the tar gene, but this usage
has now largely disappeared.] Two others of the che genes, cheC
and cheV, are in a sense usurpers of this gene symbol because the
phenotype of chemotactic incompetence only applies to certain
missense mutants, while the null phenotype is in fact non-flagellate.
These genes contribute in a very important way to the final events of
sensory transduction, at the motor, and will be discussed later.

Flagellar and Motility Genes

There are a large number of genes (ca. 35) involved in the assembly and structure of the flagellar apparatus (see Macnab and Aizawa, 1984, for a recent review). Most of these are not involved in any way with either motor rotation or motor switching between CCW and CW states, and therefore will not be discussed in the present paper, although they present extremely interesting questions regarding structure, assembly, and function. Also, there are two genes, motA and motB, that are essential for motor rotation, suggesting that their products participate in the conversion of proton potential energy into mechanical energy. No mutants in these genes have been found with abnormal switching phenotype, and so again they are not relevant to the subject of the present paper, the sensory transduction process.

EXCITATION VS ADAPTATION

The time course of the response of a cell to a stepwise stimulus consists of two phases, the first a rapid deviation of the flagellar motors from their unstimulated CCW/CW probabilities, and the second a slower return of these probabilities to close to the pre-stimulus level (Macnab and Koshland, 1972; Larsen et al., 1974). These two phases are termed the excitation phase and the adaptation phase, respectively (see, for example, Springer et al., 1979). It will be important to keep these in mind as we consider the role of the che gene products--does a given protein participate in excitation, adaptation, or both?

The Excitation Process

Knowledge regarding the primary event following transducer occupancy (either by the attractant directly, as in the case of aspartate or serine, or by an attractant-receptor complex, as in the case of maltose, ribose, or galactose) is very scanty. In spite of years of effort by a number of laboratories, the identity of "the signal" is unknown. Several putative candidates--membrane potential, Ca^{++} ion, and cyclic GMP--have not withstood the test of further experimentation. Currently, there is information on range and kinetics that restricts the possibilities regarding the signal (perhaps to be discussed in the chapter by H.C. Berg in this volume), but does not permit its identification. Although excitation occurs rapidly compared with adaptation, there is an appreciable lag (ca. 0.2 s) before the first detectable response (Segall et al., 1982), implying that there may be a chain of events rather than simple release of a small molecule followed by diffusion to the motor. We will shortly be discussing the methylation-related aspects of adaptation, but it is worth noting now that they are not necessary for excitation to occur. Absence of the methylating and

demethylating enzymes does not prevent transducers from generating
the excitation signal, although subsequent events are abnormal (Goy
et al., 1978; Stock et al., 1981; Yonekawa et al., 1983).

Adaptation and the Role of Methylation

The first biochemical clue regarding sensory transduction was
the early recognition that methionine, though at best a marginal
attractant, played a critical role in the responses to other
attractants (Adler and Dahl, 1967). The role was later narrowed to
methyl donation via S-adenosylmethionine (Armstrong, 1972; Aswad
and Koshland, 1975a). A major discovery was that the substrate for
the methylation reaction was a class of membrane proteins, the
methyl-accepting chemotaxis proteins (MCPs), and that these proteins
were products of known chemotaxis-related genes. Subsequently, it
was shown that the MCPs were, directly or indirectly, the receptors.
The kinetics of methylation and demethylation followed closely the
kinetics of adaptation--by the time methylation had approached its
new steady-state level upon attractant or repellent addition, the
cells were observed to have adapted. This led to the general assumption that the methylation system was responsible for the adaptation
phase of the tactic response. Further reinforcement for this conclusion came from the behavior of mutants defective in the methyl-
transferase (coded for by the cheR gene), which showed severely impaired adaptation (Goy et al., 1978). However, evidence that
adaptation is not entirely dependent upon the methylation system
will be discussed below.

Biochemistry of Methyl-Accepting Chemotaxis Proteins

Electrophoretic gels of radiolabelled MCPs showed a characteristic pattern of bands that suggested there might be a number of protein
species present (Springer et al., 1977). Attractant addition caused
increase of the overall extent of methylation, and also a redistribution of the label to the more mobile bands. The range of relative
mobilities of these bands corresponded to an apparent molecular
weight (M_r) range of several kilodaltons. However, extensive biochemical analysis has since revealed that the only difference between
bands is in whether or not any of a series of glutamyl residues is
methyl-esterified at its gamma-carboxyl side chain (Chelsky and
Dahlquist, 1980a; Kehry and Dahlquist, 1982). Thus, the large M_r
changes are (highly useful) artifacts of the analytical technique.
Many other examples of the deviation from a simple correlation
between M_r and true molecular weight, especially in the case of
membrane proteins, have emerged in the recent literature.

The maximum number of glutamyl residues that can be methylated
varies among the MCP classes, but a typical value is around five.
Why multiple methylation? A reasonable hypothesis would be that
whereas a single methylatable site would provide an all-or-none state

as information to the motor, multiple methylation provides the oppor-
tunity for graded information. A given transducer can only be either
empty or occupied by attractant but, if on-off events are rapid compared to the methylation-demethylation reactions, these reactions
will respond to a time-averaged extent of occupancy. An elegant
model involving successive methylations and negative feedback with
the appropriate kinetics has been presented, which yields the observed characteristics of an adapted state that at the level of motor
function is indistinguishable from the pre-stimulus state (Asakura,
1984).

A curious feature of the system is that approximately half the
methylatable sites are dormant. In other words, they are methylated
at an early stage after protein synthesis and remain methylated with
an immeasurably low turnover thereafter (Stock and Koshland, 1981).
The contribution of such a static pool remains mysterious. A possible explanation would be that these methylations are necessary for
generating the correct conformation of the protein in the membrane
and have nothing to do with their signal transmission properties,
but the argument is not very convincing. A related puzzle is that
of glutamine deamidation to glutamate, a "one-shot" reaction that
occurs shortly after synthesis (Rollins and Dahlquist, 1981; Sherris
and Parkinson, 1981). It has been suggested that this could represent vestiges of a rudimentary sensory system, with each transducer
having only a single chance to participate. This seems a dubious
explanation, because cells could only maintain sensing capability at
the rate they were synthesizing new transducers (and therefore would
become incompetent under non-growth conditions) and they could only
handle signals of a single polarity, since only one direction of
modification, glutamine to glutamate, is permitted.

Initially as a result of peptide analysis, and later as a result
of translation of the gene sequences, information has been gained
regarding the location and environment of the glutamate residues
that are methylated. They occur in two clusters, one a little over
half-way through the sequence, and the other close to the C terminus
(Krikos et al., 1983). These regions are rather highly conserved,
as might be expected since they are the sites of a common enzymatic
modification. Interestingly, some other regions within the C-terminal
half of the molecule are even more highly conserved. These could be
contributing to the enzyme binding site. Some degree of variation
in the catalytic sites would be desirable, in order to permit specificity among the MCPs with respect to the modifications to result
from a given class of stimulus. The residues occur in a characteristic pattern, separated by around seven residues and with a non-
methylatable glutamate as the immediately adjacent residue on the
N-terminal side. It has been suggested that an alpha-helical conformation would result in the presentation of the residues as a
more or less linear array along one side of the helix, and that increasing transducer occupancy would cause a successive exposure of

the residues in the array to the methyltransferase enzyme
(Terwilliger et al., 1983; Terwilliger and Koshland, 1984).

Organization of Transducers in the Membrane

The translated gene sequence yields a very strong prediction
for the organization of the molecule in the membrane (Krikos et al.,
1983; Russo and Koshland, 1983). The hydrophobic N-terminus would
form a membrane-spanning segment (which is not cleaved) from the
cytoplasm to the exterior, followed by a substantial external domain
that would include the effector binding site, then a return membrane-
spanning segment and finally the cytoplasmic domain that includes
all of the methylatable residues. The membrane-spanning segments
are only conserved at the qualitative level, as a conservation of
hydrophobicity. The proposed organization is almost inevitable on
thermodynamic grounds, and has been supported by genetic engineering
experiments that have been carried out recently. Chimeric proteins
obtained by gene fusions give the chemoeffector specificity of the
N-terminal region (A. Krikos and M. Simon, unpublished observations),
as would be expected if this contains the external domain for binding
the effector (it is the region that is least conserved of all among
the various MCPs). Another interesting result (Russo and Koshland,
1983) is that truncation of the C-terminus, even though it did not
eliminate the methylatable residues, greatly reduced the capacity
of the MCP to undergo methylation, and the capacity of the cell to
adapt. The excitation capability of the MCP, however, was unimpaired.

Communication Across the Membrane

How does a binding event on the external face of a transducer
result in generation of the excitation signal, and its cancellation
during the adaptation phase? For proteins with the organization
described in the preceding paragraph, communication across the
membrane is confined to either one or at most two single spans of
the polypeptide chain. Tere are currently two classes of model
under consideration. One can be called the "chain-pull" mechanism,
in which the consequence of effector binding is that the external
domain becomes more (or less) deeply buried, and in so doing pushes
(or pulls) on the membrane-spanning segment, causing the cytoplasmic
domain to become less (or more) deeply buried, with consequent alteration of its properties with respect to generation of the excitation
signal, and to susceptibility to methylation. Adjustment of methylation levels during the adaptation phase would then progressively
cancel the signal-generating alteration. The other class of model
invokes oligomerization of MCPs. The consequence of effector binding
would in this case be either a change in the degree of oligomerization
of the external domain, or in the orientation of the external domains
within the oligomer. By transmission of force through the membrane-
spanning segment, this in turn would cause a change in degree of
oligomerization, or orientation, of the cytoplasmic domains. There

is some evidence for tetrameric MCP complexes (Chelsky and Dahlquist, 1980), but their existence must still be regarded as uncertain.

It is worth emphasizing the sophistication of these proteins. They have multiple and quite disparate binding sites for effectors. For example, MCP II can recognize aspartate but also the periplasmic binding protein for maltose. MCP I has the remarkable ability to act as a thermometer and a pH meter, as well as a receptor for serine; it is also responsive to a variety of other repellents, by unknown mechanisms. And, in addition to their reception function, they are capable of generating an excitation signal and of progressively cancelling it. As the means by which they accomplish these multiple tasks is uncovered, they will surely become regarded as classic examples of receptor/transducer function.

Functions of the Other Chemotaxis Proteins

What are the roles of all the other che proteins? Two have been studied in considerable detail, the cheR and cheB proteins (Goy et al., 1978; Parkinson and Revello, 1978; Stock and Koshland, 1978; Clarke et al., 1980; Yonekawa et al., 1983). They are the methyltransferase and methylesterase enzymes, respectively. Loss of cheR function is accompanied by smooth-swimming phenotype, absence of methylation of MCPs, and deficient adaptation. The function of the methylation reaction can thus be seen as the enhancement of an MCP's ability to transmit a CW signal to the motor. Loss of cheB function is accompanied by opposite changes--tumbly phenotype, enhanced methylation, but again deficient adaptation. The function of the demethylation reaction can thus be seen as the diminution of an MCP's ability to transmit a CW signal to the motor. cheB function is complicated by the fact that it also mediates the glutamine deamidation reaction referred to earlier, which has parallel effects to demethylation upon the signalling properties of the MCP. Note that wherever a term such as "enhancement of an MCP's ability to transmit a CW signal" is used we could, in our present state of knowledge, equivalently say "diminution of an MCP's ability to transmit a CCW signal" since it is not really known whether the signal is a CW signal combating the motor's tendency to be in the CCW state, a CCW signal combating the tendency of the motor to be in the CW state, or whether both classes of signal might exist. A hint in this regard is the phenotype of multiple mutants lacking all classes of MCPs. Such mutants are smooth-swimming (Niwano and Taylor, 1982), suggesting that the signal generated by MCPs may be a CW signal, which is suppressed upon attractant addition and re-established as adaptation occurs.

Methylation-independent Adaptation

Although it is generally agreed that defects in methylation are associated with defects in adaptation, there is some uncertainty

as to whether adaptation requires participation of the methylation
system. For oxygen taxis, and for taxis toward phosphotransferase-
mediated sugars, it apparently does not (Niwano and Taylor, 1982).
Indeed, these responses do not employ MCPs as transducers, nor is
the extent of methylation of MCPs perturbed by such stimuli. The
nature of the sensory transduction processes in oxygen and PTS-sugar
taxis are not as well understood as that operating in taxis to amino
acids and other sugars, but there are hints that they may involve
an as yet hypothetical "phosphoryl-accepting chemotaxis protein"
or PCP, which is likely to be cytoplasmic and to involve (in the
case of the PTS sugars) some kind of interaction with the phosphorylated sugar that is the consequence of translocation of the sugar
via its specific Enzyme II (Lengeler et al., 1981). This type of
methylation-independent adaptation, as it has come to be known, is
relatively easy to accept as a parallel and independent way of
transmitting sensory information to the motors. Harder to understand is the finding that adaptation to MCP-mediated stimuli can
occur, albeit with abnormal kinetics, even in extremely tight cheR
mutants (Stock et al., 1981). This implies that even although the
MCP is presumably still generating the excitation signal at the
initial maximal level, the motor has found some other way of cancelling it. Thus, the motor itself may have to some degree the
capacity for self-homeostasis.

A Dynamic Protein Relay

     The role of the cheR and cheB proteins involves more than inter-
action with their MCP substrates. The cheR protein interacts with
the cheY protein, and the cheB protein interacts with the cheZ protein (DeFranco et al., 1979). The evidence is indirect, involving
complementation tests between E. coli and Salmonella, but it is
fairly convincing. Both the cheY and the cheZ proteins are cytoplasmic. Both also give genetic evidence, in this case of intergenic
complementation (pseudoreversion), of interactions with proteins
that are almost certainly components of the motor, namely the cheC
and cheV proteins (Parkinson et al., 1983). By interactions, we mean
to imply direct physical contact, the formation of a complex. It
does not seem likely, however, that the transducers are in physical
contact, direct or indirect, with the flagella (indeed, there is
evidence that they are not, since flagellated vesicles showed no
enrichment for MCPs over the level applying to bulk membrane;
Engstrom and Hazelbauer, 1982). The concept therefore arises of a
dynamic protein relay, with the methylation enzymes sometimes in
contact with their MCP substrates, sometimes with the cheY or cheZ
proteins, which in turn would sometimes be in contact with the
motors. The significance of these interactions is far from clear.
It is not known whether only binding events are occurring or
whether there are covalent modifications of any of the proteins or
of as yet unknown small-molecule participants. There are hints
of the involvement of nucleotides; for example, the cheY protein

binds to a Cibacron Blue column, and can be displaced by certain nucleotides (P. Matsumura, unpublished observations).

The function of the cheA and cheW proteins is not understood at all. Both appear to be important in the generation of CW information for the motor, since the mutant phenotype is smooth-swimming. cheA presents a further puzzle, because there are in fact two cheA proteins, resulting from translation from two different start sites, both in the same frame (Smith and Parkinson, 1980). The short cheA protein is thus an N-terminal truncation of the long cheA protein. The reason for this phenomenon is unknown, but the complementation behavior of mutants defective in different parts of the cheA coding region suggests that both proteins have a role in chemotaxis. Clarification of this aspect of chemotaxis may reveal novel mechanisms of control.

THE INTERFACE WITH THE MOTOR

The final components of the sensory transduction chain reside at the motor itself. We have already alluded to the fact that a few of the flagellar genes are at the interface between the sensory and motor aspects of the overall system of tactic behavior. These genes can give rise to a variety of mutant phenotypes ranging from the extreme of non-flagellate, through flagellate but paralyzed, to motile but non-chemotactic. They have therefore been ascribed different gene symbols, depending on the viewpoint of the particular study. For example, the flaAII.2 gene (described on the basis of the null phenotype) of Salmonella is also the motC gene and the cheV gene. There is another important point concerning these genes and the range of phenotypes that can accompany mutations in them. They are the exceptions to the rule that, for a given gene, deficient switching function is always in the same sense, either to a CCW bias or a CW bias. The situation is most strikingly seen in the case of cheC, where some mutants have an extreme CCW bias, while other have an extreme CW bias (Khan et al., 1978; Parkinson, 1981). This is a very revealing result, because it implies that the cheC protein is not operating on one of two parallel pathways, but at the point where these two pathways converge. These proteins, of which there are two clearly established examples, flaQ (cheC, motE) and flaAII.2 (cheV, motC) and a partially established third example, flaN (che?, motD) [gene symbols are for S. typhimurium] are thus seen to play an extraordinarily central role in the overall scheme. They are almost certainly structural components of the motor, they participate in the energy conversion mechanism, and they constitute its gear-shift mechanism or switch. In the latter function, they appear to act as a complex. The evidence for this is of the same sort that we have discussed for the cytoplasmic components of the system; all three of these genes demonstrate intergenic complementation (S.-I. Aizawa, S. Yamaguchi and R.M. Macnab, unpublished observations).

The state of the switch is determined in part autonomously, since
the motor spontaneously switches in a stochastic fashion even in
unstimulated cells (Macnab and Han, 1983; Ishihara et al., 1983).
It is also of course modulated as a result of the sensory information
generated by the MCPs and other transducers.

CONCLUSIONS

The broad outlines of a complete sensory transduction pathway
have been established, including the identification of most of the
macromolecular components. A critical feature, the excitation
signal, still has not been identified, and the manner in which the
various macromolecules participate is only understood for the
methylation modification reactions of the primary sensory transducers. Nonetheless, the system of bacterial taxis is still one
of those at the forefront of our understanding of sensory transduction processes.

REFERENCES

Adler, J., 1969, Chemoreceptors in bacteria, Science, 166:1588-1597.
Adler, J., and Dahl, M. M., 1967, A method for measuring the motility
    of bacteria and for comparing random and non-random motility,
    J. Gen. Microbiol., 46:161-173.
Adler, J., Hazelbauer, G. L., and Dahl, M. M., 1973, Chemotaxis
    toward sugars in Escherichia coli, J. Bacteriol,, 115:824-847.
Aksamit, R. R., Howlett, B. J., and Koshland, Jr., D. E., 1975,
    Soluble and membrane-bound aspartate-binding activities in
    Salmonella typhimurium, J. Bacteriol., 123:1000-1005.
Armstrong, J. B., 1972, An S-adenosylmethionine requirement for
    chemotaxis in Escherichia coli, Can. J. Microbiol., 18:1695-1701.
Armstrong, J. B., and Adler, J., 1969, Complementation of nonchemo-
    tactic mutants of Escherichia coli, Genetics, 61:61-66.
Asakura, S., 1984, A two-state model for bacterial chemoreceptor
    proteins: the role of multiple methylation, J. Mol. Biol. (in
    press).
Aswad, D. W. and Koshland, Jr., D. E., 1975, Evidence for an S-
    adenosyl methionine requirement in the chemotactic behavior of
    Salmonella typhimurium, J. Mol. Biol., 97:207-223.
Aswad, D., and Koshland, Jr., D. E., 1975, Isolation, characteriza-
    tion and complementation of Salmonella typhimurium chemotaxis
    mutants, J. Mol. Biol., 97:225-235.
Bartlett, D. H., and Matsumura, P., 1984, Identification of
    Escherichia coli Region III flagellar gene products and descrip-
    tion of two new flagellar genes, J. Bacteriol. (in press).
Berg, H. C., 1974, Dynamic properties of bacterial flagellar motors,
    Nature (London), 249:77-79.

Berg, H. C., and Anderson, R. A., 1973, Bacteria swim by rotating their flagellar filaments, Nature (London), 245:380-382.

Berg, H. C., and Brown, D. A., 1972, Chemotaxis in Escherichia coli analysed by three-dimensional tracking, Nature (London), 239: 500-504.

Boyd, A., Krikos, A., and Simon, M., 1981, Sensory transducers of E. coli are encoded by homologous genes, Cell, 26:333-343.

Boyd, A., and Simon, M., 1982, Bacterial chemotaxis, Annu. Rev. Physiol., 44:501-517.

Chelsky, D., and Dahlquist, F. W., 1980a, Structural studies of methyl-accepting chemotaxis proteins of Escherichia coli: Evidence for multiple methylation sites, Proc. Natl. Acad. Sci. USA, 77:2434-2438.

Chelsky, D., and Dahlquist, F. W., 1980b, Chemotaxis in Escherichia coli: Associations of protein components, Biochemistry, 19: 4633-4639.

Clarke, S., and Koshland, Jr., D. E., 1979, Membrane receptors for aspartate and serine in bacterial chemotaxis, J. Biol. Chem., 254:9695-9702.

Clarke, S., Sparrow, K., Panasenko, S., and Koshland, Jr., D. E., 1980, In vitro methylation of bacterial chemotaxis proteins: Characterization of protein methyl transferase activity in crude extracts of Salmonella typhimurium, J. Supramol. Biol., 13:315-328.

Dean, G. E., Macnab, R. M., Stader, J., Matsumura, P., and Burks, C., 1984, Gene sequence and predicted amino acid sequence of the motA protein, a membrane-associated protein that is required for flagellar rotation in Escherichia coli, J. Bacteriol. (in press).

DeFranco, A. L., and Koshland, Jr., D. E., 1981, Molecular cloning of chemotaxis genes and overproduction of gene products in the bacterial sensing system, J. Bacteriol., 147:390-400.

DeFranco, A. L., Parkinson, J. S., and Koshland, Jr., D. E., 1979, Functional homology of chemotaxis genes in Escherichia coli and Salmonella typhimurium, J. Bacteriol., 139:107-114.

Engstrom, P., and Hazelbauer, G. L., 1982, Methyl-accepting chemotaxis proteins are distributed in the membrane independently from basal ends of bacterial flagella, Biochim. Biophys. Acta, 686:19-26.

Goy, M. F., Springer, M. S., and Adler, J., 1978, Failure of sensory adaptation in bacterial mutants that are defective in a protein methylation reaction, Cell, 15:1231-1240.

Ishihara, A., Segall, J. E., Block, S. M., and Berg, H. C., 1983, Coordination of flagella on filamentous cells of Escherichia coli, J. Bacteriol., 155:228-237.

Kehry, M. R., and Dahlquist, F. W., 1982, The methyl-accepting chemotaxis proteins of Escherichia coli. Identification of the multiple methylation sites on methyl-accepting chemotaxis protein I, J. Biol. Chem., 257:10378-10386.

Khan, S., Macnab, R. M., DeFranco, A. L., and Koshland, Jr., D. E., 1978, Inversion of a behavioral response in bacterial chemotaxis: Explanation at the molecular level, Proc. Natl. Acad. Sci. USA, 75:4150-4154.

Kihara, M., and Macnab, R. M., 1981, Cytoplasmic pH mediates pH taxis and weak-acid repellent taxis of bacteria, J. Bacteriol., 145:1209-1221.

Kondoh, H., Ball, C. B., and Adler, J., 1979, Identification of a methyl-accepting chemotaxis protein for the ribose and galactose chemoreceptors of Escherichia coli, Proc. Natl. Acad. Sci. USA, 76:260-264.

Kort, E. N., Goy, M. F., Larsen, S. H., and Adler, J., 1975, Methylation of a membrane protein involved in bacterial chemotaxis, Proc. Natl. Acad. Sci. USA, 72:3939-3943.

Koshland, Jr., D. E., 1981, Biochemistry of sensing and adaptation in a simple bacterial system, Annu. Rev. Biochem., 50:765-782.

Krikos, A., Mutoh, N., Boyd, A., and Simon, M. I., 1983, Sensory transducers of E. coli are composed of discrete structural and functional domains, Cell, 33:615-622.

Larsen, S. H., Reader, R. W., Kort, E. N., Tso, W.-W., and Adler, J., 1974, Change in direction of flagellar rotation is the basis of the chemotactic response in Escherichia coli, Nature (London), 249:74-77.

Lengeler, J., Auburger, A.-M., Mayer, R., and Pecher, A., 1981, The phosphoenolpyruvate-dependent carbohydrate: Phosphotransferase system enzymes II as chemoreceptors in chemotaxis of Escherichia coli K12, Mol. Gen. Genet., 183:163-170.

Macnab, R. M., 1982, Sensory reception in bacteria, in: Prokaryotic and Eukaryotic Flagella, W. B. Amos and J. G. Duckett, eds., Cambridge University Press, Cambridge, pp. 77-104.

Macnab, R. M., and Aizawa, S.-I., 1984, Bacterial motility and the bacterial flagellar motor, Annu. Rev. Biophys. Bioeng., 13: 51-83.

Macnab, R. M., and Han, D. P., 1983, Asynchronous switching of flagellar motors on a single bacterial cell, Cell, 32:109-117.

Macnab, R. M., and Koshland, Jr., D. E., 1972, The gradient-sensing mechanism in bacterial chemotaxis, Proc. Natl. Acad. Sci. USA, 69:2509-2512.

Maeda, K., and Imae, Y., 1979, Thermosensory transduction in Escherichia coli: Inhibition of the thermoresponse by L-serine, Proc. Natl. Acad. Sci. USA, 76:91-95.

Matsumura, P., Silverman, M., and Simon, M., 1977, Synthesis of mot and che gene products of Escherichia coli programmed by hybrid ColE1 plasmids in minicells, J. Bacteriol. 132:996-1002.

Mesibov, R., and Adler, J., 1972, Chemotaxis toward amino acids in Escherichia coli, J. Bacteriol., 112:315-326.

Niwano, M., and Taylor, B. L., 1982, Novel sensory adaptation mechanism in bacterial chemotaxis to oxygen and phosphotransferase substrates, Proc. Natl. Acad. Sci. USA, 79:11-15.

Oosawa, K., and Imae, Y., 1983, Glycerol and ethylene glycol: Members of a new class of repellents of Escherichia coli chemotaxis, J. Bacteriol., 154:104-112.

Ordal, G. W., and Adler, J., 1974a, Isolation and complementation of mutants in galactose taxis and transport, J. Bacteriol., 117:509-516.

Ordal, G. W., and Adler, J., 1974b, Properties of mutants in galactose taxis and transport, J. Bacteriol., 117:517-526.

Parkinson, J. S., 1976, cheA, cheB and cheC genes of Escherichia coli and their role in chemotaxis, J. Bacteriol., 126:758-770.

Parkinson, J. S., 1980, Novel mutations affecting a signaling component for chemotaxis of Escherichia coli, J. Bacteriol., 142: 953-961.

Parkinson, J. S., 1981, Genetics of bacterial chemotaxis, in: Genetics as a Tool in Microbiology, S. W. Glover and D. A. Hopwood, eds., 31:265-290, Cambridge University Press, Cambridge.

Parkinson, J. S., and Hazelbauer, G. L., 1983, Bacterial chemotaxis: Molecular genetics of sensory transduction and chemotactic gene expression, in: Gene Function in Prokaryotes, Cold Spring Harbor Laboratory, pp. 293-318.

Parkinson, J. W., Parker, S. R., Talbert, P. B., and Houts, S. E., 1983, Interactions between chemotaxis genes and flagellar genes in Escherichia coli, J. Bacteriol., 155:265-274.

Parkinson, J. S., and Revello, P. T., 1978, Sensory adaptation mutants of E. coli, Cell, 15:1221-1230.

Reader, R. W., Tso, W.-W., Springer, M. S., Goy, M. F., and Adler, J., 1979, Pleiotropic aspartate taxis and serine taxis mutants of Escherichia coli, J. Gen. Microbiol., 111:363-374.

Repaske, D., and Adler, J., 1981, Change in intracellular pH of Escherichia coli mediates the chemotactic response to certain attractants and repellents, J. Bacteriol., 145:1196-1208.

Rollins, C., and Dahlquist, F. W., 1981, The methyl-accepting chemotaxis proteins of E. coli: A repellent-stimulated, covalent modification, distinct from methylation, Cell, 25:333-340.

Russo, A. F., and Koshland, Jr., D. E., 1983, Separation of signal transduction and adaptation functions of the aspartate receptor in bacterial sensing, Science, 220:1016-1020.

Segall, J. E., Manson, M. D., and Berg, H. C., 1982, Signal processing times in bacterial chemotaxis, Nature (London), 296:855-857.

Sherris, D., and Parkinson, J. S., 1981, Posttranslational processing of methyl-accepting chemotaxis proteins in Escherichia coli, Proc. Natl. Acad. Sci. USA, 78:6051-6055.

Silverman, M., and Simon, M., 1974, Flagellar rotation and the mechanism of bacterial motility, Nature (London), 249:73-74.

Slocum, M. K., and Parkinson, J. S., 1983, Genetics of methyl-accepting chemotaxis proteins in Escherichia coli: Organization of the tar region, J. Bacteriol., 155:565-577.

Slonczewski, J. L., Macnab, R. M., Alger, J. R., and Castle, A. M., 1982, Effects of pH and repellent tactic stimuli on protein methylation levels in Escherichia coli, J. Bacteriol., 152: 384–399.

Smith, R. A., and Parkinson, J. S., 1980, Overlapping genes at the cheA locus of Escherichia coli, Proc. Natl. Acad. Sci. USA, 77:5370–5374.

Springer, M. S., Goy, M. F., and Adler, J., 1977, Sensory transduction in Escherichia coli: Two complementary pathways of information processing that involve methylated proteins, Proc. Natl. Acad. Sci. USA, 74:3312–3316.

Springer, M. S., Goy, M. F., and Adler, J., 1979, Protein methylation in behavioural control mechanisms and in signal transduction, Nature (London), 280:279–284.

Stock, J. B., and Koshland, Jr., D. E., 1978, A protein methylesterase involved in bacterial sensing, Proc. Natl. Acad. Sci. USA, 75:3659–3663.

Stock, J. B., and Koshland, Jr., D. E., 1981, Changing reactivity of receptor carboxyl groups during bacterial sensing, J. Biol. Chem., 256:10826–10833.

Stock, J. B., Maderis, A. M., and Koshland, Jr., D. E., 1981, Bacterial chemotaxis in the absence of receptor carboxyl methylation, Cell, 27:37–44.

Strange, P. G., and Koshland, Jr., D. E., 1976, Receptor interactions in a signalling system: Competition between ribose receptor and galactose receptor in the chemotaxis response, Proc. Natl. Acad. Sci. USA, 73:762–766.

Taylor, B. L., 1983, Role of proton motive force in sensory transduction in bacteria, Annu. Rev. Microbiol., 37:551–573.

Terwilliger, T. C., Bogonez, E., Wang, E. A., and Koshland, Jr., D. E., 1983, Sites of methyl esterification in the aspartate receptor involved in bacterial chemotaxis, J. Biol. Chem., 258:9608–9611.

Terwilliger, T. C., and Koshland, Jr., D. E., 1984, Sites of methyl esterification and deamination on the aspartate receptor involved in chemotaxis, J. Biol. Chem., 259:7719–7725.

Tso, W.-W., and Adler, J., 1974, Negative chemotaxis in Escherichia coli, J. Bacteriol., 118:560–576.

Wang, E. A., and Koshland, Jr., D. E., 1980, Receptor structure in the bacterial sensing system, Proc. Natl. Acad. Sci. USA, 77:7157–7161.

Wang, E. A., Mowry, K. L., Clegg, D. O., and Koshland, Jr., D. E., 1982, Tandem duplication and multiple functions of a receptor gene in bacterial chemotaxis, J. Biol. Chem., 257:4673–4676.

Yonekawa, H., Hayashi, H., and Parkinson, J. S., 1983, Requirement of the cheB function for sensory adaptation in Escherichia coli., J. Bacteriol., 156:1228–1235.

Zukin, R. S., Hartig, P. R., and Koshland, Jr., D. E., 1979, Effect of an induced conformational change on the physical properties of two chemotactic receptor molecules, Biochemistry, 18:5599–5605.

PRIMARY MOLECULAR EVENTS IN ANEURAL CELL PHOTORECEPTORS

Pill-Soon Song

Department of Chemistry
Texas Tech University
Lubbock, TX 79409

INTRODUCTION

Virtually all aneural organisms exhibit varied responses to UV
and/or visible light. The spectral sensitivity of an organism is,
of course, determined by the absorption spectral characteristics of
photoreceptors and the optical bias of the organism. In this lecture,
primary molecular events in photoreception are defined to include the
perception of spectral/actinic quanta (absorption) by the photoreceptor chromophores of the organism and the resulting excitation and
relaxation processes including the primary reactions, i.e. signal
generating reactions, of the photoreceptors, as shown in Scheme 1
for photosensory transduction.

The absorption spectroscopy of photoreceptors is described in
terms of linear and cyclic conjugated chromophores, and specific
examples to illustrate the primary molecular events in selected aneural organisms, particularly dinoflagellates, Stentor coeruleus and
phytochrome-responsive algae such as Mougeotia, are discussed in this
lecture.

WHY BLOOD IS RED AND GRASS IS GREEN? or WHY CAROTENE IS YELLOW AND PHYTOCHROME IS BLUE?

Blue light is absorbed by both the linear chromophoric photoreceptors such as rhodopsin, bacteriorhodopsin and carotenoids, and
the cyclic chromophoric photoreceptors such as chlorophylls, phytochrome, phycocyanins, stentorin and flavins. The absorption spectral
characteristics of linear chromophores include strong intensity at
the first absorption band (A $\rightarrow$ B transition in the Platt notation

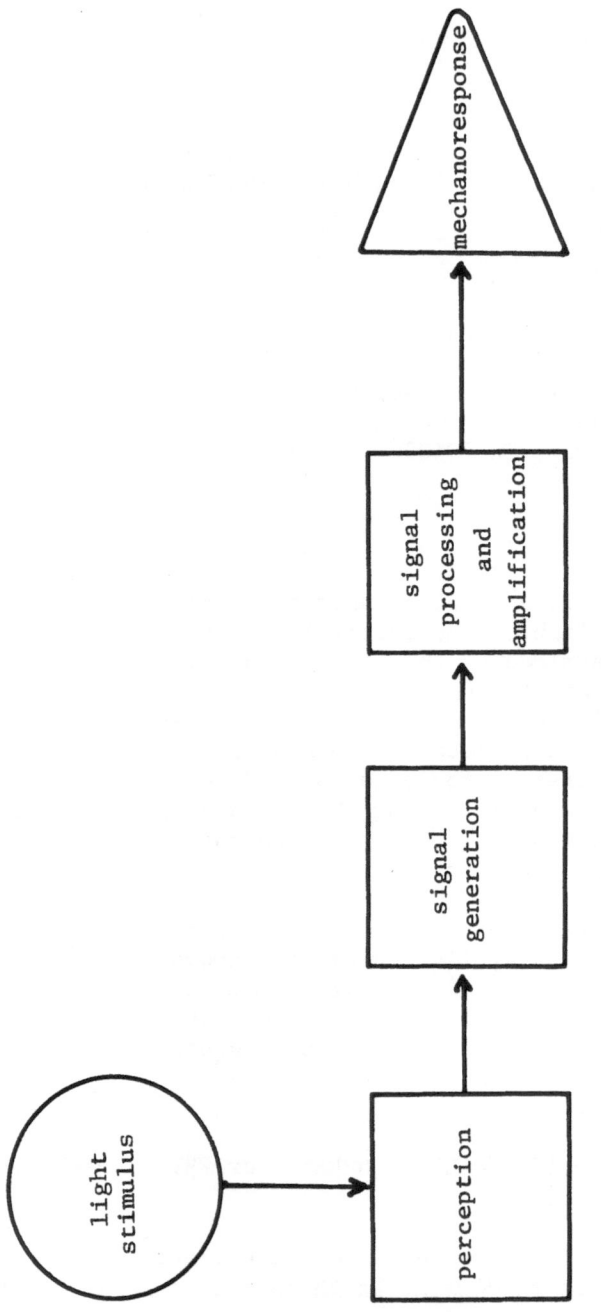

Scheme 1.

and Bu ← Ag in the C_{2h} symmetry notation), with considerably weaker absorption at shorter wavelengths (cf. the absorption spectra of β-carotene and all-<u>trans</u> retinal). Qualitatively speaking, cyclic chromophores exhibit relatively weak absorption at the first absorption band (A → $Q_{x,y}$ transition in the Platt notation), with stronger absorption at shorter wavelengths, including the so-called Soret band of porphyrins and chlorophylls in the near UV and blue region.

The above spectral "selection rules" for linear and cyclic conjugated chromophores, respectively, can be explained with the aid of "four-orbital" models in which butadiene serves as a model for all linear polyenes and benzene as a model for all cyclic polyenes (for review, see Song, 1983a). These models are schematically shown in Scheme 2.

The corresponding four-orbital model diagram for porphyrins with 18 π-electrons (18 = 4n + 2, where n = 4, to satisfy the aromaticity rule) is shown in Scheme 3. Note that the orbital angular momentum quantum numbers q = ± 4 and ± 5 are for the highest occupied and the lowest vacant orbitals, respectively.

For a linear conjugated polyene, change in orbital quantum number, Δn = odd transitions are allowed, with a strong dipole-allowed transition for the first absorption band (Δn = 1). The forbidden (thus considerably weaker in extinction than the first absorption band) transitions have Δn = even. For example, all-<u>trans</u> β-carotene shows a strong A → B band (Δn = 1) at 450 nm, whereas A → C band (Δn = 2) is barely resolvable in the near UV region. However, the

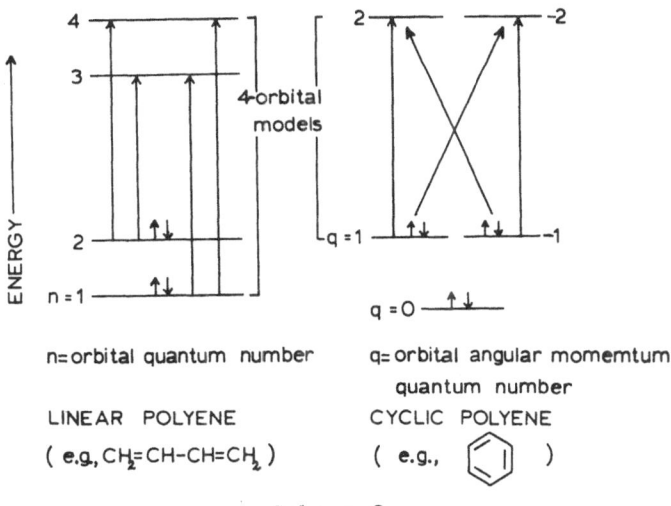

Scheme 2.

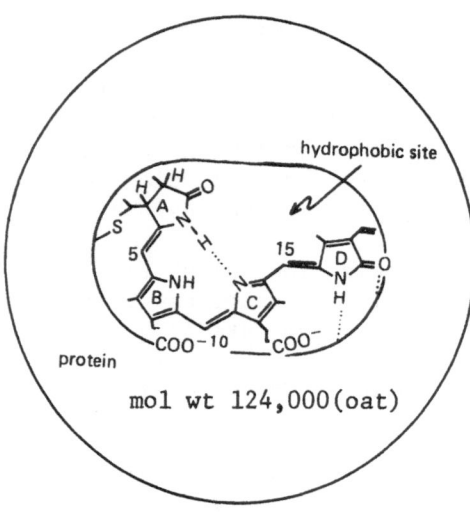

Scheme 3.

latter does get enhanced in intensity upon the loss of the center
of symmetry, thus becoming strongly dipole-allowed in, for example,
15,15'-cis β-carotene. Such a cis-enhanced band is called a "cis
peak" which occurs at 350 nm.

It is instructive to note that "bending" of the π-electron framework by trans → cis isomerization in β-carotene causes weakening and
strengthening of the first and second ("cis peak") absorption bands,
respectively. In fact, a further geometric isomerization of the
polyene chain into a cyclic or square shaped configuration will transform the typical absorption spectrum of a linear polyene (i.e. strong
first and weak second absorption bands) into a typical cyclic polyene
spectrum (i.e. weak first and strong second bands). For example, in
porphyrins, the red absorption band is due to $\Delta q = \pm 9$ transitions,
whereas the Soret band arises mainly from $\Delta q = \pm 1$ transitions (see
Song, 1983a, for review). It therefore follows from consideration of
the angular momentum conservation that the latter with the least angular momentum changes is stronger in extinction than the former. Both
calculations and qualitative predictions (cf. Hund's rule) also put
the $\Delta q = \pm 1$ transitions at a shorter wavelength region than the
$\Delta q = \pm 9$ transitions. For a thorough review of this topic, the reader
is referred to a review by the lecturer (Song, 1977, 1983a).

The above observations are qualitatively valid in describing the
absorption spectra of any linear and cyclic chromophores of photobiological importance, although quantitatively, caution must be exercised
in identifying a photoreceptor chromophore from the shape of an action
spectrum in terms of either a polyene-like or porphyrin-like molecule.

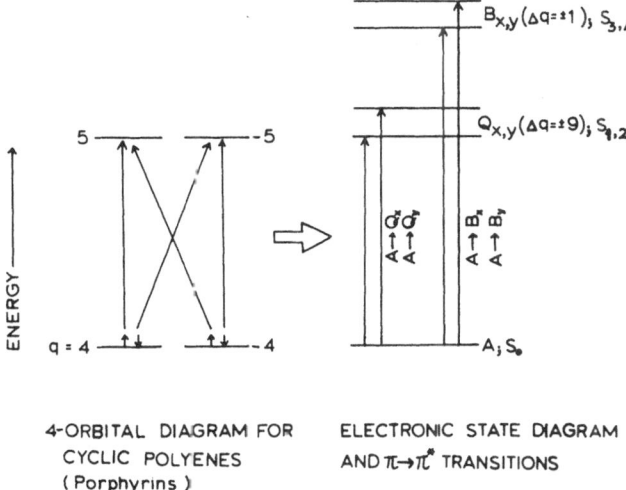

4-ORBITAL DIAGRAM FOR   ELECTRONIC STATE DIAGRAM
CYCLIC POLYENES   AND $\pi \rightarrow \pi^*$ TRANSITIONS
(Porphyrins)

Fig. 1.   A possible semi-circular conformation of the phytochrome
chromophore in its Pr form.

As an example of the utility of the above observation, phytochrome (both Pr and Pfr forms) with an open-tetrapyrrole chromophore
has been analyzed on the basis of dependence of the absorption spectral shape upon tetrapyrrole conformations (Song et al., 1979; Song,
1983a, 1984a,b). The observed absorption spectra of phytochrome are,
thus, best described by a semi-circular conformation of the chromophore, as shown in Fig. 1, which accounts for an approximate ratio
of unity for the red ($Q_{x,y}$) to Soret ($B_{x,y}$) oscillator strengths.

Similar descriptions can be extended to the absorption spectra
of other photoreceptor chromophores such as tryptophan, flavins,
chlorophylls and retinals (see Song, 1977, 1983a, for reviews).

The description of the genealogy of the excited states of linear
vs. cyclic molecules is useful for characterizing the nature of the
radiative and radiationless processes in terms of polarization direction, charge distribution, and photochemical reactivity, etc.
involved in the photoreception and its primary photoreactions. The
next section discusses the role of the lowest excited states of photoreceptor molecules in photosensory transduction.

WHY MUST PRIMARY PHOTOPROCESSES BE ULTRAFAST?

One of the essential prerequisites for the efficient operation
of a photobiological transduction process in aneural organisms is
that the photoreceptor molecule in its excited state undergoes ultrafast primary reaction(s) that lead to the sensory transduction chain.

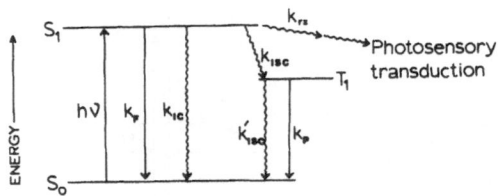

Jablonski energy level diagram
k_F ; rate constant for fluorescence, $10^8 \sim 10^9$ s^{-1}
k_{IC} ; rate constant for internal conversion, $10^7 \sim 10^{10}$ s^{-1}
k_{ISC} ; rate constant for intersystem crossing, $10^6 \sim 10^8$ s^{-1}
k_P ; rate constant for phosphorescence, $10^{-2} \sim 10^3$ s^{-1}
S_0 ; ground state
S_1 ; lowest singlet excited state
T_1 ; lowest triplet excited state

Fig. 2.   The Jablonski energy level diagram for a photoreceptor
chromophore. Various relaxation processes and their approximate values of rate constants are given.

Such primary processes, $\underline{k}_{rx}$ in Fig. 2., must be fash enough to effectively compete with other relaxation modes of the excited states.
This would ensure the high photosensitivity of photobiologically responsive organisms, namely

$$\phi_{rx} = \underline{k}_{rx}/(\underline{k}_F + \underline{k}_{IC} + \underline{k}_{ISC} + \underline{k}_{rx}) \cong 1$$

One can also argue from the above diagram (Fig. 2) that primary
photoreactions with rate constants $\underline{k}_{rx}$'s are likely to originate from
the singlet excited state, rather than from the triplet state, since
efficiency of the triplet state pathway is limited by the relatively
sluggish intersystem crossing process. This point has already been
discussed previously (Song, 1980), and evidence for the singlet state
pathway is reasonably well established in the primary photoprocess
of chlorophylls in photosynthesis (see review by Knaff, 1983), rhodopsin in vision (see review by Applebury, 1983), phytochrome in
photomorphogenesis (see review by Song, 1983b), stentorin in photomovements of Stentor coeruleus (see review by Song, 1983b) and flavins
in phototropism (Vierstra et al., 1981).

WHAT ARE PRIMARY PROCESSES OF PHOTORECEPTORS?

There are probably many different mechanisms of primary photoprocesses available from the excited states of photobiological receptor molecules. In photosynthesis, it is well established that the
primary reaction is a photoionization (see review by Parson, 1983)

$$Chl \cdot A \xrightarrow{h\nu} Chl^* \cdot A \xrightarrow{k_{rx}} Chl^{+ \cdot} A^{- \cdot}$$

where A is an acceptor molecule, namely, pheophytin in the case of the higher plant reaction center (Photosystem II).

As Dr. Chabre discusses in this lecture volume, the singlet excited rhodopsin undergoes an ultrafast conformation/ or geometric isomerism of the chromophore which is an opsin-linked, protonated cis retinal Schiff's base (see also review by Applebury, 1983).

Rhodopsin ⟶ Rhodopsin* ⟶ Bathorhodopsin,

where bathorhodopsin is the primary reaction product. The similar mechanism apparently operates in bacteriorhodopsin of Halobacterium halobium, as discussed by Dr. Spudich in this volume. These primary photoprocesses all occur within the ps time scale.

In less well-established cases of blue light responses such as phototropism, arguments have been presented for flavin as the likely photoreceptor chromophore, as opposed to carotenoids (Song and Moore, 1974; Song, 1984c). With flavins as the photoreceptor, it is most likely that the primary photoprocess is an electron transfer from an as-yet unidentified donor to the singlet excited state of flavins. Briggs and his coworkers have accumulated evidence for the possible involvement of the photoinduced electron transfer reactions between the excited flavin and b-type cytochrome in corn coleoptile photo-tropism (see review by Senger and Briggs, 1981). Earlier, Muñoz and Butler (1975) have demonstrated in a model system that a cytochrome can be photoreduced by an electron donor in the presence of flavins.

A highly speculative mechanism for the "flavin-sensitized" electron transfer as a primary photoreaction in blue-light-induced photomovement of bacteria has been proposed (Song, 1981). Unfortunately, the identity and the primary photochemical mechanism of blue light receptors remain to be established, and it is perhaps premature to conclude that photoinduced electron transfer is the primary photoprocess in all flavin photoreceptors.

In the next section, we will specifically examine the two photoreceptor systems of the lecturer's main interest, namely, stentorin of Stentor coeruleus, a unicellular ciliate, and phytochrome of plants. Here, too, the mechanisms of primary photoprocesses have not been definitely established, but it may be useful to review the information currently available regarding the possible mechanisms of the primary photoprocesses involved.

HOW DOES STENTOR PERCEIVE LIGHT STIMULUS?

The ciliary beating of Stentor coeruleus momentarily reverses from a clockwise to a counterclockwise direction in response to red light stimuli. The cell stops and then turns away from higher intensity light traps, i.e. a step-up photophobic response. In addition, the organism is able to orient itself with respect to the direction of light propagation, exhibiting negative phototaxis (see Song, 1983b, for review).

On the basis of specific inhibitory effects of protonophores and the photochemical properties of the stentorin chromophore (e.g. increase in acidity of the chromophore hydroxyl protons in the excited state, relative to the ground state), proton dissociation from the excited singlet state of stentorin has been postulated as a primary photoreaction which leads to the generation of a pH gradient across the cell membrane as an initial transduction signal (see review by Song, 1983b).

In addition to the inhibitory effects of protonophores on the photomovement of S. coeruleus noted above, light-induced membrane potentials are also quenched by the same protonophores (Song et al., 1980a, b; Song, 1981, 1983b). It is, therefore, not surprising that the photoresponse of the organism exhibits a sharp pH profile, particularly in the pH range of 5-6 (Walker et al., 1981). Although transient pH changes upon actinic stimulation of the cell have not yet been recorded, the intracellular pH does seem to decline under steady excitation of the organism with actinic wavelength light (Walker et al., 1981).

Since any isolated photoreceptor molecules are not readily assayable for their photobiological activity, it is generally difficult to demonstrate the primary photochemical reaction(s) which lead to the sensory transduction chain. Stentorin is no exception. In an attempt to ascertain whether or not proton dissociation from the excited state of stentorin serves as the primary photoprocess in Stentor coeruleus, the isolated photoreceptor proteins were examined spectrofluorometrically for their anion formation in the excited state. Both fluorescence emission spectra and ps lifetime measurements suggested that the red fluorescence of stentorin contains a significant contribution from the anionic species apparently generated during the excited state lifetime (Song et al., 1981b). An in vivo fluorescence of the organism also reflected a strong contribution of the anionic species, upon front-surface excitation of the cell suspension in a triangular cuvette (Walker, 1980; see review by Song, 1983b). However, a quantitative estimation of proton dissociation from the excited state of stentorin in vivo and in vitro depends on the angle of excitation and detection of the fluorescence between the incident beam and the surface plane of the sample cuvette (J.W. Huh and P.-S. Song, unpublished results).

The chromophores of stentorin are apparently buried within the interior space of the large oligomers, as probed by the lack of fluorescence quenching of the oligomers by potassium iodide (Song, 1981). Proton dissociation from the excited singlet state (S-OH* in Fig. 3) then proceeds within 10 ps (Song et al., 1981b). It is

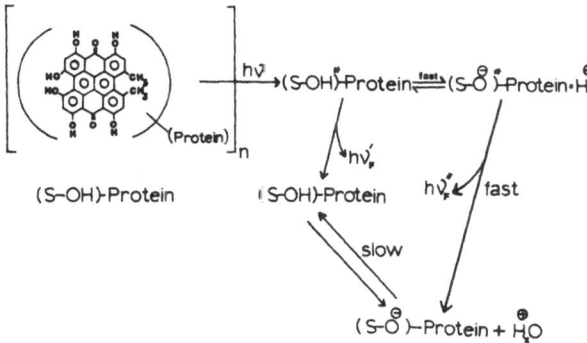

Fig. 3. Proton dissociation from the excited state of stentorin
(S-OH). See text for details.

thus possible that the oligomeric association of stentorin facilitates
the ultrafast proton transfer, via an acid-base network of amino
acid residues of the apoproteins, analogous to the charge relay network of serine proteases (Blow and Steitz, 1970). A transient proton
concentration outside of the protein oligomer is then determined by
the ratio of proton dissociation from the excited state to the sum of
proton reassociation rates of the excited and ground states of the
anionic chromophore ($S-O^-$) (Song, 1983b). In other words, the net
proton accumulation outside of the protein oligomer is inversely dependent on the rate of proton reassociation, as expected. In this
hypothetical scheme, it is assumed that a rapid conformational change
of acid-base residues, triggered by the negative charge generated by
proton dissociation in the hydrophobic core of the oligomeric protein
assembly, decreases the rate of proton reassociation in the ground
state of anionic chromophore.

Assuming that the proton dissociation from the excited stentorin
is the primary photoprocess, a working hypothesis for the photosensory transduction chain has been proposed for Stentor coeruleus
(Song, 1983b). Thus, a light-induced proton ejection from pigment
granules produces a transient pH drop in the cytosol, possibly contributing to a depolarizing receptor potential.

It must be emphasized, however, that neither the primary photoprocess nor the nature of the resulting, initial signal can be considered to be firmly established. It can only be established after
further experiments. However, the proposed mechanisms described here
serve to illustrate how proton dissociation can play an important
role in future studies.

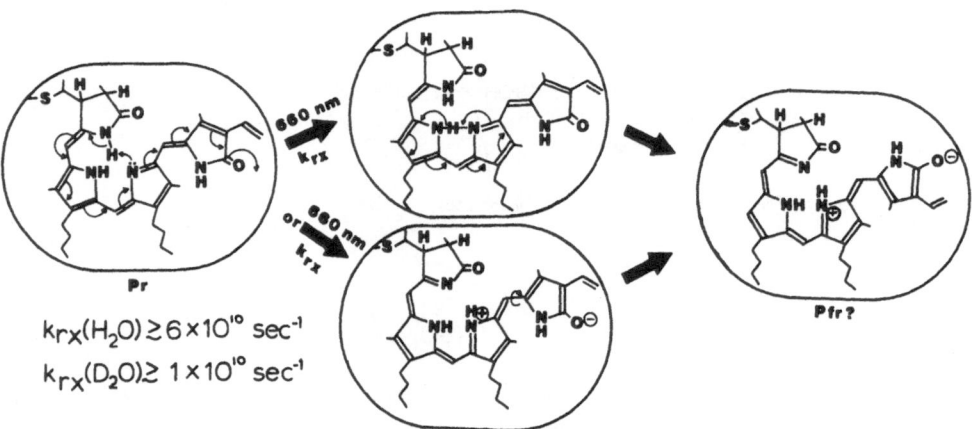

$$k_{rx}(H_2O) \gtrsim 6 \times 10^{10} \ sec^{-1}$$
$$k_{rx}(D_2O) \gtrsim 1 \times 10^{10} \ sec^{-1}$$

Fig. 4.   An intramolecular proton transfer within the central "core"
          of the tetrapyrrolic chromophore of phytochrome in the ex-
          cited singlet state. The Pfr chromophore is proposed to
          possess a semi-circular conformation of the π-electron
          network, with a configurational isomerism of the 15-16-
          bond of the Pr chromophore.

IS PROTON TRANSFER IMPORTANT IN THE PRIMARY PHOTOPROCESS OF
PHYTOCHROME?

    Phytochrome, the receptor molecule for photomorphogenesis in
plants, contains a tetrapyrrolic chromophore (see Fig. 4). The Pr
form is transformed to the physiologically active Pfr form by red
light of 660 nm. Unfortunately, the chemical structure of the latter
is not known. Based on NMR studies of the Pr- and Pfr-chromopeptides,
the primary photoprocess has been proposed to be a configurational
isomerism around the 15-16 double bond (Rüdiger et al., 1983).

    The primary photoprocess in the Pr → Pfr phototransformation of
phytochrome is ultrafast; an intermediate absorbing at 694-696 nm
can be produced after a few s of red light irradiation of Pr at 77 K
(Kendrick and Spruit, 1977; Song et al., 1981a). However, it has not
been possible to determine a deuterium isotope effect on the rate of
the primary photoprocess of phytochrome in D_2O under steady state
conditions (Song et al., 1981a). Thus, it is not possible to ascertain whether or not proton transfer is involved in the production of
the 694-696 nm intermediate. However, there is a significant deuterium isotope effect on the fluorescence lifetime of phytochrome
(Pr form), suggesting that intramolecular proton tunneling between
the tetrapyrrolic nitrogens occurs in the excited state of the chromophore (Sarkar, 1983).

If we assume that the primary photoprocess of phytochrome involves a proton transfer in the formation of the 694-696 nm intermediate at 77 K, an order of magnitude estimates of the rate constants
in water and in D_2O can be given as shown in Fig. 4. Accompanying
the lengthening of the fluorescence lifetime, the fluorescence quantum yield of phytochrome (Pr form) also increases in D_2O. Thus, the
deuterium isotope effects can be explained in terms of a slower tun-
neling/transfer of deuteron from one position [e.g. NH(D) in Ring A]
to another (e.g. Ring C or basic amino acid group).

As mentioned above, the chromophore of phytochrome undergoes a
configurational isomerization at the bridge double bond between
Ring C and Ring D. A question arises as to whether proton transfer
precedes or follows the isomerization. It is interesting to note
that ultrafast isomerization of tetrapyrroles such as bilirubin-
protein complexes occurs at room temperature, but this reaction is
effectively shut off at 77 K (Lamola, 1984). If the Pr form of
phytochrome behaves like bilirubin-protein complexes, it is then
likely that proton transfer precedes isomerization. Figure 4 illustrates how proton transfer facilitates the isomerization at the
double bond by converting it to a single bond.

CONCLUDING REMARKS

From this lecture, it is obvious that our knowledge of the
nature of the primary photoprocesses of photoreceptor molecules is
extremely limited, except for rhodopsins and chlorophylls. One of
the major difficulties in describing the primary photoprocesses of
photobiological receptors is, of course, the fact that the identities
of many of the photoreceptor chromophores are not established. Even
in those cases where the chromophores are relatively well established,
their primary photoprocesses largely remain to be elucidated. In
this lecture, the basic concepts of light absorption and the possible
role of proton transfer have been qualitatively treated, with the
hope that the student looks at an action spectrum for a photobiological response of organisms with some basis of inquisition into the
mechanism of photoreception at a molecular level.

ACKNOWLEDGEMENTS

This work was supported in part by the Robert A. Welch Foundation
(D-182), the National Science Foundation (PCM8119907) and the National
Institutes of Health (NS15426). The work on phytochrome described in
this paper was also supported by a NATO grant for International
Collaboration in Research (No. 545/83) with Prof. W. Rüdiger at the
University of Munich. I also thank Mr. Rabi K. Prusti for his
drawing of the original Figures in this chapter.

REFERENCES

Applebury, M. L., 1983, Light and the primary molecular processes
     of vision, in: "Topics in Photobiology," H. O. Kim, and P.-S.
     Song, eds., Jeju National University, Jeju, pp. 195-205.
Blow, D. M., and Steitz, T. A., 1970, X-ray diffraction studies of
     enzymes, Annu. Rev. Biochem., 39:63-100.
Kendrick, R. E., and Spruit, C. J. P., 1977, Phototransformation of
     phytochrome, Photochem. Photobiol., 26:201-214.
Knaff, D. B., 1983, Recent advances in photosynthetic electron trans-
     port, in: "Topics in Photobiology," H. O. Kim, and P.-S. Song,
     eds., Jeju National University, Jeju, pp. 209-221.
Lamola, A. A., 1984, The effects of environment on photophysical
     processes of bilirubin, in: "Optical Properties and Structure
     of tetrapyrroles," G. Blauer and H. Sund, eds., W. de Gruyter,
     Berlin, in press.
Muñoz, V., and Butler, W. L., 1975, Photoreceptor pigment for blue
     light response in Neurospora crassa, Plant Physiol., 55:421-426.
Parson, W. W., 1983, Photosynthesis, in: "Biochemistry," G. Zubay,
     ed., Addison-Wesley, Reading, MA, pp. 409-435.
Rüdiger, W., Thümmler, F., Cmiel, E., and Schneider, S., 1983, Chromo-
     phore structure of the physiologically active form (Pfr) of
     phytochrome, Proc. Natl. Acad. Sci. USA, 80:6244-6248.
Sarkar, H. K., 1983, Hydrodynamic properties and molecular topography
     of phytochrome, Ph.D. dissertation, Texas Tech University,
     Lubbock, TX.
Senger, H., and Briggs, W. R., 1981, The blue light receptor(s):
     Primary reactions and subsequent metabolic changes, Photochem.
     Photobiol. Rev., 6:1-38.
Song, P.-S., 1977, UV-visible spectroscopy of bioorganic molecules,
     in: Annu. Reports (B), G. P. Moss, ed., The Chemical Society,
     London, 74:18-44.
Song, P.-S., 1980, Primary photophysical and photochemical reactions:
     Theoretical background and general introduction, in: "Photore-
     ception and Sensory Transduction in Aneural Organisms," F. Lenci
     and G. Colombetti, eds., Plenum Press, New York, pp. 189-210.
Song, P.-S., 1981, Photosensory transduction in Stentor coeruleus and
     Related Organisms, Biochim. Biophys. Acta, 639:1-29.
Song, P.-S., 1983a, The electronic spectroscopy of photoreceptors
     (other than rhodopsin), Photochem. Photobiol. Rev., 7:77-139.
Song, P.-S., 1983b, Protozoan and related photoreceptors: Molecular
     aspects, Annu. Rev. Biophys. Bioengin., 12:35-68.
Song, P.-S., 1984a, Phytochrome, in: "Adv. Plant Physiology," M. B.
     Wilkins, ed., Pitman Books, London, pp. 354-379.
Song, P.-S., 1984b, The molecular model of phytochrome deduced from
     optical probes, in: "Optical Properties and Structure of Tetra-
     pyrroles," G. Blauer and H. Sund, eds., W. de Gruyter, Berlin,
     in press.

Song, P.-S., 1984c, Photophysical aspects of blue light receptors: The old question (flavins versus carotenoids) re-examined, in: "Blue Light Effects in Biological Systems," H. Senger, ed., pp. 75-80, Springer-Verlag, Berlin.

Song, P.-S., Chae, Q., and Gardner, J. G., 1979, Spectroscopic properties and chromophore conformations of the photomorphogenic receptor: Phytochrome, Biochim. Biophys. Acta, 476:479-495.

Song, P.-S., Häder, D.-P., and Poff, K. L., 1980a, Step-up photophobic response in the ciliate Stentor coeruleus, Arch. Microbiol., 126:181-186.

Song, P.-S., Häder, D.-P., and Poff, K. L., 1980b, Phototactic orientation by the ciliate, Stentor coeruleus, Photochem. Photobiol., 32:781-786.

Song, P.-S., Kim, I. S., and Poff, K. L., 1981a, Primary photoprocesses of undegraded phytochrome excited with red and blue light at 77°K, Biochem. Biophys. Acta, 635:369-382.

Song, P.-S., Walker, E. B., Auerbach, R. A., and Robinson, G. W., 1981b, Proton release form Stentor photoreceptors in the excited states, Biophys. L., 35:551-555.

Song, P.-S., and Moore, T. A., 1974, On the photoreceptor pigment for phototropism and phototaxis: Is a carotenoid the most likely candidate?, Photochem. Photobiol., 19:435-441.

Vierstra, R. D., Poff, K. L., Walker, E. B., and Song, P.-S., 1981, Effect of xenon on the excited states of phototropic receptor flavin in corn seedling, Plant Physiol., 67:996-998.

Walker, E. B., 1980, Photosensory transduction in Stentor coeruleus, Ph.D. dissertation, Texas Tech University, Lubbock, TX.

Walker, E. B., Yoon, M., and Song, P.-S., 1981, The pH dependence of photosensory responses in Stentor coeruleus and model system, Biochim. Biophys. Acta, 634:289-308.

METHODS FOR THE MICROSPECTROSCOPIC INVESTIGATION OF

PHOTORECEPTIVE STRUCTURES

P. A. Benedetti

Istituto di Biofisica
Consiglio Nazionale delle Ricerche
Pisa, Italy 56100

The investigation of functional and distributional properties
of cellular structures is often based on extractive procedures,
followed by biochemical or spectroscopic assays performed on the
materials obtained. The techniques involved in the extraction tend
to modify, to some extent, the nature of the components under examination. The drawback is especially marked in the case of photoreceptive organelles which originally contain labile photopigments
confined in delicate microstructures. For instance, extraction
processes on phototactic protozoa have been attempted with limited
results, both due to the physiological "stress" imposed in the
process and to the difficulty in achieving a satisfactory separation
of the materials.

A relatively more sophisticated approach tending to overcome
the cited limitations is represented by the in vivo microspectroscopy, applied directly on individual organisms. The methods derive
from the association of techniques used in the spectroscopy of
macroscopical preparations, together with those of light microscopy,
and have been demonstrated largely non-destructive to the organisms
(Francon, 1961; Piller, 1977; Liebman, 1973; Von Sengbusch and
Thaer, 1973). As far as the optical properties are of interest,
the possibility exists, in this way, of performing a variety of
spectroscopic and structural studies on intact specimens.

It has to be noted, however, that a certain amount of power is
sent to the specimen with the analyzing beam, and any irradiation
process can, in principle, induce photomodifications of the
physiological state. Consequently, care must be used in order to
control such photoeffects, should they verify. In general, this
means that the irradiation doses have to be kept at a minimum, and

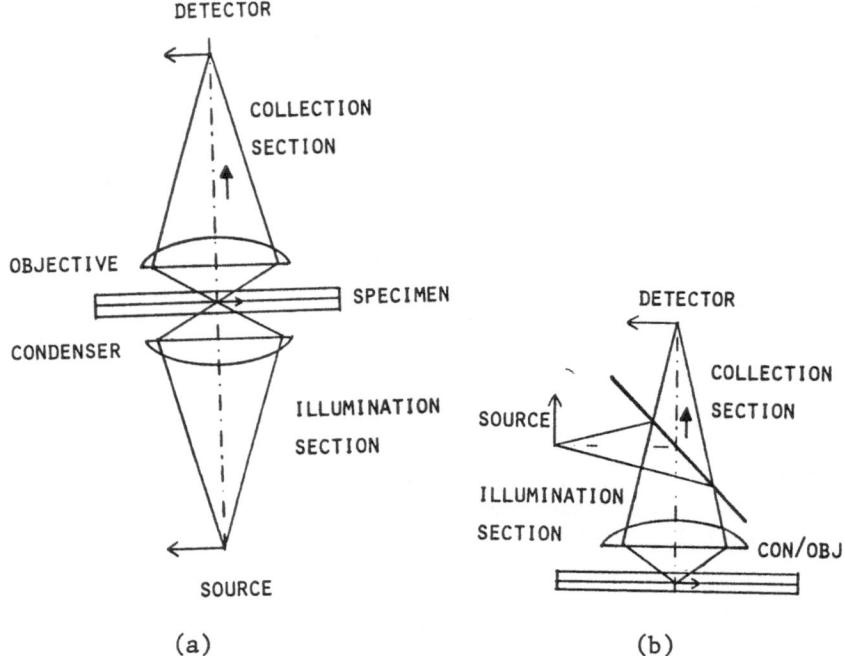

Fig. 1. Optical diagrams typical in microspectroscopy.
         (a) Microphotometry and <u>trans</u> illumination microfluorometry;
         (b) Epi-illumination microfluorometry.

low-noise techniques are to be employed to provide acceptable accuracy in the measured data.

    Moreover, it is worth noting that the characteristics of com-
mercially available microphotometers are not adequate for most photopigment investigations; therefore, the majority of the work, in
this field, has been carried out, in the past, with the aid of
systems specifically designed and constructed in the laboratory
(e.g. Liebman, 1973; Mac, 1964; Wolken, 1968). On the other hand,
even if suitable instruments are scarce, the advances of modern
technology, especially in the fields of optoelectronics and automatic computers, are to the advantage of those willing to undertake
the construction of systems for wide purpose microspectroscopy,
including photoreceptor and photopigment analysis in single living
cells.

GENERAL REQUIREMENTS

    Maximally relevant to the field of photoreceptor studies are
the absorption micro(spectro)photometry and the emission or excitation micro(spectro)fluorometry. Moreover, since the spatial

distribution of optically differentiated substances or the temporal changes are often of interest, provisions are generally introduced capable of furnishing spatial descriptions of the specimen, as well as recordings of the evolution in time of the optical characteristics.

The optomechanical arrangement of an instrument designed for microspectroscopy can be ideally viewed by consisting of three principal components: the illumination section, the specimen and the detection section. The convenient specialization of the cited components, in a particular measurement, and the way of treating the gathered data, serve to afford the performance desired in a variety of spectroscopic investigations.

A Classical Example: Simple Microdensitometry

Before looking more explicitly at the details of the techniques, let us examine the relatively simplified problem of the punctual, single-wavelength, absorption microphotometry or microdensitometry (Fig. 1a). With the aid of the illumination section, serving to concentrate a spot of light onto the microscopical preparation, the radiation transmitted is measured by means of a detector included in the light collection section. If two measurements are made, pointing into the specimen and successively into a blank outer zone, following Lambert-Beer's law, absorbance A, regarding the illuminated portion of the specimen and the wavelength employed, is known as

$$A = \log (I/T)$$

where I and T, respectively, represent the illumination and the transmitted intensities.

The basic arrangement described is not far removed from those employed in more complicated measurements. In fact, (1) Spectral measurements can be obtained introducing filtering components along the optical path, (2) Spatial distribution studies can be performed with the aid of techniques for specimen displacement or beam displacement or deflection at some level in the optics, (3) Fluorescence studies can be carried out if convenient wavelength separation parts are introduced in the illumination and collection sections which can eventually correspond, in part, in the epi-illumination configuration.

OPTICAL ASPECTS

Let us now analyze the optical components in more detail, together with the principles involved in the measurements. As previously introduced, we distinguish:

(1)   The illumination section
      (a) Source
      (b) Intermediate optics
      (c) Condenser

(2)   The specimen

(3)   The collection section
      (a) Objective
      (b) Intermediate optics
      (c) Detector

Absorption Microspectroscopy

In the case just described regarding absorption spectroscopy
(Fig. 1a), the illumination section serves to illuminate a selected
portion of the specimen, using a radiation of desired spectral properties. This section acts as an "inverted" microscope, in the
sense that a macroscopic image due to the source is formed after
demagnification onto the specimen plane, with the aid of a convenient intermediate optics and condenser. The combined power of the
lenses corresponds to the demagnification factor. Actually, factors
of the order of 1000 are realizable and spots having effective diameters down to the order of the wavelength can be formed, when
operating near the diffraction limit. Different illumination methods
can be employed. The Kohler method is preferred for low stray
light; "critical" illumination, on the other hand, is more efficient
in terms of power and directly realizable by imaging a compact arc.

The specimen is usually mounted between thin transparent slides.
Oil immersion can be employed and provisions for centering the
specimen should be introduced in the supporting stage in order to
select the portions to be analyzed. Focus depth is reduced, in
high resolution work, due to the large angular aperture of illumination to be employed.

The collection section closely resembles that of a conventional
microscope in that the light coming from the specimen is sent to the
photodetector instead of reaching the eye of an observer. Intermediate optical parts serve to image the rear focal plane of the
objective onto the detector. If the beam is maintained stationary
during the measurement, spatial filtering, before detection, can be
employed with the purpose of matching the sensitive area to that of
the illuminated zone of the specimen for improved rejection of the
stray light.

Fluorescence Microspectroscopy

Methods for fluorescence measurements are essentially based
on one of the two following configurations:

(1) Trans-illumination
(2) Epi-illumination

Whilst in the first case, apart from the convenient introduction of spectrally selective components, the opto-mechanical set-up resembles the one described for absorption microphotometry (Fig. 1a), in the second case (Fig. 1b), part of the optics is common to both the illumination and the detection sections, and the source and the detector are placed in conjugate positions, with respect to a dichroic mirror serving to separate the incident and the emitted radiation. Spatial filtering considerations equally apply as above, even if some simplification is connected to the fact that most of the optics serves to accomplish both the illumination and the light collection tasks.

SPECTRAL MEASUREMENTS

Studies regarding how the optical characteristics depend on the wavelength are fundamental in spectroscopy. In the actual case of microscopic spectroscopy, fairly familiar methods of wavelength selection are employed, i.e. filters and monochromators. However, some specificity exists in the way of choosing and matching their characteristics in ways suitable for the particular application. First, the light efficiency is of concern. Most biological and markedly photopigment spectral shapes are relatively broad, so as not to require narrow bandwidths in the devices employed for wavelength selection. This results in direct advantages in terms of throughput. Filters, being compact and cheaper, may be used principally in positions where fixed characteristics are sufficient or blocking properties are requested. Continuously variable filters and dispersion of diffraction monochromators, however, are necessary in the cases in which wavelength "scanning" has to be realized and spectral recordings consequently produced. Fast spectral changes and flexible selection of wavelengths in a sequence are also required in many experiments and specially designed, programmable monochromators are often useful (Benedetti and Evangelista, in preparation). Photomultiplier cooling is generally important to decrease the noise due to the dark current (Benedetti and Grassi, 1980). Finally, multichannel simultaneous detection of several wavelengths offers unique performance in experiments involving the study of faint fluorescent signals. These methods are based on the combined use of polychromators and intensified camera-tubes or solid-state photoarrays, in association with convenient control and processing techniques (Benedetti et al., 1983).

SPATIAL MEASUREMENTS

Particularly important in the investigation of structural parameters of the cells is the possibility of obtaining quantitative

spatial descriptions of the optical properties of the specimen
under examination. This requirement is covered with the aid of
techniques for positional scanning, the role of which can be distinguished into a couple of principal cases:

    (1) Test-reference scanning
    (2) Area scanning

Test-reference scanning or dual-beam techniques are at the
base of "corrected" measurements, both in the case of absorption
studies and in fluorescence. In fact, the optical density calculation would require the knowledge of source, instrumental and detector efficiency versus the wavelength. As previously described,
by Lambert-Beer's formula, this corresponds to the corrected light
measurements, in a dual-beam technique, performed by means of the
alternate positioning of the analyzing spot onto a reference (I) and
a test (T) location of the specimen. Similarly, in fluorescence
measurements, one can obtain values corrected against the background
emission or the dark signal by systematic subtraction of the reference data from the test one.

Moreover, area scanning and associated control and automatic
processing techniques give the possibility of systematically ex-
ploring a desired portion of the specimen in order to produce a
"map" or an "image" which is not actually the image of the specimen
in the optical sense, but the spatial description of a potentially
complex set of spectroscopic and distributional properties, often
removed definitely from a conventionally obtained image. To be more
concrete, let us cite two examples: (1) Mapping of the relative
abundance of some substance, as calculated by linear deconvolution,
after that the point optical densities are measured at selected
wavelengths for which the specific absorbances are known, and (2)
Spatial distribution of absorbing structures laying in a plane which
is internal to a thick specimen. This is accomplished taking advantage of the high focal discrimination intrinsic in the point
illumination of the specimen at large angular apertures

Positional-scanning Methods

The detection of positional differences and spatial distributions involves the introduction of specific devices capable of performing the cited scanning operation. Methods employed in microspectroscopy, for that purpose, play a central role in the complex
performance of the instrumentation.

Some of the units which are commercially available (Leitz,
1973; Zeiss, 1973) adopt the principle depicted in Fig. 2a. Object
scanning (or stage scanning) requires relatively straightforward
additions to the design of the microscope in order to render the
conventional instrument capable of performing quantitative

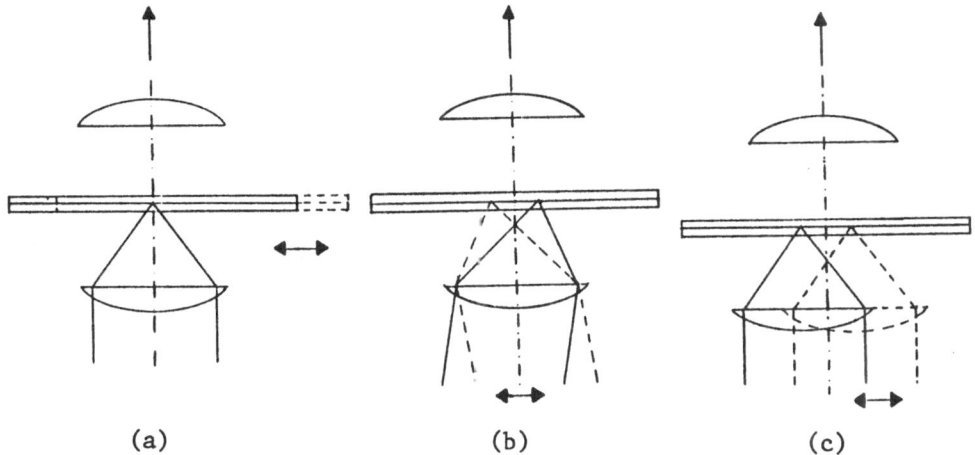

Fig. 2. Positional scanning methods:
(a) Object (stage) scanning;
(b) Source scanning;
(c) Moving-condenser scanning.

measurements. The solution is also to be considered very correct, in the optical sense, since the rays remain stationary, the sole moving optical component being the specimen itself.

Unfortunately, in practice, the method is markedly slow and also offers poor path flexibility due to the heavy masses to be displaced. For instance, test-reference positioning at reasonable speeds is unfeasible in this way, rendering the technique not adaptable to high-sensitivity photopigment work on samples of living material or in liquid suspension.

In order to overcome some of the cited limitations, some authors have introduced techniques using so-called "source-scanning" methods, in the form of dual sources (e.g. MacNichol, 1964) or beam deflection (e.g. Vickers, 1972). The optical situation is depicted in Fig. 2b which substantiates the presence of angular changes of the optical axis of illumination with respect to the plane of the specimen. The photometric uniformity is consequently affected and, although in this case the sample is held stationary, difficulties remain in the set-up of opto-mechanical solutions offering speed and path flexibility.

An alternative technique permitting both fast test-reference and area scanning while maintaining fair optical accuracy was introduced with the moving-condenser method (Benedetti et al., 1976b). In this case (Fig. 2c), the condenser is displaced onto a plane parallel to specimen one. All the optical parts except for the light-weight condensing lenses are held stationary and the accuracy is not substantially impaired with respect to the object

scanning method. A system, based on the condenser-scanning principle and currently employed in various micro-spectroscopic investigations (Brunori et al., 1974; Antonini et al., 1978; Benedetti et
al., 1980; Antonini et al., 1982) and high-resolution photostructure
analysis (Benedetti and Checcucci, 1975; Benedetti et al., 1976a;
Benedetti and Lenci, 1977; Nultsch and Benedetti, 1978; Nultsch et
al., 1983), will be described in the Appendix.

TIME MEASUREMENTS

A variety of cases exists in which the time evolution of
spectroscopic properties or positional changes have to be detected
and monitored. A detailed description of cases and techniques is
probably beyond the scope of this presentation, but since some of
the applications are of special interest in the field of photoreceptor microspectroscopy, a list of relevant examples seems
appropriate:

    Spectroscopic time-changes:
        Fluorescence signal decay
        Flash photolysis
        Bleaching experiments, etc.

    Positional time-changes:
        Photomovement of structures
        Circadian rhythms
        Protoplasm movements, etc.

The experimental methods to be employed are correspondingly
various and require the flexible configurability of the optomechanical part of the instrumentation. Moreover, specialized electronics,
laser techniques, sophisticated data processing and other advanced
methods are usually required by the increasing complexity of the
measurements.

APPENDIX: A PRACTICAL IMPLEMENTATION

With the purpose of presenting a more concrete illustration of
an actual instrument employing some of the principles introduced
above, a brief description will be given of the principal characteristics and applications related with a complete system designed,
constructed and utilized at the Istituto di Biofisica of the CNR
in Pisa. The system, which is composed of optical, mechanical,
electronic and computing parts, offers versatility and performance
adequate to carry out high spatial resolution and high sensitivity
measurements in various microspectroscopic applications within the
visible spectral range.

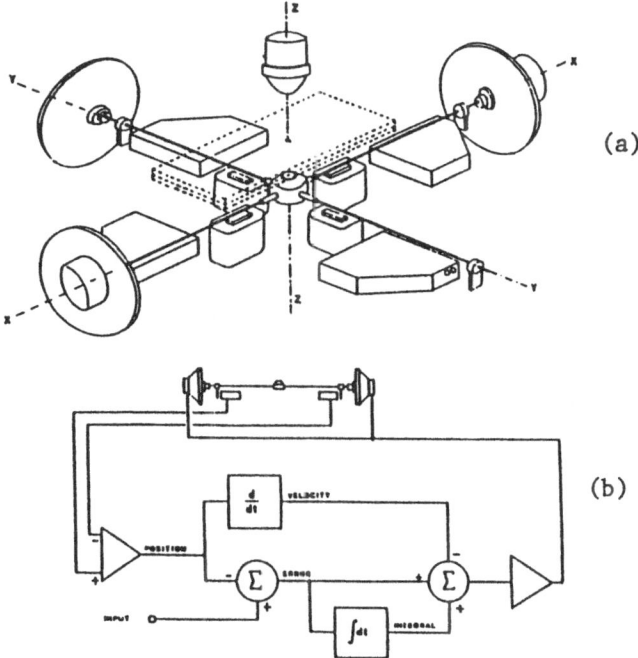

Fig. 3.   The electromechanical, moving-condenser transducer:
          (a) Set-up of the parts;
          (b) Electrical diagram (one axis.

The positional scanning is achieved by means of an electro-mechanical transducer (Fig. 3a) employing the moving-condenser principle (Benedetti et al., 1976b).   The unit is servo assisted (Fig. 3b) and covers an area of 500 x 500 micrometers square, the accuracy being better than 0.25 micrometers.   Fine focusing is also electrically driven.   The spectral measurements are guaranteed by a fast monochromator (Fig. 4), in which a power galvanometer actuates the rotation of a holographic grating.   The approximate bandwidth is 8 nm and, depending on the application, the unit nay be used for source filtering (Fig. 4a) or detector filtering (Fig. 4b).   Both the positioning and the wavelength are driven by means of voltages and the speed of response is such that arbitrary values can be addressed in times shorter than 10 ms.

As depicted in Fig. 5, the instrument can be flexibly con-figurated for measurements of different nature which also depend on the appropriate programming of the processing equipment used in association (Fig. 6).

Among the principal measuring capabilities, we mention the following:

(1) Absorption studies (Fig. 5a): Corrected spectra of the optical density, maps of transmittance, component detection, etc.

(2) Photostimulation studies (Fig. 5b): Selective irradiation of structures, photomodification, etc.

(3) Fluorescence studies, scanning (Fig. 5c) and multichannel (Fig. 5d): Emission and excitation spectra in the trans- and in the epi-illumination configuration, emission maps, etc.

(4) Transient studies of the optical and of the distributional properties during the measurements cited in the preceding points.

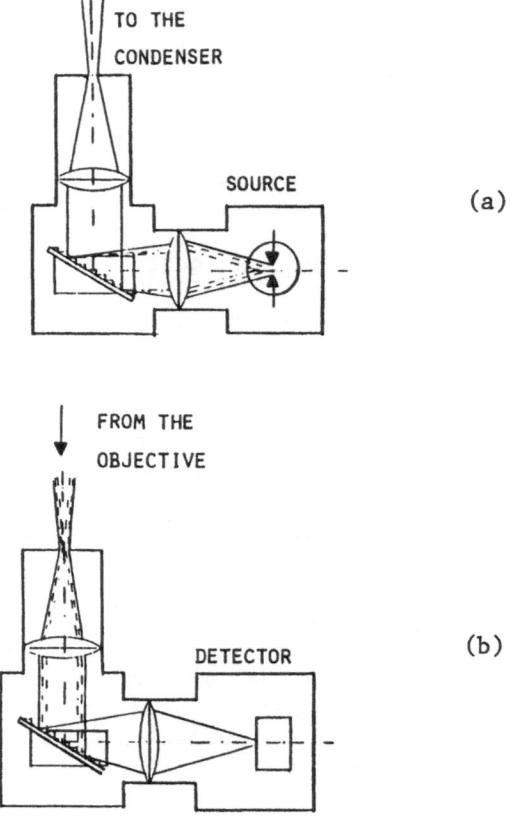

Fig. 4.   The fast-scanning galvanometer monochromator
          (a) Use in microspectrophotometry
          (b) Use in microspectrofluorometry

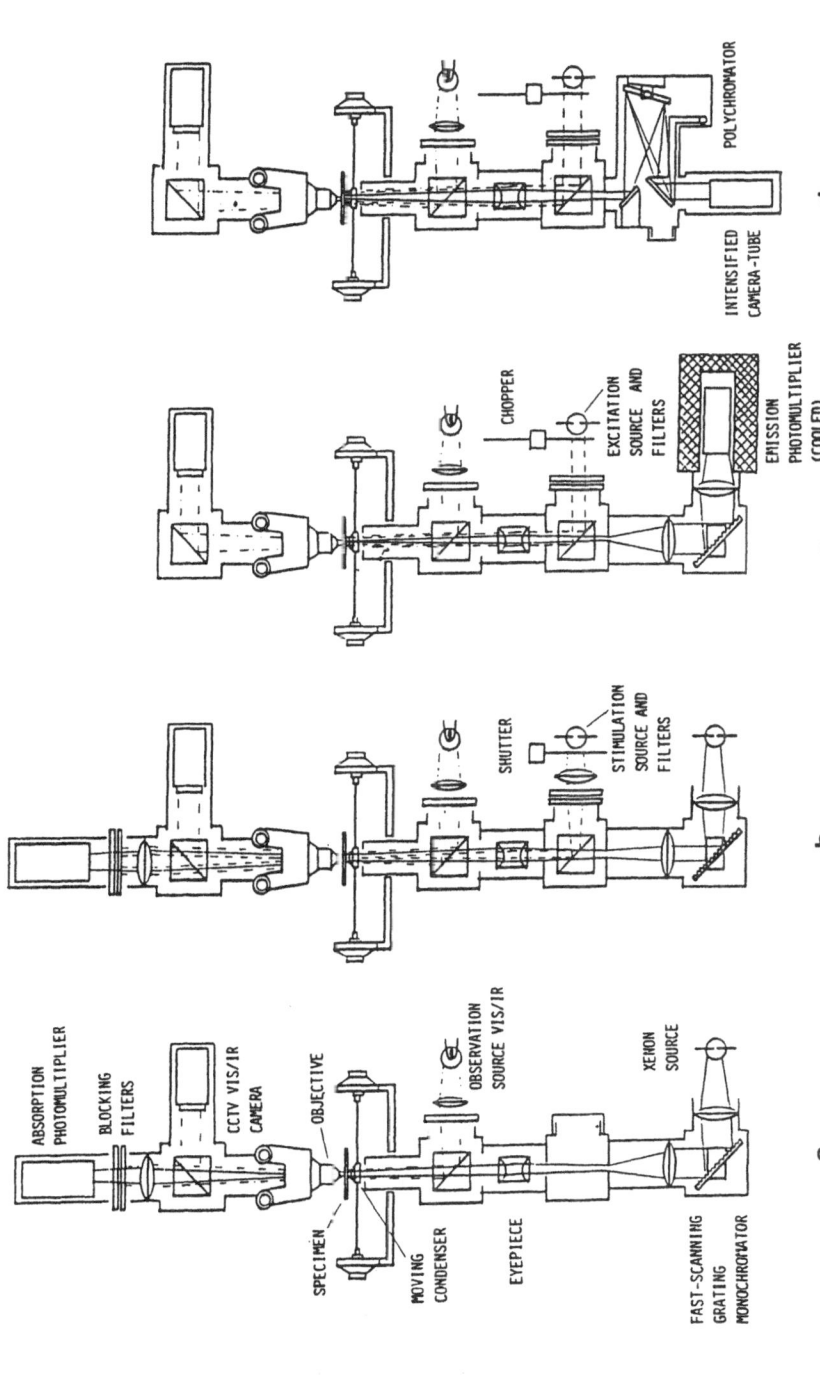

Fig. 5. Optical configurations in moving-condenser, microspectroscopic applications: (a) Micro-spectrophotometry; (b) as (a), with photostimulation; (c) epi-illumination microspectro-fluorometry; (d) as (c), with multi-channel emission detection.

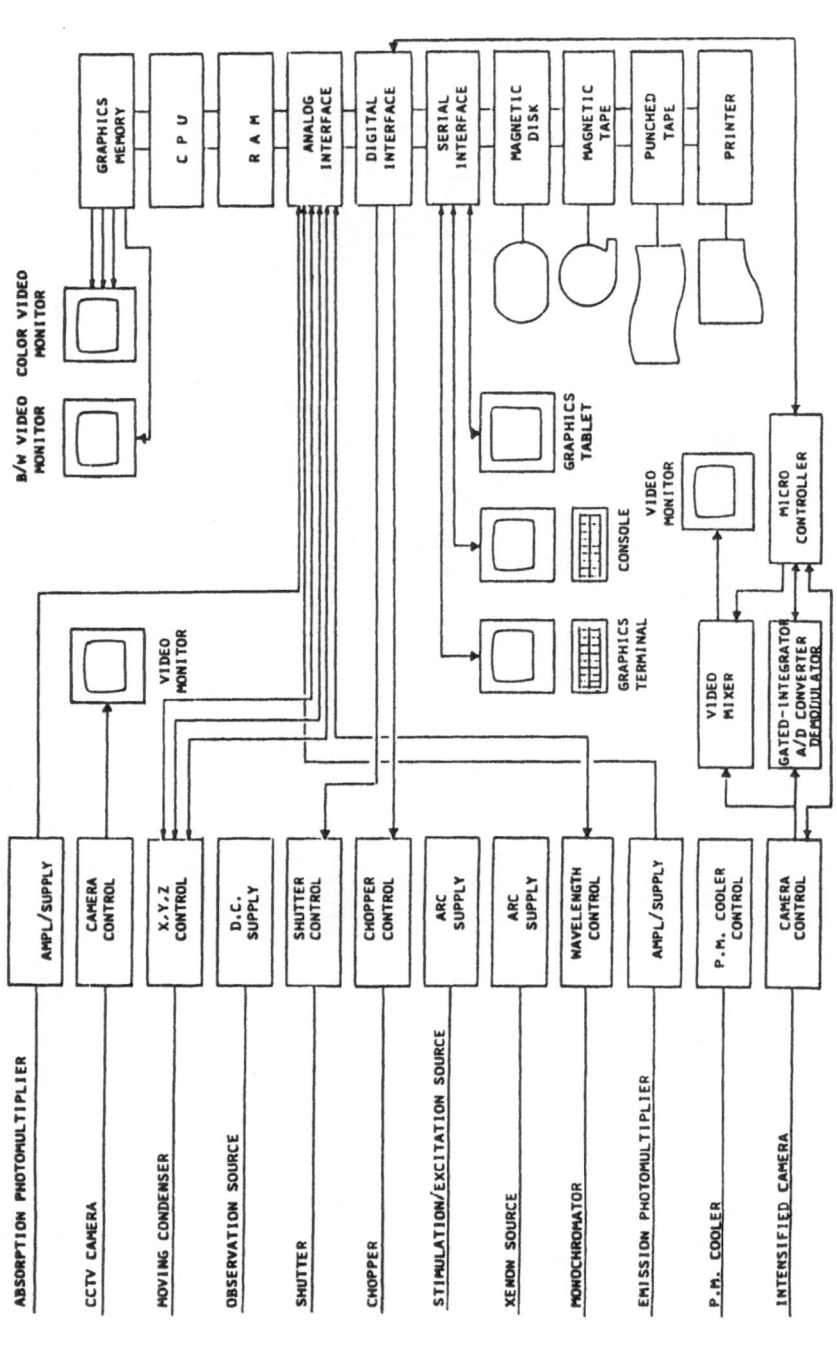

Fig. 6. Electronic and computer parts associated to the optical configurations depicted in Fig. 5.

Typical applications specifically belonging to the field of photoreception spectroscopy are:

(1) Photoreceptor pigment absorption spectra (Benedetti et al., 1976a).

(2) Photoreceptor pigment spectra of fluorescence emission (Benedetti and Lenci, 1977).

(3) Photomovement of cellular structures (Nultsch and Benedetti, 1978).

(4) Photobleaching experiments (Nultsch et al., 1983).

ACKNOWLEDGEMENTS

The author wishes to acknowledge the contributions of G. Bianchini, P. Coltelli, V. Evangelista, S. Grassi and P. Gualtieri who participated in this activity within the program "Image Micro-spectroscopy (IMSS)."

REFERENCES

Antonini, E., Brunori, M., Giardina, B., Benedetti, P. A., Bianchini, G., and Gressi, S., 1978, Single-cell observations of gas reactions and shape changes in normal and sickling erythrocytes, Biophys. J., 1001:187.

Antonini, E., Brunori, M., Giardina, B., Benedetti, P. A., Grassi, S., and Gualtieri, P., 1982, Kinetics of reactions within single erythrocytes: Studies by microspectrophotometry, Haematologia, 15(1):3.

Benedetti, P. A., Bianchini, G., Checcucci, A., Ferrara, R., Grassi, S., and Percival, D., 1976a, Spectroscopic properties and related functions of the stigma measured in living cells of Euglena gracilis, Arch. Microbiol., 111:73.

Benedetti, P. A., Bianchini, G., and Chiti, G., 1976b, Fast-scanning microspectroscopy: An electrodynamic moving-condenser method, Appl. Opt., 15:2554.

Benedetti, P. A., and Checcucci, 1975, Paraflagellar body (PFB) pigments studied by fluorescence microscopy in Euglena gracilis, Plant. Sci. Lett., 4:47.

Benedetti, P. A., Coltelli, P., Evangelista, V., and Gualtieri, P., 1983, Sistema per la misura ottica multicanale di spettri e immagini deboli, Atti del VI Congr. SIBPA - VII Congr. GNCB, Camogli (I) 6-8 Ott.

Benedetti, P. A., Dini, F., Grassi, S., Gualtieri, P., and Pintus, N., 1980, DNA determination in complex structures using Image Microspectroscopy (IMSS), VI Intl. Histoch. and Cytoch. Congr., Brighton, UK (Aug).

Benedetti, P. A., and Evangelista, V., 1985, Modular grating-mono-
    chromator with rapid voltage-programming of the wavelength,
    in preparation.

Benedetti, P. A., and Grassi, S., 1980, Thermoelectric photocathode
    cooler allowing rapid interchange of end-on photomultipliers,
    Appl. Opt., 19:192.

Benedetti, P. A., and Lenci, F., 1977, In vivo microspectrofluorome-
    try of photoreceptor pigments in Euglena gracilis, Photochem.
    Photobiol., 26:316.

Brunori, M., Giardina, B., Antonini, E., Benedetti, P. A., and
    Bianchini, G., 1974, Distribution of the haemoglobin components
    of trout blood among the erythrocytes: Observations by single-
    cell spectroscopy, J. Mol. Biol., 86:165.

Francon, M., 1961, "Progress in Microscopy," Vol. X, Pergamon Press,
    Oxford.

Leitz MPV Brochure 620-18, 1973, Leitz Wetzlar, W. Germany.

Liebman, P. A., 1973, Microspectrophotometry of visual receptors,
    in: "Biochemistry and Physiology of Visual Pigments,"
    H. Langer, ed., Springer Verlag, Berlin.

MacNichol, Jr., E. F., 1964, Three-pigment color vision, Sci. Am.,
    48, December.

Nultsch, W., and Benedetti, P. A., 1978, Microspectrophotometric
    measurements of the light-induced chromatophore movements in
    a single cell of the brown alga Dyctyota dichotoma, Z.
    Pflanzenphysiol., 87:173.

Nultsch, W., Benedetti, P. A., and Gualtieri, P., 1983, Microspectro-
    photometric investigations of photobleaching in Anabaena
    variabilis cells and heterocysts and its prevention by sodium
    azide, Z. Pflanzenphysiol., 111:327.

Piller, H., 1977, "Microscope Photometry," Springer Verlag, Berlin.

Vickers M85/M86 Brochure VM85/86/1, 1972, Vickers Inst., York, UK.

Von Sengbusch, G., and Thaer, A., 1973, Some aspects of instrumenta-
    tion and methods as applied to fluorometry at the microscale,
    in: "Fluorescence Techniques in Cell Biology," A. Thaer and
    M. Sernetz, eds., Springer Verlag, Berlin.

Wolken, J. J., Forsberg, R., Gallik, G., and Florida, R., 1968,
    Rapid recording microspectrophotometer, Rev. Sci. Instrum.,
    39:1734.

Zeiss SMP Brochure 41-820, 1973, Zeiss Oberkochen, W. Germany.

COMPUTER-AIDED STUDIES OF PHOTOINDUCED BEHAVIORS

Donat-P. Häder

Fachbereich Biologie-Botanik
Philipps-Universität
D-3550 Marburg
W. Germany

1. INTRODUCTION

With the availability of inexpensive and versatile microcomputers, a number of tedious and time-consuming tasks can be facilitated or expedited. This survey is intended to summarize some computer techniques and applications to enhance the study of orientation of microorganisms with respect to light.

Photomovement can be analyzed by two basically different approaches. The behavior can be studied using population techniques which assay the result of movement of a large number of organisms (Häder, 1979). On the other hand, the tracks of individuals can be followed and their velocity and angular deviation from the light source can be analyzed. In order to ensure statistical significance, a large number of track elements needs to be studied involving time-consuming measurement of distances and angles; thus, computer-aided data requisition could be of great help. In addition, the selection of individual organisms (in a microscope or on a video screen) by a human operator could be biased by a non-random preference. Even in "double blind" experiments, organisms moving towards or away from a light source could be selected preferentially over those moving in different directions. This problem can be solved by digitizing the video image of moving cells and analyzing the tracks in a computer. Either all organisms are assayed simultaneously or individuals are selected by a truly random process.

The second useful application is the development of mathematical models. The movement of both individuals and populations can be modeled by choosing appropriate parameters. The advantage is two-fold. First, once a model has been developed successfully, individual stimuli can be varied and their effect can be tested and compared with the natural response of real organisms. Secondly, the cumulative result of several, simultaneously-applied stimuli can be studied in a much more complex way than a lab experiment would allow. Thus, the ecological situation in the natural environment of the organism can be simulated.

2. COMPUTER-AIDED DATA ACQUISITION

Computers can be applied on several levels of automation for data acquisition: Image recognition and movement analysis, data input and analysis of relevant parameters (i.e. angles and distances) and finally statistical treatment of the measurement values.

2.1 Video Digitization and Pattern Recognition

The most ambitious task during automatic computer-aided data acquisition is the pattern recognition of moving objects. The video image from the camera is digitized using a video A/D converter and appropriate converter hardware (Fig. 1). The image is stored in an array of points (called pixels) in a row and column matrix. Common array sizes are 128 x 128 to 2048 x 2048 pixels. Each pixel can have a certain number of gray levels attributed. The upper limit of resolution in size and gray level is given by the available memory space and velocity of transfer and calculation. The digitization can be accomplished either by sequentially addressing each data point or by a flash converter. While the former method requires several seconds to minutes for each video frame, the second allows video digitization in real time.

In order to recognize organisms in an image, a number of restrictions need to be considered: The contrast between background and organisms should be high enough and the background brightness level should be uniform. Ultimately, it is desirable to find a threshold in brightness to set the background to one level and the organisms to another, i.e., to use only two gray levels.

Debris should be carefully avoided in the microscope image, since otherwise algorithms need to be incorporated to discriminate organisms from other objects by size and/or shape. Several methods have been developed to recognize objects and to decide which pixel belongs to the object. First, the outline could be traced by following the dark/light border. All pixels inside the outline are part of the organism. This method works well for organisms with rounded shapes, but not necessarily for long cells or objects of variable shape with concave edges. Second, the "blow-up" technique starts at the first pixel found during a line by line search. The next neighbors of this pixel are assayed. Around each of the

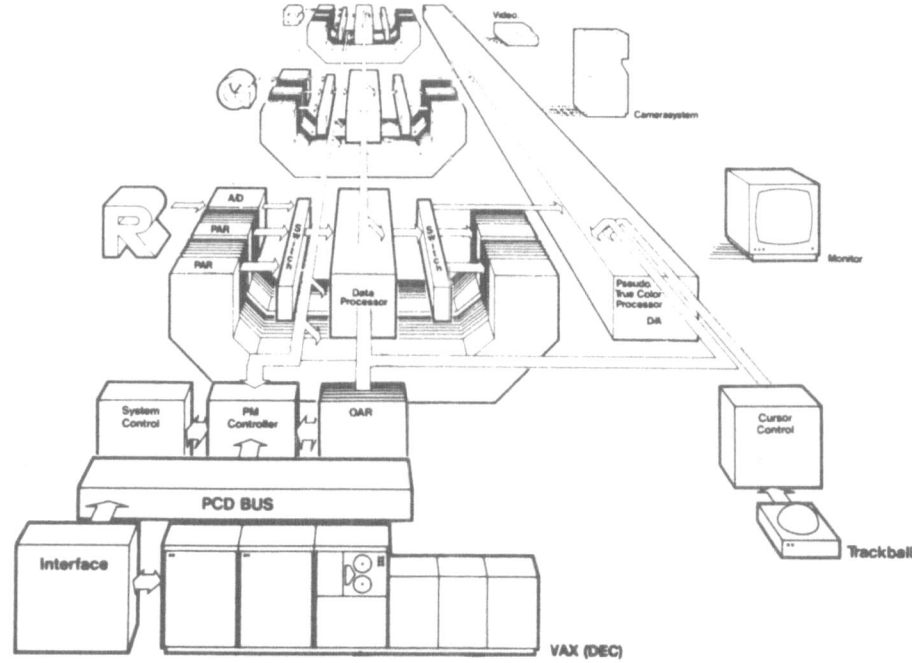

Fig. 1. Video image digitizer with video camera, video recorder, monitor, A/D converters for three colors connected to a host computer. Different gray levels can be displayed in pseudo colors. Courtesy by VTE, Herrsching.

pixels, found to belong to the organism, a new shell is analyzed until the whole area has been covered. Then the center of the area can be computed which defines the position of the cell on the screen. This procedure is continued until all organisms have been found on the screen and is then repeated iteratively in fixed time intervals. The cell centers are then connected to produce a calculated track of the organisms. Special precaution has to be taken to handle situations where organisms cross each others path or touch each other. The algorithms must guarantee that the computer follows the correct path. Another problem arises when organisms leave or enter the plane of focus. The easiest solution is to abandon controversial tracks. In addition, non-motile organisms can be filtered out.

2.2 Angle and Distance Calculation

In case of a computerized video digitization, the track information is already available in machine readable form. If the necessary hardware is not available, the track information can be

retrieved semiautomatically (Häder, 1981; Lipson and Häder, 1984);
the tracks of organisms are copied onto acetate sheets overlying
the video monitor. Instead of manually measuring the angles and
distances the organisms have moved in, given time intervals the data
can be transferred into the computer using a simple device consisting
of precision potentiometers linked to A/D converters (Fig. 2).

Another possibility is to use a bitpad where beginning and end
of each track segment are marked (Häder, 1981). Computer programs
have been developed to extract angles and distances from these

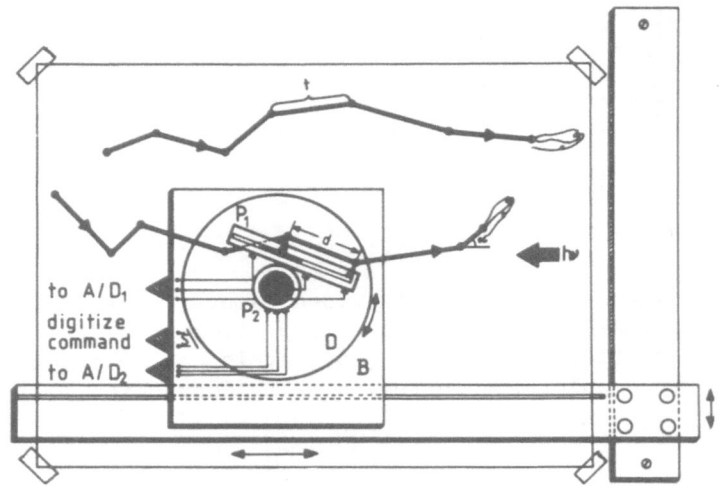

Fig. 2. Experimental set-up for data acquisition from a hardcopy
         of tracks of moving microorganisms. A rotary potentiometer
         (P_2) is mounted on a transparent plate (B), which is moved
         in the plane with the help of a commercial drafting machine.
         On the shaft of the rotary potentiometer, a transparent disk
         (D) is mounted, which carries a linear potentiometer (P_1).
         During operation the linear potentiometer is oriented paral-
         lel to the track segment; the output from the rotary poten-
         tiometer then accounts for the angle of deviation with the
         light direction indicated by the heavy arrow. The markers
         of the linear potentiometer are adjusted to point to the
         beginning and end of the track segment to determine the
         distance (d) traveled during the constant time interval (t).
         The analog voltages from the two potentiometers are sent to
         two independent A/D converters. Each time the potentiometers
         are adjusted to a track segment, the switch is closed mo-
         mentarily to instruct the computer to input the data (Lipson
         and Häder, 1984).

cartesian coordinates. The angles can be sorted in linear or circular histograms. In some cases, these histograms are clear enough to demonstrate an obvious positive or negative phototaxis. Other organisms produce more obscure histograms (Häder et al., 1981; Colombetti et al., 1982). In these cases, a statistical treatment of the data is necessary.

2.3 Statistical Treatment of Directional Data

An easy way of evaluating the statistical significance of an oriented movement is the Raleigh test (Batschelet, 1965, 1981):

$$\bar{r} = \sqrt{\left(\frac{\Sigma \cos\alpha}{N}\right)^2 + \left(\frac{\Sigma \sin\alpha}{N}\right)^2}$$

An $\bar{r}$-value near 0 indicates a random orientation while a value near 1 demonstrates an almost perfect orientation. The mean direction in which the population moves is calculated by

$$\theta = \arccos \frac{\Sigma \cos\alpha}{N \, \bar{r}}$$

This test is only valid for unimodal distributions. Populations which show simultaneous positive and negative phototaxis as the green flagellate Euglena gracilis (Lenci et al., 1983) or bimodal phototaxis as the pseudoplasmodia of some strains of the slime mold Dictyostelium discoideum (Fisher and Williams, 1981 a,b) or multi-modal phototaxis as the amoebae of some Dictyostelium strains (Häder et al., 1983) do not yield significant r-values. Another way of describing the degree of phototactic orientation is by using k-values (Mardia, 1972) which are derived from $\bar{r}$-values. Other common tests in directional statistics are the V-test, the U-test and the χ^2-test. The latter can be used when the number of groups (K) in a unimodal distribution is known:

$$\chi^2 = \sum_{N=1}^{K} \frac{(o_i - e_i)^2}{e_i}$$

where e_i is the expected and o_i the observed number of organisms in the group.

2.4 Fourier Analysis and Smoothing

An elegant way of interpreting angular data and extracting relevant information is by fast Fourier Analysis (Emerson, 1980; Kincaid and Schneider, 1983).

Fourier series describe continuous periodic functions and are based on a discrete summation of an infinite number of harmonic

terms. The transform of the original function is an infinite number
of coefficients with the increasing frequencies and appropriate
phase shifts. The function x(t) can be transformed into the Fourier
transform X(f) by

$$X(f) = \int_{-\infty}^{\infty} x(t) \exp (i\ 2\pi\ ft)\ df$$

The application for directional data starts with the assumption
that the original histogram, which may have a rather complex form,
can be approximated by superimposing sine waves of increasing frequencies (Häder and Lipson, in press). These individual components
have a certain amplitude and phase (Fig. 3). The Fourier analysis
algorithm uses the original histogram and produces a transform of
the data (Fig. 4). The transform is simply a different representation
of the original data and shows the dominant frequency components with
their corresponding phase. This technique allows to analyze the
modality of a distribution, i.e. in how many directions the organisms
move. The phase indicates the direction of movement.

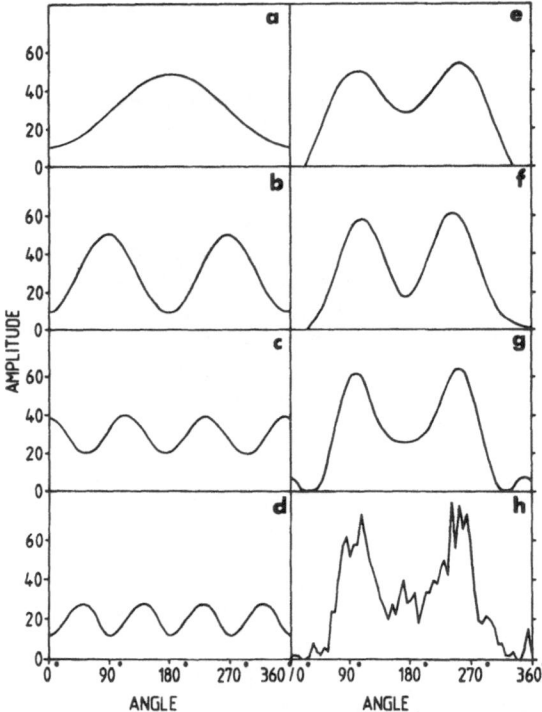

Fig. 3.   Reconstruction of a histrogram (h) by subsequently adding
          components with frequencies 1 to 4 (a-d) with appropriate
          phases and amplitudes showing the result of each step (e-g).
          The histogram represents a bimodal distribution of
          Dictyostelium discoideum slugs strain HO 596 to lateral
          white light of 3.3×10^{-2} Wm^{-2} (Häder and Lipson, in press).

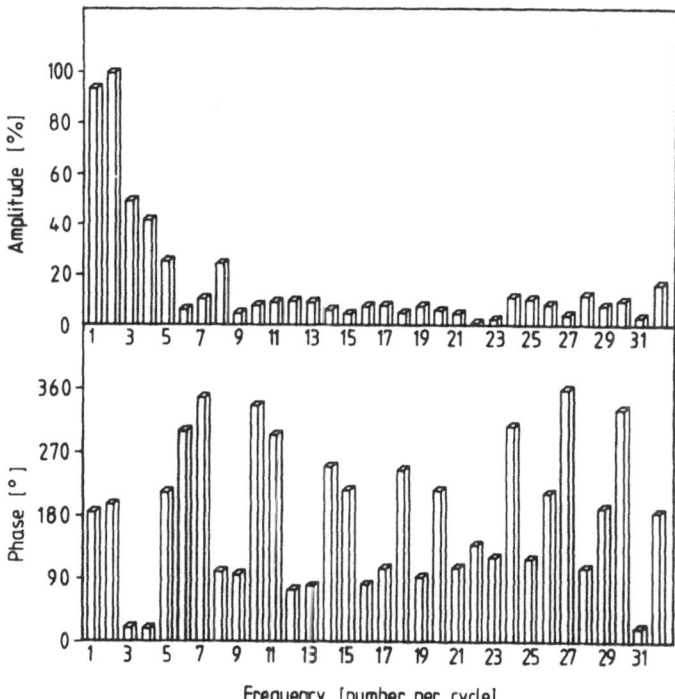

Fig. 4.   Fourier transformation of the histogram shown in Fig. 3 h
with amplitude (top) and phase (bottom) of the first 32
frequencies (abscissa) (Häder and Lipson, in press).

The next step is a smoothing operation. In the frequency domain
certain components are chosen; either low frequency components (low
path filtering) or components exceeding a certain amplitude are selected. On this subset a Fourier synthesis is carried out which is
the inverse of the Fourier analysis and reconstructs a smoothed
histogram

$$x(t) = \int_{-\infty}^{\infty} X(f) \exp (-i\, 2\pi\, ft)\, dt$$

3. MATHEMATICAL MODELS

The aim of mathematical models is to simulate the behavior of
individual organisms or populations in a non-stimulated situation
and under the influence of external stimuli. Examples of models for
population movements are the chemotactic behavior of bacteria
(Rosen, 1975; Odell and Keller, 1976) and the phototactic orientation
of algae (Kiknadze et al., 1974; Burkart and Häder, 1980). Two examples of mathematical simulations of individual organisms are discussed below.

3.1 Simulation of Photophobic Responses of the Blue-Green Algae, Phormidium

The blue-green alga (Cyanobacteria = Cyanophyta), Phormidium uncinatum, moves by a yet unknown mechanism. Upon leaving a light field, it undergoes a photophobic response (Diehn et al., 1977), which, in this organism, consists of a reversal of movement. The current direction of movement is dictated by an electrical gradient between the front and rear ends (Murvanidze and Glagolev, 1982). During a photophobic response, this electrical gradient is reversed.

The mathematical simulation is based on a discrete dynamic system characterized by recurrent equtions (Häder and Burkart, 1982). Each cell in a trichome x_i is defined by an inside negative electrical potential Q at a given time t

$$Q\,(x_i,t) = Q_D(x_i,t) + Q_L(x_i,\ t) + Q_P(x_i,t)$$

The overall potential is the sum of the resting potential of the cell Q_D plus a light-induced potential Q_L plus a potential induced during a phobic response. The light-induced potential results from the vectorial transport of protons from the cytoplasm into the thylakoids (intracellular, closed vesicles) by the activity of plastoquinone during the photosynthetic electron transport.

When the front cells of a filament enter a dark area, the proton gradient breaks down. The resulting polarization of the cytoplasm causes calcium channels in the cytoplasma membrane to open. This allows a massive influx of calcium along a previously established gradient, which depolarizes the cells even further. Thus, the front of the filament is less negative than the rear portion and the filament reverses its direction of movement.

In the model, the electrical potential (consisting of the above-mentioned components) is calculated for each cell of a filament at a given time and repeated at fixed time intervals. The speed and direction of the filament is calculated from the overall potential. The position of the filament is plotted graphically at predefined time intervals showing the electrical potential of each cell (Figs. 5 and 6). The model predicted that the number of cells entering a dark area before the filament reversed depends on the total number of cells in a filament: long filaments take longer to reverse than short ones. The same quantitative data have been found in vivo when the behavior of filaments leaving a light field was studied microscopically.

The model has been used to study the photobehavior of Phormidium under various experimental conditions. All predicted results have been verified by microscopical observations in vivo. Filaments are not 'fooled' by a very small shaded area. They slow down for some

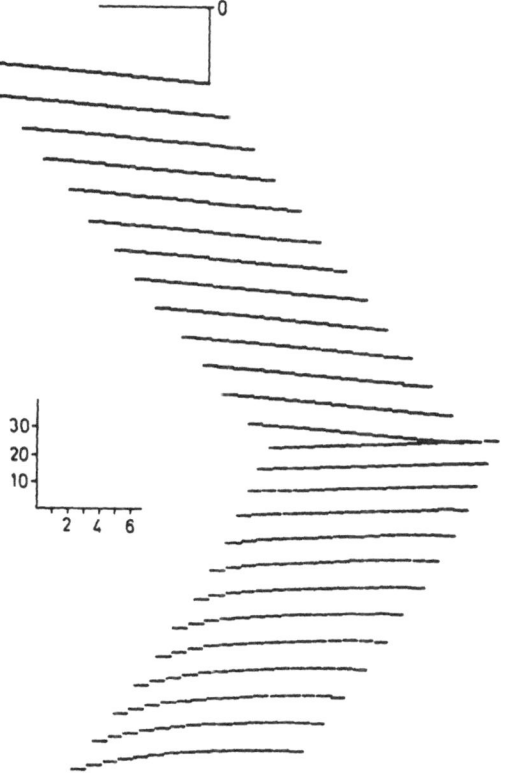

Fig. 5. Computer simulation of a photophobic reversal induced in a
         trichome of 15 cells by a decrease in light intensity
         simultaneously in all cells. One unit of the x-axis corre-
         sponds to 1 mV. The initial negative potential of each
         cell is given by the distance on the y-axis between the
         horizontal bar and the position of each cell. The trichome
         is drawn every third cycle with a vertical offset of 10 mV.
         An offset on the x-axis indicates the distance travelled
         between three cycles (Häder and Burkart, 1982).

time, but before they reverse the front cells have entered the next
bright field and the potential builds up again, so that no reversal
occurs. Filaments also do not reverse when up to 70% of the rear
end are darkened, while a darkening of the front 10% is sufficient
to trigger a photophobic response.

    Recently a new response type was predicted by the computer
model (Häder and Burkart, 1982). The simulation postulated a step-
up photophobic response to occur when the organism enters a very
bright field. Such behavior has never been found in cyanobacteria

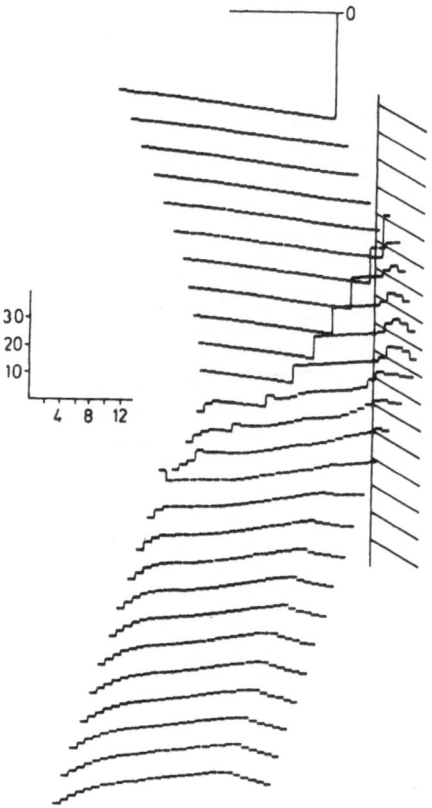

Fig. 6.   Simulated phobic response in a trichome of 28 cells induced
          by entering a dark field (hatched area).   Details as in
          Fig. 3 (Häder and Burkart, 1982).

with the exception of an <u>Oscillatoria</u> species (Schmidt, 1923). This
response could also be induced in <u>Phormidium</u> at higher light intensities (Table 1). At intermediate intensities, an oscillation could be
observed. The filaments reversed the direction of movement when
entering the bright field. But then the new front end entered a
dark field which also caused a reversal.

     The direction of movement can be reversed experimentally even
in the dark by applying an external electrical field. A DC field
reversed the internal electrical gradient along the length of the
filament. This behavior is also found in the model when an external
component $Q_{EX}(x_i,n)$ is added to the dark potential. Furthermore,
the behavior found under application of the calcium ionophore A23187
has been simulated. As <u>in vivo</u> the filaments oscillate in their

Table 1. Step-up Photophobic Responses at the Border
Between Dark and Light (Laser Irradiation
Converted to lx) in vivo and in the Computer
Simulation.

Illumination (lx)	Behavior	
	Simulation	In vivo
200 000	+	+
100 000	+	+
70 000	+	+
50 000	+	+,(a)
20 000	a	a
10 000	a	a
5 000	a	a
2 000	–	–,a
1 000	–	–

+, Step-up response.
–, No response.
a, Alternating step-up and step-down responses (after
Häder and Burkart, 1982).

direction of movement at intermediate molar concentrations. This
may be due to an oscillation between the Ca^{2+} influx and the
balancing calcium pump activity.

3.2 Simulation of Dictyostelium Pseudoplasmodia Movement

The multicellular stage of the cellular slime mold Dictyostelium
discoideum is called pseudoplasmodium or slug. Slugs are motile and
respond to a variety of external stimuli. In lateral light they
respond positive phototactically (Bonner et al., 1950; Francis, 1964),
even at an extreme low illuminance of $> 10^{-5}$ lx (Poff and Häder,
1984). The action spectrum differs from that of the unicellular
amoebae (Häder and Poff, 1979 a,b) which show both positive and
negative phototaxis.

The light direction is detected by a lens effect (Francis, 1964;
Häder and Burkart, 1983a). Since the refractive index of slugs
(n = 1.369) is significantly higher than that of the surrounding air,
parallel light is focussed onto the rear surface. Fisher and
Williams (1981 a,b) have postulated that a low molecular weight substance (termed slug turning factor, STF) is produced in a higher
concentration on the (brighter) distal side than on the proximal

side. The slug responds negative chemotactically to this self-produced and excreted substance so that it moves towards the light source. In addition, <u>Dictyostelium</u> slugs display an extreme sensitivity towards temperature differences (Poff and Skokut, 1977; Whitaker and Poff, 1980). The organisms detect temperature differences of 1/1000°C over their width and respond maximally to 0.2°C/cm.

The mathematical model considers a slug as a compartment on a plane; it is characterized by a sensor potential on the right and left sides. The difference between the two sensor potentials $\Delta S(t)$ results from the sum of several components adjusted by appropriate factors, which may be functions rather than simple numerical terms.

$$\Delta S(t) = w_1 \, \Delta S_R(t) + w_2 \, \Delta S_T(t) + w_3 \, \Delta S_P(t)$$

The first term $\Delta S_R(t)$ simulates the random temperature fluctuations in the agar surface on which the slug moves. These variations in the order of 10^{-4} to 10^{-3}°C over the width of the slug cause random deviation from the path. The factor $\Delta S_T(t)$ describes the influence of an externally applied uniform temperature gradient, which can be defined in either of two directions. The effective component across the slug is calculated from the orientation of the slug in this gradient by $|\vec{T}|\sin\alpha$. Finally, ΔS_P defines the sensor potential difference resulting from a laterally impinging parallel light beam. The effective intensity difference across the slug is derived from the deviation of the slug from the light beam and the intensification of light by the lens effect.

Any differences detected between the two sensors result in a deviation from the straight path the organism otherwise follows. At a time t the slug is described by a position vector x(t) in a Cartesian coordinate system. The movement is described by a direction vector v(t), the absolute value of which characterizes the speed.

Figure 7 shows the simulated movement of 20 slugs in the presence of a light beam (10^{-3} lx) impinging from above (a). The angular deviation from the light beam is analyzed in small track segments and a histogram is calculated (Fig. 7b). By appropriate selection of the adjustment factor w_3 describing the signal strength difference on the two sensors in dependence of the light intensity I

$$w_3 = a \, (\log I)^2 + b$$

the directedness $\bar{r}$ of the simulated movement can be calculated from the illuminance of the impinging lateral light (Fig. 8). Similar curves have been calculated for the dependence of directedness on the steepness of a temperature gradient (Häder and Burkart, 1983b).

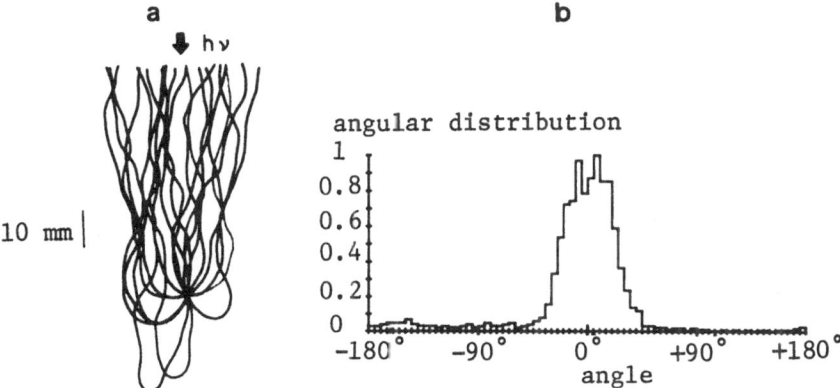

Fig. 7. Graphical presentation of a simulated movement of Dictyostelium discoideum slugs in the presence of 10^{-3} lx white
light stimulus impinging in the plane of movement (arrow).
The statistical analysis in the form of a linear histogram
(b) shows the distribution of angles in 72 sectors of 5°
each as a percentage of the highest value. The $\bar{r}$-value
has been calculated as 0.98, and the angle of deviation
(θ) as 12.66° ± 11.26° (Häder and Burkart, 1983).

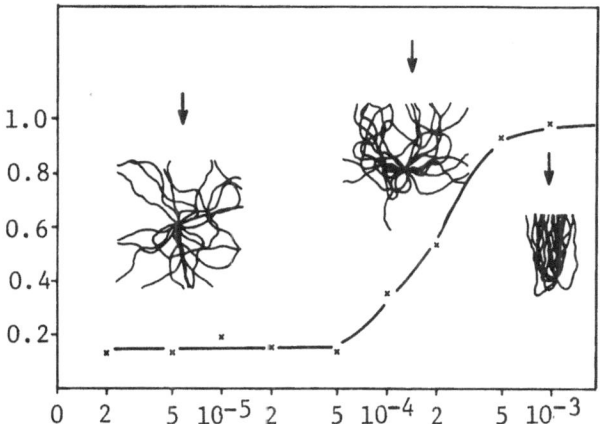

Fig. 8. Dependence of the directedness of simulated movement ($\bar{r}$,
ordinate) on the illuminance of the impinging lateral light
(in lx, abscissa). The insets show a pattern of orientation
at 5×10^{-6}, 2×10^{-4}, and 10^{-3} lx. The light direction
is indicated by arrows (Häder and Burkart, 1983).

   In the presence of both a light stimulus and a temperature
gradient applied perpendicular to each other, the computer simulation predicts a movement in the direction of the vectorial resultant.
These experiments are a first step on the way to the understanding
of the behavior under multiple simultaneous stimulation which is the
normal situation in the natural environment of organisms.

   Fisher and Williams (1981a) have demonstrated that under certain
conditions some strains of Dictyostelium do not move directly towards
the light source but deviate to the left and right of it (diaphoto-
taxis). The resulting bimodal distribution can be simulated by the
computer model by simply changing the function for w_3 (Fig. 9).
Likewise, the STF hypothesis (see above) has been studied. If the
slugs turn away from that side which produces a higher STF concentration (caused by a light or temperature gradient), one would
expect the slugs to avoid areas of higher STF concentrations in the

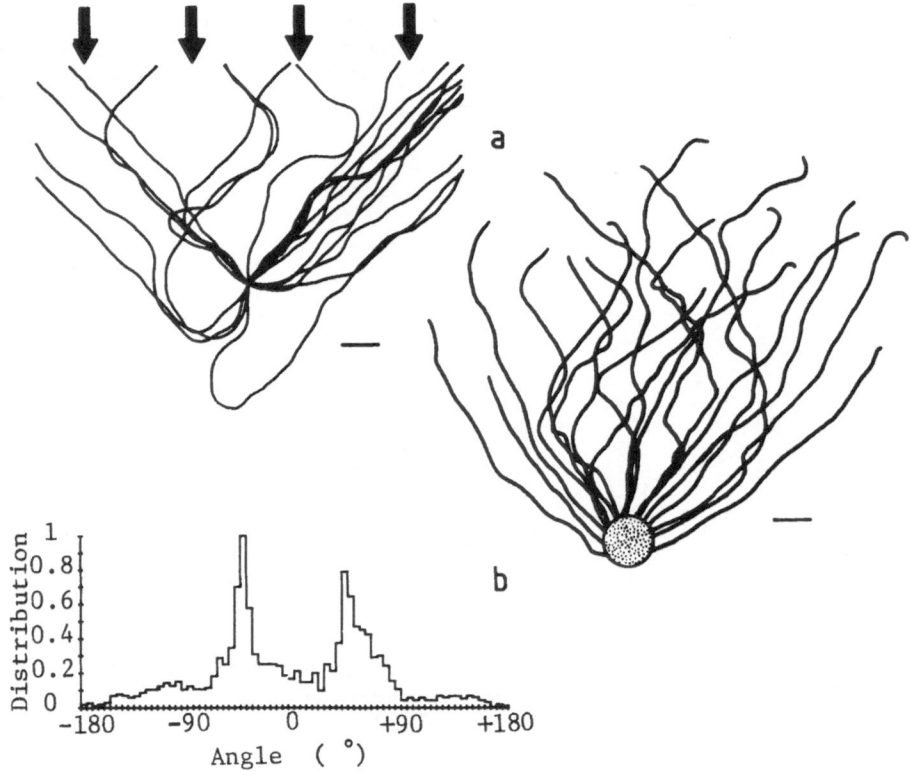

Fig. 9. Simulated movement patterns of a bimodal mutant in lateral
         white light (a. 5 x 10^{-4} lx from the top, arrows) and
         linear histogram of the angular distribution of track seg-
         ments (b) in comparison to the movement patterns in vivo
         (c). Bars represent 10 mm (Häder and Burkart, 1984).

agar surface. Indeed, slugs are repelled by strips saturated with a crude exudate from cells even when attracted by a light stimulus (Fisher, 1981). This behavior could be simulated by defining a field of appropriate STF concentration in the plane of movement (Häder and Burkart, 1984).

CONCLUSIONS

A number of mathematically oriented projects can be facilitated using the capabilities of computers; other tasks could not be performed without. In photomovement, the data acquisition can be enhanced by either a fully automatic video analysis of the tracks of organisms or by computer-assisted assessment of movement direction and velocity. Statistical tests geared at directional analysis can be performed as well as Fourier transformations with subsequent smoothing of the resultant histograms.

Mathematical models of the movement of microorganisms and their responses to external stimuli can be constructed using an iterative calculation of parameters as a function of time. The resulting simulation can be compared to the behavior in vivo and allows valuable predictions of the responses to multiple stimuli which occur under natural conditions in the habitat of the organisms too complex to be duplicated in laboratory experiments.

ACKNOWLEDGEMENTS

This work was supported in part by grants from the Deutsche Forschungsgemeinschaft (Ha 985/5-4) and the Bundesminister für Forschung und Technologie (KBF 57). The author thanks I. Herrmann, J. Meyer-Wegener, R. Möller and S. Wölk for skillful technical assistance.

REFERENCES

Batschelet, E., 1965, Statistical methods for the analysis of problems in animal orientation and certain biological rhythms, in: "Animal Orientation and Navigation," S.F. Galles, K. Schmidt-Koenig, G. J. Jacobs, R. F. Belleville, eds., NASA, Washington.
Batschelet, E., 1981, "Circular Statistics in Biology," Academic Press, London.
Bonner, J. T., Clarke, Jr., W. W., Neely, Jr., Ch. L., and Slifkin, M. K., 1950, The orientation to light and the extremely sensitive orientation to temperature gradients in the slime mold Dictyostelium discoideum, J. Cell Comp. Physiol., 36:149-158.

Burkart, U., and Häder, D.-P., 1980, Phototactic attraction in light
    trap experiments: A mathematical model, J. Math. Biol., 10:
    257-269.
Colombetti, G., Häder, D.-P., Lenci, F., and Quaglia, M., 1982,
    Phototaxis in Euglena gracilis: Effect of sodium azide and
    triphenylmethyl phosphonium ion on the photosensory trans-
    duction chain, Curr. Microbiol., 7:281-284.
Diehn, B., Feinleib, M., Haupt, W., Hildebrand, E., Lenci, F., and
    Nultsch, W., 1977, Terminology of behavioral responses of
    motile microorganisms, Photochem. Photobiol., 26:559-560.
Emerson, P. H., 1980, Fast Fourier transform, fundamentals and
    applications, Creat. Comp., July:58-63.
Fisher, P. R., 1981, Orientation Behaviour by Dictyostelium
    discoideum slugs, Thesis, University of Queensland.
Fisher, P. R., and Williams, K. L., 1981a, Bidirectional phototaxis
    by Dictyostelium discoideum slugs, FEMS Microbiol. Lett., 12:
    87-89.
Fisher, P. R., and Williams, K. L., 1981b, Activated charcoal and
    orientation behavior by Dictyostelium discoideum slugs, J.
    Gen. Microbiol., 126:519-523.
Francis, D. W., 1964, Some studies on phototaxis of Dictyostelium,
    J. Cell. Comp. Physiol., 64:131-138.
Häder, D.-P., 1979, Photomovement, in: "Encyclopedia of Plant
    Physiology," New Series Vol. 7, W. Haupt and M. E. Feinleib,
    eds., Springer Verlag, Berlin, Heidelberg, pp. 268-309.
Häder, D.-P., 1981, Computer-based evaluation of phototactic
    orientation in microorganisms, EDV Med. Biol., 12:27-30.
Häder, D.-P., and Burkart, U., 1982, Enhanced model for photophobic
    responses of the blue-green alga, Phormidium uncinatum,
    Plant Cell Physiol., 23:1391-1400.
Häder, D.-P., and Burkart, U., 1983a, Mathematical simulation of
    Dictyostelium pseudoplasmodia movements, Math. Bioscience,
    67:41-57.
Häder, D.-P., and Burkart, U., 1983b, Optical properties of Dictyo-
    stelium discoideum pseudoplasmodia responsible for phototactic
    orientation, Exp. Mycol., 7:1-8.
Häder, D.-P., and Burkart, U., 1984, Movement of Dictyostelium
    pseudoplasmodia in vivo and in a mathematical simulation,
    Plant Cell Physiol., 25:1-10.
Häder, D.-P., Claviez, M., Merkl, R., and Gerisch, G., 1983, Response
    of Dictyostelium discoideum amoebae to local stimulation by
    light, Cell Biol. Int. Rep., 7:611-616.
Häder, D.-P., Colombetti, G., Lenci, F., and Quaglia, M., 1981,
    Phototaxis in the flagellates, Euglena gracilis and Ochromonas
    danica, Arch. Microbiol., 130:78-82.
Häder, D.-P., and Lipson, E. D., 1984, Fourier analysis of angular
    distribution, in press.
Häder, D.-P., and Poff, K. L., 1979a, Light induced accumulation of
    Dictyostelium discoideum amoebas, Photochem. Photobiol., 29:
    1157-1162.

Häder, D.-P., and Poff, K. L., 1979b, Photodispersal from light traps by amoebas of Dictyostelium discoideum, Exp. Mycol., 3:121-131.

Kiknadze, G. S., Esakow, R., Schubirow, W. L., and Tageewa, S. W., 1974, Applications of statistical physic methods to the study of phototaxis, Doklady, 219:220-223.

Kincaid, D. T., and Schneider, R. B., 1983, Quantification of leaf shape with a microcomputer and Fourier transform, Can. J. Bot., 61:2333-2342.

Lenci, F., Colombetti, G., and Häder, D.-P., 1983, Role of flavin quenchers and inhibitors in the sensory transduction of the flagellate, Euglena gracilis, Curr. Microbiol., 9:285-290.

Lipson, E. D., and Häder, D.-P., 1984, Video data acquisition for movement responses in individual organisms, Photochem. Photobiol., 39:437-441.

Mardia, K. V., 1972, "Statistics of Directional Data," Academic Press, London.

Murvanidze, G. V., and Glagolev, A. N., 1982, Electrical nature of the taxis signal in cyanobacteria, J. Bacteriol., 152: 239-244.

Odell, G. M., and Keller, E. F., 1976, Traveling bands of chemotactic bacteria revisited, J. Theoret. Biol., 56:243-247.

Poff, K. L., and Häder, D.-P., 1984, An action spectrum for phototaxis by pseudoplasmodia of Dictyostelium discoideum, Photochem. Photobiol., 39:433-436.

Poff, K. L., and Skokut, M., 1977, Thermotaxis by pseudoplasmodia of Dictyostelium discoideum, Proc. Natl. Acad. Sci. USA, 74:2007-7010.

Rosen, G., 1975, Analytical solution to the initial-value problem for traveling bands of chemotactic bacteria, J. Theoret. Biol., 49:311-321.

Schmidt, G., 1923, Das Reizverhalten künstlicher Teilstücke, die Kontraktilität and das osmotische Verhalten der Oscillatoria jenensis, Jb. Wiss. Bot., 62:328-419.

Whitaker, B. D., and Poff, K. L., 1980, Thermal adaptation of thermosensing and negative thermotaxis in Dictyostelium, Exp. Cell Res., 128:87-93.

SENSORY TRANSDUCTION IN HALOBACTERIUM

Eilo Hildebrand and Angelika Schimz

Institut für Neurobiologie
Kernforschungsanlage Julich
D-5170 Jülich
W. Germany

1. INTRODUCTION

Bacteria can sense and integrate outside stimuli and adapt to new environmental conditions, thereby revealing a simple mode of behavior. By a temporal sensing mechanism, bacteria orient indirectly in a biased three-dimensional random walk which finally leads the cells to accumulate in favorable surroundings or to avoid unfavorable areas. Chemosensing in eubacteria has been studied in a number of laboratories and is the sensory system most advanced in biochemical understanding (Koshland, 1980; Berg, 1985; Macnab, 1985).

The archaebacterium Halobacterium halobium responds to light stimuli (Hildebrand and Dencher, 1975) and to chemical stimuli (Schimz and Hildebrand, 1979), and its sensory systems seem to be organized in conceptual outline rather similarly to the organization of chemosensory systems in eubacteria. Photobehavior of H. halobium seems to be highly suitable to analyze certain steps in the transduction chain, e.g. the response regulation and signal integration, since light stimuli can be applied and removed almost instantaneously, which allows to study events with a high resolution in time. The analysis of behavioral responses to selected stimulation programs together with biochemical studies and the use of genetically defective strains appears as an appropriate tool to delineate the sensory transduction chain in bacteria.

2. SWIMMING BEHAVIOR AND LIGHT-CONTROLLED RESPONSES

Halobacterium halobium is polarly flagellated and swims in both directions of its long axis. From time to time the cells reverse their swimming direction due to a change of flagellar rotation from clockwise (CW) to counterclockwise (CCW) or vice versa (Alam and Oesterhelt, 1984). Most cells from the logarithmic growth phase carry flagella on one pole only, while in the stationary phase the bacteria become bipolarly flagellated (Alam and Oesterhelt, 1984). The time intervals between spontaneous reversals, normally about 10 s, are not significantly different, i.e. the cells swim equally long in both directions (so far proved only for cells from the stationary phase). External stimuli bias this pattern of behavior. Increase of yellow-green light or decrease of UV light suppresses the spontaneous reversals for a certain time and is therefore regarded as an attractant stimulus (Fig. 1). Decrease of yellow-green light or increase of UV

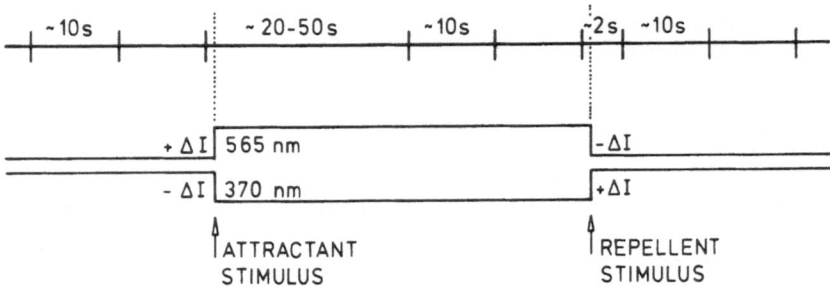

Fig. 1.   Scheme of spontaneous and light-controlled swimming behavior
           of Halobacterium halobium. Each vertical bar indicates a
           reversal of the swimming direction. For details, see text.

light, on the other hand, causes the next reversal to occur earlier, within a few seconds, and is, therefore, called a repellent stimulus (Hildebrand and Dencher, 1975; Spudich and Stoeckenius, 1979; Hildebrand and Schimz, 1983a) (Fig. 1). After a single response, i.e. one increased or decreased interval, the cells return to their normal frequency of spontaneous reversals, i.e. they adapt to constant light conditions. This behavior has so far been observed only upon step-like changes of light intensity. In a light intensity gradient, we would expect a similar mode of behavior.

Many of the results described in the following were obtained by behavioral analysis. Once behavioral responses have been defined, their evaluation can be readily used as a quantitative method to study certain events in sensory transduction.

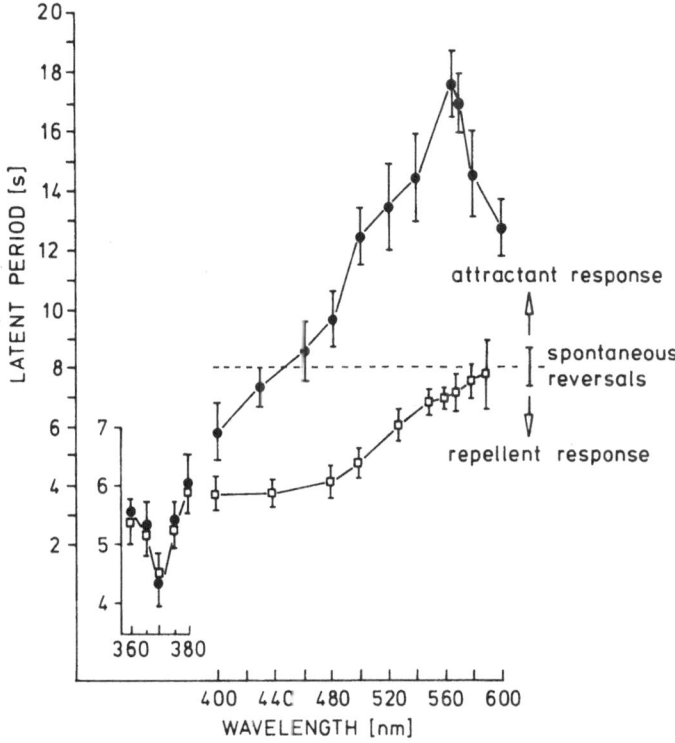

Fig. 2. Action spectra of photosensory behavior of Halobacterium
halobium obtained upon step-up stimuli of constant quantum
flux at each wavelength. Because of the different sensitivity in the UV and in the visible range, 1 x 10^{12} photons
x $mm^{-2}s^{-1}$ were used between 360 and 380 nm, and 3.4 x 10^{14}
between 400 and 600 nm. Photosensory activity is given as
the latent period between the onset of the stimulus (applied
2 s after a spontaneous reversal) and the subsequent reversal. PS 370 and PS 565 fully active (●); PS 370 alone,
before PS 565 had developed (□). Each point is the mean
± SEM of 30 measurements. (After Hildebrand and Schimz,
1983b.)

3. SENSORY PHOTOSYSTEMS

     An action spectrum measured with a step-up stimulus of constant
quantum flux at each wavelength shows a repellent response in the UV
with a maximum at 370 nm and an attractant response in the visible
range with a maximum at 565 nm (Fig. 2). In contrast, a step-down
stimulus leads to an attractant response in the UV and to a repellent
response in the visible. Sensitivity maxima were found at the same
wavelengths (Hildebrand and Schimz, 1983a). Because of the opposing
effects of UV and visible light, two sensory photosystems, PS 370 and

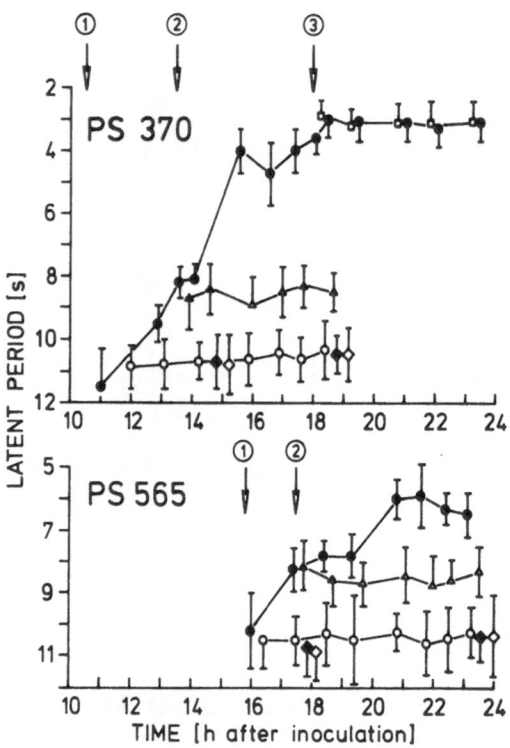

Fig. 3. Consecutive formation of PS 370 and PS 565 in Halobacterium
halobium, and blocking of their synthesis by puromycin.
Photosensory activity is given as the latent period between
the onset of a stimulus (applied 2 s after a spontaneous
reversal) and the subsequent reversal. Stimuli: Step-up
stimulus of 2.3×10^{13} photons $\times$ $mm^{-2}s^{-1}$ at 370 nm or step-
down stimulus of 9.5×10^{14} photons $\times$ $mm^{-2}s^{-1}$ at 565 nm.
Arrows indicate the time of puromycin addition. Without
puromycin (●); after addition of puromycin at t_1 (O), t_2
(△), t_3 (□); time between spontaneous reversals without
puromycin (◆) and after puromycin addition at t_1 (◊). Each
point is the mean ± SEM of 30 measurements. (After Hildebrand and Schimz, 1983b.)

PS 565, have been assumed (Hildebrand and Dencher, 1975). PS 370 is
about 50 times more sensitive than PS 565 (Hildebrand and Dencher,
1975).

     It may be expected that there is a certain spectral overlap of
the photosystems. The action spectra measured so far may, therefore,
be difference spectra (Spudich and Stoeckenius, 1979).

Bacteria from a newly inoculated culture do not respond to
light. The sensitivity to UV light occurs prior to the sensitivity
to visible light (Hildebrand and Dencher, 1975; Schimz and Hildebrand,
1979; Hildebrand and Schimz, 1983b). The reappearance of PS 370 and
PS 565 can be stopped with puromycin (Hildebrand and Schimz, 1983b),
which is an effective blocker of protein biosynthesis in halobacteria
(Schimz, 1981). Puromycin does not inhibit the activity of partially
or fully resynthesized photosystems (Fig. 3). These results indicate that proteins which are directly involved in the process of
sensory recognition or transduction are degraded and synthesized de
novo in a growing culture.

Blocking the resynthesis of PS 565 at the right moment allows
to measure the action spectrum of PS 370 separately (Fig. 2). Moreover, a mutant strain, ET-15-7, could be isolated, which lacks photosensory activity in the visible range but shows a normal action
spectrum in the UV range (Hildebrand and Schimz, 1983b). The maximum
of PS 370 is neither influenced by the presence of PS 565 nor by the
presence of carotenoids which act as accessory pigments in PS 370
(Hildebrand and Schimz, 1983b; Dencher and Hildebrand, 1979). Furthermore, PS 565, which so far could be measured only in the presence
of PS 370, does not change its maximum during its reappearance. We,
therefore, conclude that the sensitivity maxima at 370 nm and 565
nm will reflect the absorption maxima of the underlying sensory pigments (Hildebrand and Schimz, 1983b).

4. PHOTORECEPTOR PIGMENTS

Both sensory photosystems require retinal as the chromophore;
therefore, their pigments are regarded as "rhodopsins," i.e. retinal-
protein complexes. When the biosynthesis of retinal is biochemically
blocked by nicotine, the bacteria do not respond to light stimuli
(Dencher and Hildebrand, 1979). Also, mutant strains which are not
able to synthesize retinal lack photosensory activity (Spudich and
Bogomolni, 1983b). Photosensory activity of such cells reappears
when retinal is added to the cell suspension (Dencher and Hildebrand,
1979; Spudich and Bogomolni, 1983b). When the synthesis of endogen-
eous retinal is blocked, photosensory activity can also be restored
with retinal analogues (Sperling and Schimz, 1980; Schimz et al.,
1982, 1983). This allows us to substitute the naturally-occurring
retinal by several different retinal chromophores. When retinal is
substituted by analogues, action spectra of both photosystems show
a characteristic wavelength shift. $Retinal_2$ shifts the action spectra
both of PS 565 (Fig. 4) and PS 370 (not shown) to longer wavelengths,
whereas 13-ethyl and 13-propyl retinal cause a shift to shorter wavelengths. In addition, PS 565 shows a secondary peak, when retinal
analogues are used (Figs. 4 and 5).

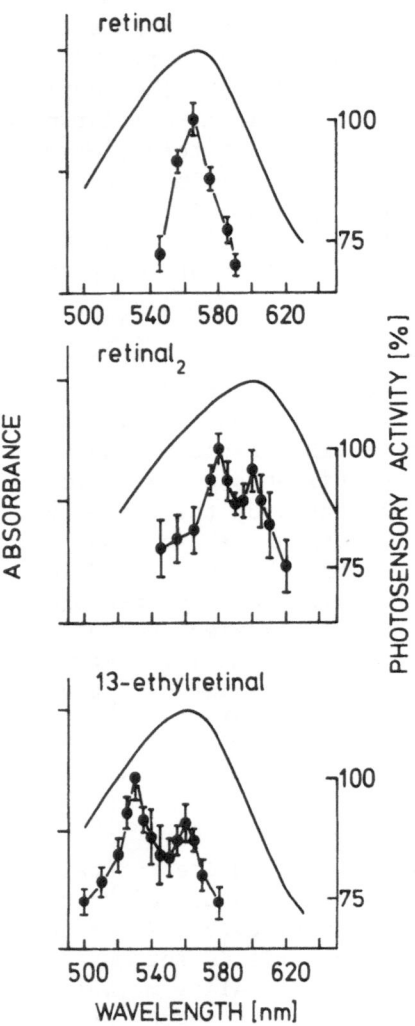

Fig. 4. Absorption spectra of BR and action spectra of PS 565 after
         reconstitution of retinal pigments in the BR-containing
         strain R_1L_3 by retinal or analogues. Cells were grown in
         nicotine to block the synthesis of endogeneous retinal.
         Reconstitution of the pigments was done in vivo. Absorption
         spectra of BR (upper curves) were measured with suspensions
         of purified purple membranes and normalized with respect to
         their maxima. Action spectra (lower curves) were measured
         with step-down stimuli of 2.5×10^{14} photons x $mm^{-2}s^{-1}$ at
         each wavelength. The maxima of photosensory activity were
         normalized to 100%. Each point is the mean ± SEM of 20 –
         30 measurements. Propyl retinal (not shown) causes about
         the same shift as ethyl retinal. (After Schimz et al.,
         1983).

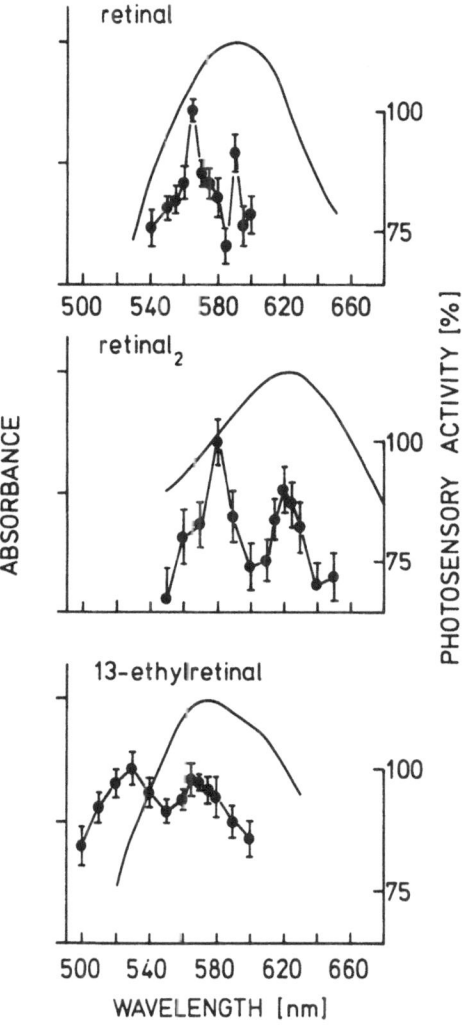

Fig. 5. Absorption spectra of envelope vesicles and action spectra
         of PS 565 obtained from a BR-deficient strain, L-33, after
         reconstitution of retinal pigments by retinal or analogues.
         Absorption spectra (upper curves) were measured with vesi-
         cles prepared from retinal-containing cells, which were
         bleached in the presence of hydroxylamine to remove retinal,
         and then reconstituted by addition of retinal or analogues.
         The absorption spectra obtained under these conditions may
         be combined spectra of HR and sR. Action spectra (lower
         curves) were measured as described in Fig. 4. Propyl
         retinal (not shown) causes about the same shift as ethyl
         retinal. (After Schimz et al., 1983.)

Retinal is also the chromophore of the light-energy converting pigments bacteriorhodopsin (BR) (Stoeckenius et al., 1979) and halorhodopsin (HR) (Lindley and MacDonald, 1979; Matsuno-Yagi and Mukohata, 1980; Lanyi and Weber, 1980) and of the "slowly cycling rhodopsin" (sR) (Spudich and Bogomolni, 1983b; Bogomolni and Spudich, 1982). In mutant strains which lack BR or HR, both sensory photosystems are present (Hildebrand and Schimz, 1983a; Spudich and Spudich, 1982). This shows that BR and HR are not required for photosensing. Therefore, the photosensory pigments must be different from the light-energy converting ones. PS 370 and PS 565 thus can be regarded as genuine sensory systems which are not functionally coupled to the conversion of light into cellular energy.

The retinal analogues mentioned above can substitute retinal in all known "rhodopsins" of Halobacterium, but the resulting wavelength shifts are different for each pigment. The secondary peak in the action spectra of BR-containing strains (e.g. R_1L_3), measured after substitution of retinal by analogues, matches the absorption peak of the corresponding BR (Fig. 4). Consequently, the BR-deficient mutant ET-15 lacks this secondary peak and shows only the main peak which is identical in all strains investigated so far (Schimz et al., 1983). With normal retinal as the chromophore, the absorption maximum of BR (in the light-adapted state at 568 nm) is too close to the main activity peak at 565 nm to be separated in the action spectrum (Fig. 4).

Action spectra of another BR-deficient mutant, L-33, show a different secondary peak at longer wavelengths than in R_1L_3, which is probably caused by another retinal pigment (Fig. 5). With the normal retinal as the chromophore, this secondary peak is found at about 590 nm (Schimz et al., 1983; Traulich et al., 1983) and matches fairly well the absorption maximum of sR found at 587 nm (Spudich and Bogomolni, 1983b).

The secondary peak in the action spectrum of PS 565 in L-33 was formerly believed to be caused by HR (Schimz et al., 1983; Traulich et al., 1983) for the following reasons: L-33 contains high amounts of HR (Lanyi, 1982), the absorption maximum of which was reported to be at 588 nm at that time (Lanyi and Weber, 1980), but is now found at 578 nm (Spudich and Bogomolni, 1983b). The synthesis of the protein component of HR seems to depend on the presence of retinal (Spudich and Bogomolni, 1983a). Since, for behavioral experiments, we used nicotine cells from the stationary growth phase where little protein biosynthesis occurs, one cannot expect to obtain sufficient amounts of HR upon addition of retinals. Actually, we could not restore uncoupler-enhanced H^+ uptake under these conditions (unpublished results) which further supports the finding that HR cannot be reconstituted in retinal-free cells in the stationary growth phase. BR and sR, however, can be reconstituted under these conditions (Spudich and Bogomolni, 1983a; Sumper et al., 1976). We, therefore, conclude from the action spectra shown in Figs. 4 and 5 that BR and probably also sR can contribute to the photosensory activity of PS 565 while the possible role of HR in photosensing is not yet known.

Spudich and Bogomolni proposed sR and its intermediate in the photocycle, S_{373}, to act as the sensory receptors of PS 565 and PS 370, respectively (Bogomolni and Spudich, 1982; Spudich, 1985). We rather assume that sR, like BR, may act as an accessory pigment in PS 565 or through a parallel pathway, and that the main sensory pigments of both photosystems have not yet been detected biochemically or spectroscopically.

One has to consider that probably few receptor molecules per cell could be sufficient to mediate photobehavior.

5. INFLUENCE OF BACKGROUND LIGHT

As schematically drawn in Fig. 1, Halobacterium detects temporal changes of light intensity ($\Delta I/dt$), but its behavioral pattern is not influenced by the actual level of constant light, i.e. the frequency of spontaneous reversals is not changed by different background light.

One would expect, however, that the sensitivity to light stimuli, i.e. step-up or step-down changes of intensity, should depend on background light. Background light of moderate intensity was found not to reduce the sensitivity of either photosystem as compared to complete darkness or far red background light outside the absorption range of rhodopsins in Halobacterium (Tables 1 and 2). High intensities of background light reduce the photosensory activity. PS 370 is maximally inhibited by background light in the UV around 370 nm and also in the yellow-green range (Table 1). PS 565 is most efficiently inhibited by background light in the yellow-green range around 565 nm and by blue or UV light (Table 2).

Some of these results are consistent with the assumption of two separate photopigments as well as with the idea that two forms of a single (photochromic) pigment undergoing a photocycle mediate the different responses to UV and visible light as proposed by Wagner (1984) and Spudich (Bogomolni and Spudich, 1982; Spudich, 1985). The fact, however, that yellow-green light of high intensity reduces the response to UV stimuli can hardly be explained by the latter hypothesis. According to this hypothesis, yellow-green light should increase the amount of the UV-absorbing intermediate and thus lead to an increased sensitivity of PS 370.

Inhibition of either photosystem by background light which is maximally absorbed by its pigment can be easily explained as a saturation phenomenon at the receptor level, e.g. pigment conversion

Table 1. Photosensory Activity of PS 370 of <u>Halobacterium</u>
         <u>halobium</u>, Strain WS, at Different Background
         Light.

BACKGROUND λ, intensity ($h\nu \times mm^{-2}s^{-1}$)	STIMULUS [a] λ, intensity ($h\nu \times mm^{-2}s^{-1}$)	RESPONSE latent period [b] ($\bar{x} \pm$ SEM)	(n)
FAR RED cut off 780 nm	NONE	8.7 ± 0.9	(20)
	STEP-UP 370 nm (6×10^{13})	3.0 ± 0.3	(20)
	STEP-DOWN 370 nm (6×10^{13})	14.6 ± 1.7	(30)
NONE ($<10^{-6}$ μW × cm^{-2})	STEP-UP 364 nm (2.3×10^{13})	3.5 ± 0.3	(33)
UV λ_{max} 370 nm, HW 12 nm (5.4×10^{12})	STEP-UP 370 nm (1.6×10^{13})	2.1 ± 0.1	(30)
(2.1×10^{13})	STEP-UP 370 (1.6×10^{13})	5.7 ± 0.9	(30)
YELLOW cut off 515 nm I = 0.03	STEP-UP 370 nm (0.8×10^{13})	2.5 ± 0.1	(30)
I = 1.0	STEP-UP 370 nm (0.8×10^{13})	4.8 ± 0.6	(30)

[a] delivered 2 s after a spontaneous reversal

[b] time between stimulus and subsequent reversal (s)

or bleaching. Inhibition of PS 565 by intense UV light and inhibition of PS 370 by intense yellow-green light may indicate that signals
from two photoreceptors compete with each other at a certain site of
sensory integration in the transduction chain (Strange and Koshland,
1976). In summary, most of our results, together with the successful
separation of PS 370 from PS 565, lead us to favor the assumption
of two distinct photosensory pigments in <u>Halobacterium</u>.

Table 2. Photosensory Activity of PS 565 of <u>Halobacterium</u>
<u>halobium</u>, Strain WS, at Different Background
Light.

BACKGROUND λ, intensity ($h\nu \times mm^{-2}s^{-1}$)	STIMULUS [a) λ, intensity ($h\nu \times mm^{-2}s^{-1}$)	RESPONSE latent period [b) ($\bar{x} \pm SEM$) (n)
FAR RED cut off 780 nm	NONE	11.3 ± 1.3 (30)
	STEP-UP 565 nm (8.3×10^{14})	23.7 ± 1.8 (20)
	STEP-DOWN 565 nm (1.0×10^{15})	2.8 ± 0.2 (20)
YELLOW-GREEN λ_{max} 566 nm, HW 18 nm (2.2×10^{13})	STEP-UP 565 nm (8.3×10^{14})	29.6 ± 3.1 (30)
(5.2×10^{14})	STEP-UP 565 nm (8.3×10^{14})	15.3 ± 1.8 (30)
BLUE λ_{max} 401 nm, HW 20 nm (3.4×10^{13})	STEP-UP 565 nm (8.3×10^{14})	25.7 ± 3.4 (30)
(1.1×10^{14})	STEP-UP 565 nm (8.3×10^{14})	19.5 ± 3.3 (30)
UV λ_{max} 370 nm, HW 30 nm (8.5×10^{13})	NONE	9.9 ± 0.8 (20)
	STEP-UP 565 nm (1.0×10^{15})	2.0 ± 0.3 (20)
	STEP-DOWN 565 nm (1.0×10^{15})	23.5 ± 1.0 (20)

a) delivered 2 s after a spontaneous reversal

b) time between stimulus and subsequent reversal (s)

In one of the strains of <u>Halobacterium</u> so far investigated,
"wild type" strain WS, we found that intense UV background light
causes an "inverse" response: A step-up stimulus at 565 nm leads
to a repellent response, while a step-down stimulus causes an

attractant response (Table 2). This effect could be explained by
the assumption of a photochromic pigment (Spudich, 1985). However,
under certain experimental conditions, we observed similar effects
also at other kinds of background light when low stimulus intensities
were used (unpublished). Therefore, other mechanisms might be responsible for these "inverse" effects.

6. RESPONSE REGULATION AND SIGNAL INTEGRATION

The mechanisms which control the directional change of flagellar
rotation are presently not known. A model proposed by Koshland (1980)
for bacterial chemotaxis assumes a "response regulator" X, which is
enzymatically produced and degraded, and, at a critical concentration,
induces a conversion of the flagellar motor from CCW to CW. In order
to look more closely to the mechanism of response regulation in
Halobacterium, it seemed promising to apply light stimuli at different
moments between two spontaneous reversals and to measure how the
probability for the occurrence of the next reversal might be changed.

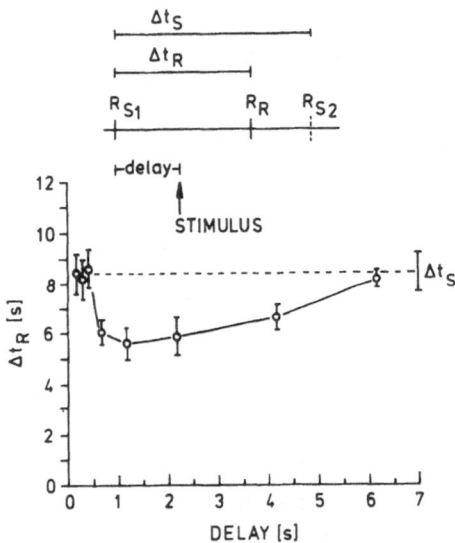

Fig. 6. Effectiveness of repellent stimuli at different moments
during the interval between two reversals. With different
delays after a spontaneous reversal (R_{S_1}), the cells were
stimulated by a step-like intensity increase of 6×10^{12}
photons $\times mm^{-2}s^{-1}$ at 370 nm. Δt_R, the time between R_{S_1} and
the subsequent reversal, R_R, reflects the effectiveness of
the stimulus to raise the probability for the next reversal
to occur. The response can be regarded as a repellent response as long as Δt_R is shorter than Δt_S, the time between
two spontaneous reversals. Each point is the mean ± SEM
of 30 measurements.

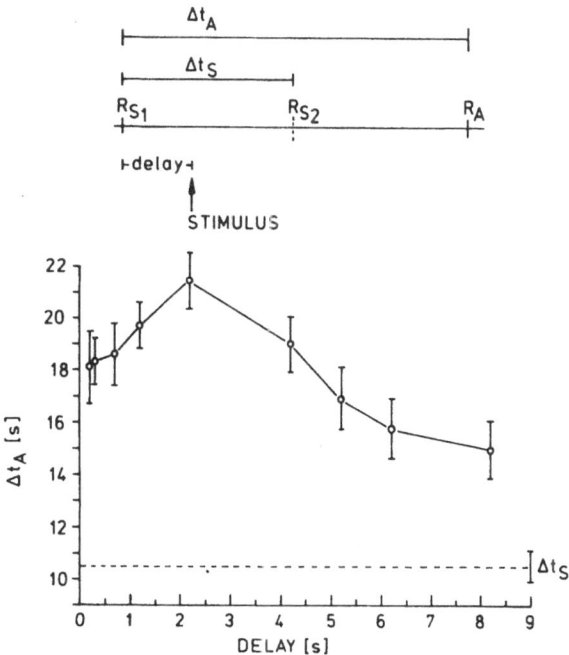

Fig. 7. Effectiveness of attractant stimuli at different moments
during the interval between two reversals. With different
delays after a spontaneous reversal (R_{S_1}), the cells were
stimulated by a step-like intensity decrease of 6 x 10^{12}
photons x $mm^{-2}s^{-1}$ at 370 nm. Δt_A, the time between R_{S_1} and
the subsequent reversal, R_A, reflects the effectiveness of
the stimulus to lower the probability for the next reversal
to occur. The response can be regarded as an attractant
response as long as Δt_A is longer than Δt_S, the time between two spontaneous reversals. Each point is the mean
± SEM of 30 measurements.

    We found that the probability for the next reversal to occur
strongly depends on the moment at which a repellent or an attractant
stimulus is delivered. Shortly after a spontaneous reversal, repellent stimuli do not change the probability for the next reversal
at all (Fig. 6). Even a 10-fold intensity of the stimulus could not
change the average time to the next reversal. This "refractory
period" lasts for about 400 ms. With further delay, repellent stimuli decrease the interval to the next reversal. This effect has its
maximum at a delay of about 1 s, thereafter it becomes weaker again,
i.e. the probability for the occurrence of a reversal first increases
significantly and then approaches the original value the closer the
stimulus comes to the next expected spontaneous reversal (Fig. 6).

Attractant stimuli, on the other hand, lead to an increase of the time to the next reversal at any time. No refractoriness can be observed in this case. The effect increases during the first 2 s following a reversal; it decreases with further increase of the delay until the stimulus approaches the next spontaneous reversal (Fig. 7).

The effectiveness of both attractant and repellent stimuli to alter the probability for the occurrence of a reversal is greatest when they are applied 1 - 2 s after the preceding reversal. These results give experimental support to the hypothesis of a periodically changing level of a "regulator," a cellular oscillator, which is altered in opposite directions by attractant and repellent stimuli (submitted for publication).

It has been reported that Halobacterium can integrate signals from PS 370 and PS 565 and also signals from photosensory and chemosensory inputs (Spudich and Stoeckenius, 1979). Simultaneous stimulation by attractant and repellent stimuli through both photosystems showed that opposite signals cancel each other at any time between two reversals, even during the refractory period for repellents. This indicates that repellent stimuli are recognized even during the refractory period and that signal integration may take place in a step prior to response regulation.

7. ADAPTATION AND REVERSIBLE METHYLATION OF MEMBRANE PROTEINS

As mentioned earlier, Halobacterium, after a single response to a stimulus, resumes its normal frequency of spontaneous reversals. There is no difference in sensitivity whether a stimulus is applied after a spontaneous reversal or after a stimulus-dependent one. That means that the cells quickly adapt to constant light conditions.

In chemosensory systems of eubacteria, it has been shown that reversible methylation of membrane proteins is essential for adaptation (Koshland, 1980). Also, in Halobacterium, we found that attractant stimuli, which may be either light stimuli or chemical stimuli, are accompanied by methylation of membrane proteins of a molecular weight of about 60 000 daltons. Like in eubacteria the methyl donor is S-adenosyl methionine. Repellent stimuli, on the other hand, lead to a removal of methyl groups from these proteins (Schimz, 1981, 1982; Bibikov et al., 1982). According to the assumption made for eubacteria, we regard this reversible covalent modification of membrane proteins as a feed-back mechanism which leads to adaptation.

Methylation and demethylation reactions in Halobacterium are influenced by calcium in opposite directions. Methylation is enhanced by lowering the Ca^{2+} concentration, whereas increasing Ca^{2+} concentrations activate demethylation (Schimz, 1982).

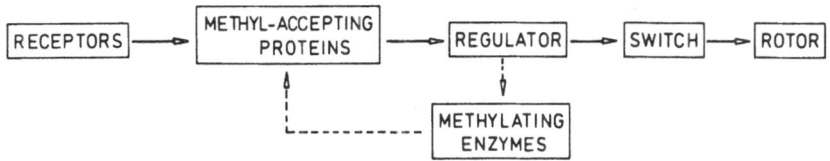

Fig. 8. Scheme of signal transduction (solid lines) and adaptation
(dashed lines) in Halobacterium halobium.

Removal of Ca^{2+} in the experimental saline suppresses reversals
of the swimming direction in Halobacterium (Hildebrand, 1980;
Murvanidze and Glagolev, 1981), while high concentrations of Ca^{2+} in
the presence of an ionophore are reported to cause frequent spontaneous reversals (Murvanidze and Glagolev, 1981). Calcium, therefore,
is thought to be essential for directional changes of flagellar rotation in Halobacterium. Since Ca^{2+} acts on methylation and demethylation in a corresponding manner, we assume that Ca^{2+} may also be
involved in the process of adaptation.

8. WORKING HYPOTHESIS FOR SENSORY TRANSDUCTION IN HALOBACTERIUM

According to the general scheme established for chemosensory
transduction in eubacteria (Koshland, 1980), we assume a causal chain
as presented in Fig. 8.

Our present knowledge is put together in a working hypothesis
shown in Figs. 9 and 10. Upon excitation, the receptor pigment is
assumed to interact with a methyl-accepting protein (cf. Koiwai and
Hayashi, 1979; Zukin et al., 1979), which is also supposed to be the
site of signal integration. As a result of this interaction, the
level of regulator, X, changes in opposite directions upon attractant
(Fig. 9) and repellent stimuli (Fig. 10). The nature of the regulator
is not yet known. It might be the activity of an allosteric enzyme,
the turnover of small transmitter molecules like cyclic nucleotides,
and/or the concentration of a cation. Since calcium seems to play
an important role in the sensory pathway and in adaptation, we consider it possible that calcium itself is at least part of the regulator. The level of X may determine the conformational state of the
switch which in its active form causes a reversal of flagellar rotation. It has been suggested earlier that the conformational change
of the switch may be controlled by calcium (Ordal and Fields, 1977).

Adaptation to constant light conditions is achieved by covalent
modifications, i.e. methylation or demethylation of the methyl-
accepting proteins. The activities of both the methyl-transferase
and esterase are changed by the change in Ca^{2+} concentration, which
leads to a new level of methylation. Reversible methylation counter-
acts the effect of excitation on the Ca^{2+} (X) level and resets the

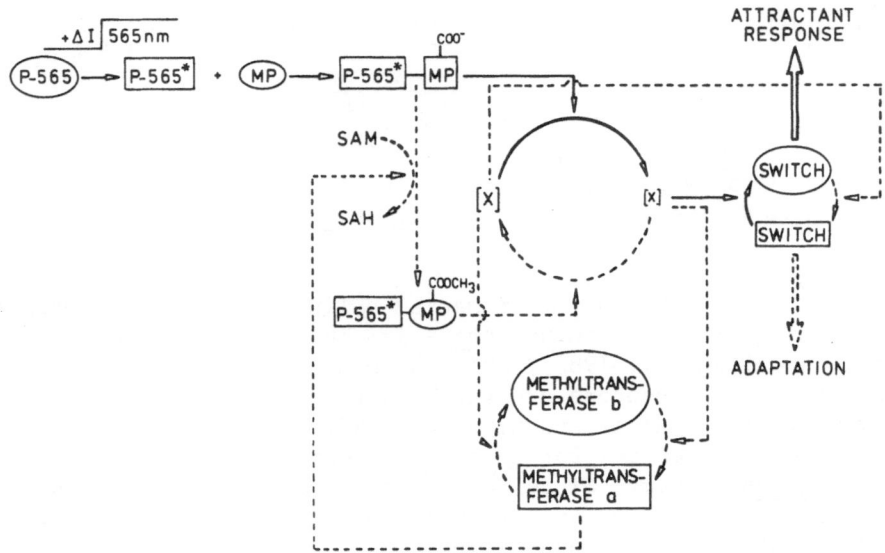

Fig. 9. Working hypothesis for the mechanism of signal transduction
        (solid lines) and adaptation (dashed lines) upon an
        attractant stimulus (light increase at 565 nm). For details,
        see text.

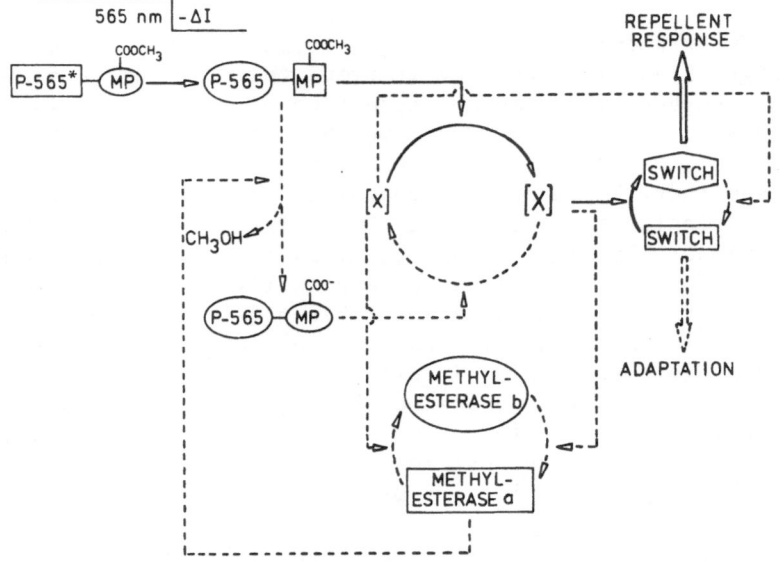

Fig. 10. Working hypothesis for the mechanism of signal transduction
         (solid lines) and adaptation (dashed lines) upon a repellent
         stimulus (light decrease at 565 nm, i.e. removal of an
         attractant). For details, see text.

Ca^{2+} concentration, thereby returning the system to a state equal to that before excitation. The altered level of methylation can be regarded as a memory (cf. Koshland, 1980).

Some steps in these schemes are rather speculative and further work on the molecular mechanisms is necessary for a better understanding of photosensory transduction in Halobacterium.

ACKNOWLEDGEMENTS

We thank Regina König, Ronald Backbier, Dieter Grammig and Helmut Erkens for valuable technical help, and the Deutsche Forschungsgemeinschaft (SFB 160) for financial support.

Thanks to a generous gift of J. L. Spudich, we were able to repeat some of our experiments on the influence of background light with the mutant strain Fl x 3 (see Spudich and Spudich, 1982). The sensitivity of PS 370 at far red background light was found to be the same as in strain WS as shown in Table 1. The response to stimulation of PS 565 at UV background light was also "inverse" as shown in Table 2. We want to stress the point, however, that these results could be well interpreted by the assumption of a photochromic pigment as proposed by Spudich (1985).

REFERENCES

Alam, M., and Oesterhelt, D., 1984, Morphology, function and isolation of halobacterial flagella, J. Mol. Biol., 176:459-475.

Berg, H. C., 1985, Physics of bacterial chemotaxis, in: "Sensory Perception and Transduction in Aneural Organisms," G. Colombetti, F. Lenci and P.-S. Song, eds., Plenum Press, New York.

Bibikov, S. I., Baryshev, V. A., and Glagolev, A. N., 1982, The role of methylation in the taxis of Halobacterium halobium to light and chemo-effectors, FEBS Lett., 146:255-258.

Bogomolni, R. A., and Spudich, J. L., 1982, Identification of a third rhodopsin-like pigment in phototactic Halobacterium halobium, Proc. Natl. Acad. Sci. USA, 79:6250-6254.

Dencher, N. A., and Hildebrand, E., 1979, Sensory transduction in Halobacterium halobium: Retinal protein pigment controls UV-induced behavioral response, Z. Naturforsch., 34c:841-847.

Hildebrand, E., 1980, Comparative discussion of photoreception in lower and higher organisms. Structural and functional aspects, in: "Photoreception and Sensory Transduction in Aneural Organisms, F. Lenci and G. Colombetti, eds., Plenum Press, New York, pp. 319-340.

Hildebrand, E., and Dencher, N., 1975, Two photosystems controlling behavioural responses of Halobacterium halobium, Nature, 257: 46-48.

Hildebrand, E., and Schimz, A., 1983a, Photosensory behavior of a
    bacteriorhodopsin-deficient mutant, ET-15, of Halobacterium
    halobium, Photochem. Photobiol., 37:581-584.

Hildebrand, E., and Schimz, A., 1983b, Consecutive formation of
    sensory photosystems in growing Halobacterium halobium,
    Photochem. Photobiol., 38:593-597.

Koiwai, O., and Hayashi, H., 1979, Studies on bacterial chemotaxis.
    IV. Interaction of maltose receptor with a membrane-bound
    chemosensing component, J. Biochem., 86:27-34.

Koshland, Jr., D. E., 1980, "Bacterial Chemotaxis as a Model Behav-
    ioral System," Raven Press, New York.

Lanyi, J. K., 1982, Spectrophotometric determination of halorhodopsin
    in Halobacterium halobium membranes, in: "Methods in Enzymol-
    ogy," Vol. 88, L. Packer, ed., Academic Press, New York, pp.
    439-443.

Lanyi, J. K., and Weber, H. J., 1980, Spectrophotometric identifica-
    tion of the pigment associated with light-driven primary sodium
    translocation in Halobacterium halobium, J. Biol. Chem., 255:
    243-250.

Lindley, E. V., and MacDonald, R. E., 1979, A second mechanism for
    sodium extrusion in Halobacterium halobium: A light-driven
    sodium pump, Biochem. Biophys. Res. Commun., 88:491-499.

Macnab, R. M., 1985, Biochemistry of sensory transduction in bacteria,
    in: "Sensory Perception and Transduction in Aneural Organisms,"
    G. Colombetti, F. Lenci and P.-S. Song, eds., Plenum Press,
    New York.

Matsuno-Yagi, A., and Mukohata, Y., 1980, ATP synthesis linked to a
    light-dependent proton uptake in a red mutant strain of Halo-
    bacterium lacking bacteriorhodopsin, Arch. Biochem. Biophys.,
    199:297-303.

Murvanidze, G. V., and Glagolev, A. N., 1981, Calcium ions regulate
    reverse motion in phototactically active Phormidium uncinatum
    and Halobacterium halobium, FEMS Microbiol. Lett., 12:3-6.

Ordal, G. W., and Fields, R. B., 1977, A biochemical mechanism for
    bacterial chemotaxis, J. Theoret. Biol., 68:491-500.

Schimz, A., 1981, Methylation of membrane proteins is involved in
    chemosensory and photosensory behavior of Halobacterium halo-
    bium, FEBS Lett., 125:205-207.

Schimz, A., 1982, Localization of the methylation system involved in
    sensory behavior of Halobacterium halobium and its dependence
    on calcium, FEBS Lett., 139:283-286.

Schimz, A., and Hildebrand, E., 1979, Chemosensory responses of
    Halobacterium halobium, J. Bacteriol., 140:749-753.

Schimz, A., Sperling, W., Ermann, P., Bestmann, H. J., and Hildebrand,
    E., 1983, Substitution of retinal by analogues in retinal pig-
    ments of Halobacterium halobium. Contribution of bacteriorho-
    dopsin and halorhodopsin to photosensory activity, Photochem.
    Photobiol., 38:417-423.

Schimz, A., Sperling, W., Hildebrand, E., and Köhler-Hahn, D., 1982,
    Bacteriorhodopsin and the sensory pigment of the photosystem 565
    in Halobacterium halobium, Photochem. Photobiol., 36:193-196.

Sperling, W., and Schimz, A., 1980, Photosensory retinal pigments in Halobacterium halobium, Biophys. Struct. Mech., 6:165-169.

Spudich, E. N., and Spudich, J. L., 1982, Control of transmembrane ion fluxes to select halorhodopsin-deficient and other energy transduction mutants of Halobacterium halobium, Proc. Natl. Acad. Sci. USA, 79:4308-4312.

Spudich, J. L., 1985, Color-sensing by phototactic Halobacterium halobium, in: "Sensory Perception and Transduction in Aneural Organisms," G. Colombetti, F. Lenci, and P.-S. Song, eds., Plenum Press, New York.

Spudich, J. L., and Bogomolni, R. A., 1983a, Spectral and chemical discrimination of hR and sR, Biophys. J., 41, 21a.

Spudich, J. L., and Bogomolni, R. A., 1983b, Spectroscopic discrimination of the three rhodopsin-like pigments in Halobacterium halobium membranes, Biophys. J., 43:243-246.

Spudich, J. L., and Stoeckenius, W., 1979, Photosensory and chemosensory behavior of Halobacterium halobium, Photobiochem. Photobiophys., 1:43-53.

Stoeckenius, W., Lozier, R. H., and Bogomolni, R. A., 1979, Bacteriorhodopsin and the purple membrane of halobacteria, Biochim. Biophys. Acta, 505:215-278.

Strange, P. G., and Koshland, Jr., D. E., 1976, Receptor interactions in a signalling system: Competition between ribose receptor and galactose receptor in the chemotaxis response, Proc. Natl. Acad. Sci. USA, 73:762-766.

Sumper, M., Reitmeier, H., and Oesterhelt, D., 1976, Biosynthesis of the purple membrane of halobacteria, Angew. Chem. Int. Ed. Engl., 15:187-194.

Traulich, B., Hildebrand, E., Schimz, A., Wagner, G., and Lanyi, J. K., 1983, Halorhodopsin and photosensory behavior in Halobacterium halobium mutant strain L-33, Photochem. Photobiol., 37: 577-580.

Wagner, G., 1984, Blue light effects in halobacteria, in: "Blue Light Effects in Biological Systems," H. Senger, ed., Springer Verlag, Berlin, Heidelberg, pp. 48-54.

Zukin, R. S., Hartig, P. R., and Koshland, Jr., D. E., 1977, Use of a distant reporter group as evidence for a conformational change in a sensory receptor, Proc. Natl. Acad. Sci. USA, 74:1932-1936.

COLOR-SENSING BY PHOTOTACTIC HALOBACTERIUM HALOBIUM

John Spudich

Albert Einstein College of Medicine
1300 Morris Park Avenue
Bronx, NY 10461

PHOTOSENSORY BEHAVIOR

Halobacterium halobium, a flagellated bacterium which grows in near saturated brine, is phototactic: the cells are attracted to long wavelength visible light (the "red" light response) and repelled by shorter wavelength light (the "blue" light response) (Hildebrand and Dencher, 1975). In the absence of photostimuli, the motility pattern is a three-dimensional random walk resulting from short periods of swimming interrupted by spontaneous reversals which re-orient the cell's swimming direction. Light intensity gradients bias the random walk by modulating the frequency of reversals (Spudich and Stoeckenius, 1979). These spatial gradients are sensed by the cell as time-dependent changes in light intensity incurred during its translational motion. Abrupt temporal changes in light intensity mimic the spatial gradients, causing changes in reversal probability, and can be used to assay the cell's sensory responses. For example, decreases in blue light intensity or increases in red light intensity are interpreted by the cell as movement in a favorable direction, and therefore these stimuli cause a period of low reversal probability ("smooth" swimming). This response is transient: the cells adapt to the new light intensity and resume their prestimulus behavior reversal probability[*].

---

[*]The type of photosensitive motility response exhibited by bacteria would be called "photoklinokinesis" rather than "phototaxis" by some investigators, who use the latter term in a more restrictive sense (see discussion in Spudich and Stoeckenius, 1979).

Vitamin A aldehyde (retinal), which forms the chromophore of
visual pigments, is also required for photosensory reception in
halobacterial phototaxis. This is clear because cells chemically
(Dencher and Hildebrand, 1979) or genetically (Spudich and Bogomolni,
1983) blocked in retinal synthesis lack phototaxis responses to both
red and blue stimuli and full sensitivity is restored by retinal
addition. Two retinal-containing pigments, bacteriorhodopsin (bR)
and halorhodopsin (hR), function as light-driven ion transporters in
H. halobium (Stoeckenius and Bogomolni, 1982). Both of these pigments absorb in the region of attractant (red) sensitivity and on
this basis were initially considered as candidates for the attractant
receptor. Mutants lacking both bR and hR were found to exhibit normal
color-discriminating phototaxis (Spudich and Spudich, 1982) indicating
at least one additional retinal pigment must exist in this organism.

PHOTOCHEMISTRY OF SLOW-CYCLING RHODOPSIN (sR)

We have detected a third photoreactive retinal-containing
membrane protein in H. halobium (Bogomolni and Spudich, 1982). Like
bR and hR, photoexcitation of the third pigment is followed by a sequence of thermal steps which returns the molecule to its original
state. This cyclic process (or "photocycle") has a half-time of
~0.8 s at 23°, in contrast to the faster (~10 ms) photocycles of bR
and hR. The slow-cycling rhodopsin (sR) does not appear to mediate
active ion transport (Spudich and Spudich, 1982; Bogomolni and
Spudich, 1982) and is the only spectroscopically detected retinal
pigment in phototactic strains which lack bR and hR (Spudich and
Bogomolni, 1983; Bogomolni and Spudich, 1982). Addition of retinal
to a mutant blocked in retinal synthesis generates sR and phototaxis
with similar kinetics and retinal concentration dependence (Spudich
and Bogomolni, 1983; Ehrlich et al., 1984). These properties and
its absorbance in the region of attractant (red) sensitivity
(Spudich and Bogomolni, 1983) make sR a likely candidate for the
attractant photoreceptor.

The repellent photoreceptor, on the other hand, poses a dilemma
because no separate pigment active in the region of repellent (blue)
sensitivity is detectable by sensitive flash photolysis analysis
(detection limit 50 molecules/cell, Bogomolni and Spudich, unpublished). Recent experiments conducted by Roberto Bogomolni and
myself show that a solution to this dilemma can be found within the
sR photocycle. The repellent receptor is expected to absorb maximally near 370 nm according to phototaxis action spectra (Hildebrand
and Dencher, 1975; Traulich et al., 1983) and sR photoexcitation
generates in μs an intermediate species with maximal absorbance at
373 nm (S_{373}) (Bogomolni and Spudich, 1982). Because of its fast
rise and slow decay, S_{373} accumulates in significant amounts in
light absorbed by sR_{587} producing a photostationary state mixture
of the two species: sR_{587} and S_{373}.

We considered that sR in its S_{373} form might mediate the repellent responses. If this were correct, no repellent responses
should occur in the absence of S_{373} and repellent sensitivity should
appear when S_{373} is generated by illumination of sR_{587}. The key to
testing these predictions was to construct an optical arrangement
which permitted us to observe the cells swimming behavior (i) under
conditions in which sR was not photoexcited, and (ii) upon selective
photoexcitation of sR_{587} and S_{373} in a quantitative manner.

The absorption spectrum of sR_{587} is so broad that any visible
light will generate S_{373}. Because of this all published measurements
of phototaxis responses have been made in the presence of S_{373},
generated by the light used to observe the bacteria. To avoid exciting sR_{587}, we observed the cells with light of wavelengths >750
nm using an infrared sensitive RCA camera (Model 2000), and tracked
the cells on a video monitor. This light is beyond the absorption
range of sR_{587} and therefore no S_{373} is generated. Under these
conditions, strong UV light which previously has been found to be
strongly repellent caused no change in reversal probability.
Further, addition of red background light which generates S_{373}
sensitizes the cells to the repellent stimulus, causing nearly
every cell to reverse swimming direction. Quantitation of swimming
reversals in a variety of stimulus regimes shows a strict correlation
between S_{373} photoexcitation and the repellent response, and sR_{587}
photoexcitation and the attractant response (Spudich and Bogomolni,
submitted).

SIGNAL TRANSDUCTION BY SLOW-CYCLING RHODOPSIN (sR)

As shown in the photoreaction pathways on the left (Fig. 1),
photoexcitation of the dark form of sR (sR_{587}) generates a short-
lived species (S_{680}) which decays rapidly into S_{373}. The S_{373}

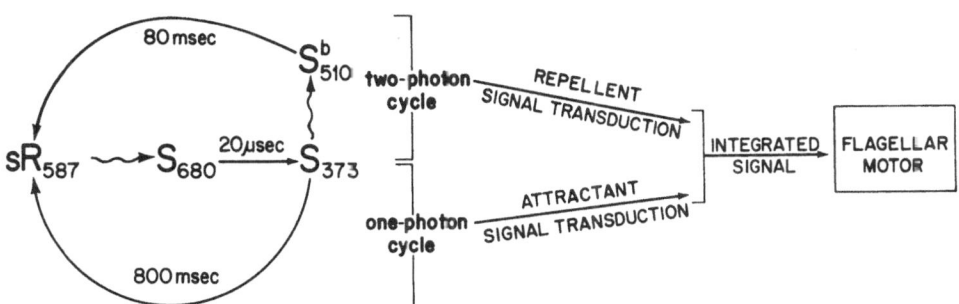

Fig. 1. Mechanism of color-sensing based on the dual role of sR as
         attractant and repellent photosensory receptor. The Figure
         summarizes the photoreactions of sR and their proposed
         coupling to the flagellar motor.

form can return to the sR_{587} form by either of two paths: a purely
thermal relaxation or a photoinduced path. If S_{373} does not absorb
a photon, it relaxes thermally to sR_{587} completing the "one-photon"
photocycle. During its transit through the one-photon cycle, the
sR molecule generates an attractant signal to the flagellar motor.
If S_{373} absorbs a photon, it generates the intermediate S^b510, which
does not occur in the thermal path of S_{373} decay. The relaxation
of S^b510 to sR_{587} completes a cycle containing two photon absorption
steps. After absorption of the second photon in the "two-photon"
cycle, sR generates a repellent signal to the flagellar motor.

Spectral discrimination by sR can be understood as an interplay
between the two-photon cycle discussed above and the one-photon
cycle generated by photoexcitation of sR_{587} and its subsequent
thermal steps. The one-photon reaction generates an attractant
signal, which is transmitted to the flagellar motor by an attractant
signal transduction pathway. Increases (decreases) in this signal
suppress (induce) flagellar reversals. These effects occur in the
absence of any background illumination. In contrast, the two-photon
cycle generates a repellent signal, transmitted and amplified separately from the attractant signal, but eventually integrated before
or at the site of action on the flagellar motor (Spudich and
Stoeckenius, 1979). Increases (decreases) in this signal induce
(suppress) flagellar reversals. Both induction and suppression of
reversals are transient responses; i.e. the cell adapts to the
signals. The net result is that the cell's motility apparatus responds to changes in the numbers of sR molecules undergoing the one-
photon and two-photon cycles.

Previous studies have shown that the cells are at least an
order of magnitude more sensitive to repellent (blue) light than to
attractant (red) light intensity changes (Hildebrand and Dencher,
1975; Spudich and Stoeckenius, 1979). According to our interpretation, this sensitivity difference must be conferred by amplification
step(s) beyond the photoreceptor, because there can never be more
repellent receptor molecules (S_{373}) than the maximum number of molecules available for attractant reception (sR_{587}), and from flash
photolysis measurements we know their photochemical quantum yields
are nearly equal. Therefore, the repellent signal from the two-
photon cycle must be at least an order of magnitude more effective
in inducing reversals than the one-photon cycle is in suppressing
them.

The relative amplification in the two signaling pathways appears
to depend on the specific conditions of cell culture growth (Spudich
and Bogomolni, unpublished). Measurements of sensitivities at different stages of growth indicate the cells maintain a high constant
sensitivity to blue light in exponential and stationary phase, while

increasing the amplification in the attractant transduction pathway
as they progress into stationary phase (Sundberg and Spudich, unpublished). Variation in the amplification may explain the apparent
consecutive formation and "separation" of attractant and repellent
sensitivities under various conditions (Hildebrand and Schimz,
1983).

The absorption spectrum of S_{373} matches closely the ~ 370 nm
action spectrum peak reported for phototaxis repulsion (Hildebrand
and Dencher, 1975; Traulich et al., 1983). However, the absorption
spectrum of sR_{587} matches some but not all of the published action
spectra for phototaxis attraction. The attractant response was
reported to have a single peak at ~ 565 nm in wild-type (Hildebrand
and Dencher, 1975), but two peaks: one at ~ 565 nm and the other
at ~ 590 nm in a bR-deficient strain (Traulich et al., 1983). More
studies are needed to understand the variations in action spectra.
The action spectra for bR⁻hR⁻ double mutants (Spudich and Spudich,
1982) would be especially valuable, since these are free of possible
complicating effects of photoexcitation of bR and hR during attractant stimulation. Further discussion of this appears in Spudich
(1984, in press).

The color-sensing mechanism described here depends on the existence of two spectrally distinct signaling forms of the same receptor. This could be a unique property of the sensory rhodopsin
of halobacteria. Alternatively, the sR mechanism may reflect an
inherent potential of retinal/protein complexes -- photoreversibility
of photoconversion products of the retinal/opsin chromophore --
important to the evolution of the rhodopsin family as photosensory
receptors. Even retinal pigment photoproducts which are too short
lived to accumulate under physiological conditions (such as the K,
L, and M intermediates formed from bR and the early photoproducts
formed from mammalian rod rhodopsin) are photoreversible. Inverte-
brate rhodopsins, like sR and like the non-retinal plant chromoprotein phytochrome, have photocycles sufficiently slow to cause significant photostationary state accumulation of their photoreversible
conformations. The sR mechanism, preserved in the archaebacterium,
H. halobium, may have provided a primitive color-sensing capability,
which in some organisms was maintained in the evolution of visual
systems, while in others was replaced by multiple receptor mechanisms
with increased color-resolution.

ACKNOWLEDGEMENTS

During this work the author was supported by NIH grant GM27750,
Jane Coffin Childs Memorial Fund Project Grant No. 360, and an
Irma T. Hirschl Trust Career Scientist Award.

REFERENCES

Bogomolni, R. A., and Spudich, J. L., 1982, Identification of a
    third rhodopsin-like pigment in phototactic Halobacterium
    halobium, Proc. Natl. Acad. Sci. USA, 79:6250-6254.
Dencher, N. A., and Hildebrand, E., 1979, Sensory transduction in
    Halobacterium halobium: Retinal protein pigment controls
    UV-induced behavioral response, Z. Naturforsch., 34:841-847.
Ehrlich, B. E., Schen, C. R., and Spudich, J. L., 1984, Bacterial
    rhodopsin monitored with fluorescent dyes in vesicles and in
    vivo, J. Memb. Biol., in press.
Hildebrand, E., and Dencher, N. A., 1975, Two photosystems controlling
    behavioural responses of Halobacterium halobium, Nature, 257:
    46-48.
Hildebrand, E., and Schimz, A., 1983, Consecutive formation of sensory
    photosystems in growing Halobacterium halobium, Photochem.
    Photobiol., 38:593-597.
Spudich, E. N., and Spudich, J. L., 1982, Control of transmembrane
    ion fluxes to select halorhodopsin-deficient and other energy-
    transduction mutants of Halobacterium halobium, Proc. Natl.
    Acad. Sci. USA, 79:4308-4312.
Spudich, J. L., 1984, Genetic demonstration of a sensory rhodopsin in
    bacteria, in: "Information and Energy Transduction in Biological
    Membranes," E. Helmreich, L. Bolis, and H. Passow, eds., Alan
    R. Liss, Inc., New York (in press).
Spudich, J. L., and Bogomolni, R. A., 1983, Spectroscopic discrimi-
    nation of the three rhodopsinlike pigments in Halobacterium
    halobium membranes, Biophys. J., 43:243-246.
Spudich, J. L., and Stoeckenius, W., 1979, Photosensory and Chemo-
    sensory behavior of Halobacterium halobium, J. Photobiochem.
    Photobiophys., 1:43-53.
Stoeckenius, W., and Bogomolni, R. A., 1982, Bacteriorhodopsin and
    related pigments of halobacteria, Annu. Rev. Biochem., 52:587-
    616.
Traulich, B., Hildebrand, E., Schimz, A., Wagner, G., and Lanyi,
    J. K., 1983, Halorhodopsin and photosensory behavior in
    Halobacterium halobium mutant strain L-33, Photochem. Photobiol.,
    37:577-579.

BEHAVIORAL STUDIES OF FREE-SWIMMING PHOTORESPONSIVE ORGANISMS

Mary Ella Feinleib

Department of Biology
Tufts University
Medford, MA 02155

Photomovement in microorganisms provides us with tractable
model systems for studying sensory transduction, while also offering
a delightful variety of behaviors to examine. Behavior can be
studied at several levels. Once we are familiar with the phenomenology of a photoresponse, we can use that behavior to ask questions
about the response mechanism. Information can be obtained from determining action spectra, from examining responses of selected mutants, and from studying "behavior" of isolated cell parts, such as
flagella. After reviewing the various types of photomovement found
in microorganisms, this chapter will focus on behavioral studies in
the biflagellated alga Chlamydomonas.

GENERAL FEATURES OF PHOTOMOVEMENT

Types of Photoresponse

Photomovement "strategies" in microorganisms show a diversity
of evolutionary solutions to the same problem: how to move into an
optimally illuminated region of the environment. The mechanisms for
detecting that optimal region are more complex in the case of light
stimuli than in the case, for example, of chemical stimuli. (See
Macnab, this volume.) The spatial concentration gradient of a
chemical stimulus is identical with the direction of that stimulus.
Thus, movement toward the region of highest concentration coincides,
in effect, with movement toward the source of the chemical. On the
other hand, the spatial intensity[*] gradient of light does not

---

[*]The common term "intensity" will be used in this chapter in place
 of the more correct term "fluence rate."

necessarily coincide with the direction of light. Thus, movement
toward the region of highest light intensity may not be the same
as movement toward the light source.

Accumulation of microorganisms in an illuminated region (a
"light trap") often occurs as the result of a behavior that has
nothing to do with the perception of light direction; namely a photophobic response (Diehn et al., 1977). In a typical photophobic response, a sudden change in light intensity elicits a transient change
in the activity of the organism. In many microorganisms, the photophobic response includes a brief cessation of forward movement (stop
response), followed by a change in the direction of movement. The
response pattern is "programmed" with respect to the morphology of
the organism and is independent of light direction. Photophobic
response may be elicited by an increase (step-up) in light intensity
or by a decrease (step-down) in intensity. A response typically
lasts for a few seconds, after which the organism becomes adapted to
the new light level and resumes normal movement.

One can readily see how "step-down photophobic responses" can
lead to photoaccumulation. If every organism in the population
changes direction when it crosses the border from a lighted area
into a dark area, whereas none shows a direction change upon entering
the light, the organisms will accumulate in the light. Conversely,
"step-up photophobic responses" can lead to photodispersal from a
light trap.

By contrast, phototaxis is a movement that is oriented with
respect to the direction of the stimulus light. An organism may
move toward the light source (positive phototaxis) or away from it
(negative phototaxis). The distinction between light direction and
intensity gradient has been illustrated dramatically by observing
the behavior of a flagellate in a converging light beam, under conditions known to elicit positive phototaxis (Buder, 1915; Halldal,
1959). The cells swam toward the light source, but away from the
region of highest light intensity.

A third type of photomovement merits only brief mention here.
In photokinesis, the steady-state rate of movement is controlled by
the absolute magnitude of the light intensity. Because light acts
as an energy source for photokinesis, this phenomenon is not,
strictly speaking, a photosensory response (Nultsch, this volume).

Photophobic Responses

In bacteria, photomovement is characterized by photophobic responses. The effectiveness of this strategy can be seen in
Thiospirillum, a photosynthetic sulfur bacterium (Buder, 1915;
Pfennig, 1968; Nultsch, 1975). At one pole of this corkscrew-shaped
cell is a bundle of flagella which rotates rigidly as a unit,

generating both torque and thrust. (See Berg, this volume.) When
the flagellar bundle extends out from the pole, it functions as a
"push propeller," causing the cell to rotate about its longitudinal
axis, while also propelling it forward. When Thiospirillum encounters
a sudden decrease in light intensity, as happens when it crosses a
border from light to dark, the flagellar bundle flips over like an
umbrella being opened over the pole of the cell. The flagellar
bundle reverses its sense of rotation, now acting as a "pull propeller," and the cell reverses its direction of movement. Repeated
step-down photophobic responses of this type lead to a dense accumulation of cells in a light trap.

The photoresponses of Halobacterium are discussed in detail in
this volume, by Hildebrand and by Spudich. This bacterium shows
spontaneous reversals of direction during normal swimming. Light
may increase the frequency of reversals or may cause their suppression, depending on whether the stimulus is in the "attractant" or the
"repellent" range of the spectrum, and on whether it is step-up or
step-down. For example, a step-up of light intensity in the repellent
wavelength range (ca. 370 nm) causes an increase in frequency of reversals, and can therefore lead to dispersal of cells from a light
trap (Hildebrand, 1983).

Various photophobic response patterns are found in blue-green
algae and in eukaryotic microorganisms (Nultsch, this volume; Diehn,
this volume; section on Chlamydomonas, below). Despite their diversity, all photophobic responses require that the organism collect
the same kind of information about the stimulus; namely, it must be
able to detect a temporal change in light intensity.

Perception of Light Direction in Phototaxis

The distinction between photophobic response and phototaxis
becomes significant in those organisms that display both kinds of
behavior, most notably the motile algae. Photophobic responses and
phototaxis both occur in gliding algae, including blue-greens,
diatoms, and desmids, as well as in flagellated algae (Nultsch and
Häder, 1979; Feinleib, 1980; Colombetti and Lenci, 1983; Häder, 1984;
Diehn, this volume). These two kinds of behavior have also been
reported in some ciliates; e.g. Stentor (Song et al., 1983; Song,
this volume), and in the cellular slime-mold Dictyostelium, in both
the amoeboid and the pseudoplasmodial stages (Poff and Loomis, 1973;
Häder and Poff, 1979 a and b; Poff, this volume).

Unlike photophobic response, phototaxis demands that the organism have a means of perceiving light direction, by detecting either
a spatial gradient or a temporal gradient of light absorption. An
absorption gradient may be established by a variety of mechanisms,
including refraction and attenuation (by absorption or scattering)
(Feinleib, 1980; Poff, this volume). In the hypothetical examples

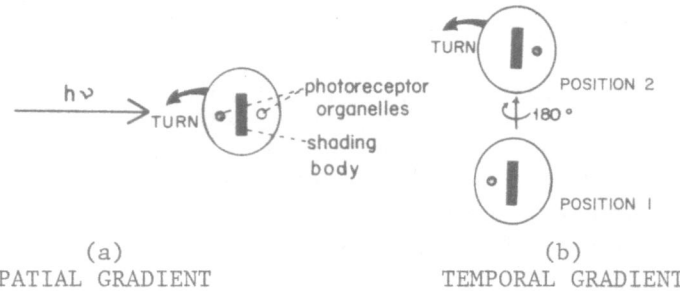

(ONE-INSTANT MECHANISM)   (TWO-INSTANT MECHANISM)

(a)   (b)
SPATIAL GRADIENT   TEMPORAL GRADIENT

Fig. 1.   Examples of two fundamental mechanisms for detecting light
          direction in a hypothetical cell (shown in longitudinal
          section). (a) Spatial gradient or one-instant mechanism.
          (b) Temporal gradient or two-instant mechanism. In both
          examples shown, a light-absorption gradient is established
          by attenuation. The photoreceptor organelles are shaded
          via absorption by another pigmented structure in the cell,
          a "shading body" (oriented perpendicular to the plane of
          the section). The arrows marked "turn" indicate the
          initial direction of response in positive phototaxis. See
          text for further details (after Feinleib, 1980).

shown in Fig. 1, the photoreceptor organelles are shaded via absorption by another pigmented structure in the cell.

     A simple hypothetical case of a spatial-gradient or "one-instant"
mechanism is shown in Fig. 1a (Feinleib, 1975, 1980). Two photoreceptor "organelles" are separated by a screen of shading pigments
(a shading body), oriented perpendicular to the plane of the organelles. If a stimulus beam enters from the left, the photoreceptor
organelle on the left side of the cell absorbs more light than the
organelle on the right side. To account for positive phototaxis in
this scheme, one can postulate that the cell always turns toward the
side where the photoreceptor pigment absorbs maximally--in this case,
the left side.

     The essential feature of a one-instant mechanism is that light
direction is perceived by comparing the light absorbed in two photoreceptive regions of the cell at one instant in time--essentially
instantaneously. It is doubtful that any flagellate uses a one-
instant mechanism for phototactic orientation, but this type of
mechanism presumably occurs in some gliding organisms, e.g. Anabaena
(Nultsch, this volume).

A rapidly moving organism such as a flagellate may well perceive
light direction using a temporal-gradient or "two-instant" mechanism.
There are various possible versions of this model. The hypothetical
case depicted in Fig. 1b corresponds to the orientation mechanism
generally believed to occur in Euglena (Diehn, 1973, 1979; Lenci and
Colombetti, 1978). A two-instant mechanism for detecting light direction requires only one photoreceptive point ("organelle"), but it
demands that some optical asymmetry be built into the structure of
the organism. The photoreceptor organelle may be asymmetrically
located relative to other light-absorbing cell components, as in the
case shown. Alternatively, it may be asymmetrically shaped or it may
be asymmetrically positioned in a transparent cell which refracts
light (Feinleib, 1980; Poff, this volume).

The essential feature of a two-instant mechanism, in any version,
is that light direction is perceived by comparing the light absorbed
in one photoreceptive region of the cell at two instants in time. In
continuous unilateral light, the cell obtains two successive absorption measurements by changing its position with respect to the
light source.

The hypothetical cell in Fig. 1b rotates about its longitudinal
axis as it swims forward from position 1 to position 2. This kind of
movement is, in fact, characteristic of flagellates. As the cell
rotates, the shading body periodically comes between the light source
and the photoreceptor, modulating the light reaching the photoreceptor. (Accordingly, this scheme is often referred to as the
"Periodic-shading" mechanism.) Positive phototaxis can be accounted
for in this model by postulating that, whenever the cell perceives a
step-down signal, it turns toward its shading-body side. When the
cell is in position 1, the photoreceptor faces the light source; no
step-down signal occurs. As the cell rotates into position 2, however, the shading body enters into the light path, attenuating the
light reaching the photoreceptor. In response to this step-down
signal, the cell turns about its lateral axis, toward its shading-
body side--which also happens to be the side facing the light source.

Typically, a cell would perform a series of "course corrections"
of this type (one per rotation about its longitudinal axis), until
it is heading toward the light source. When the axis of the swimming
path is aligned with the stimulus-beam axis, no further modulation of
the light occurs; the photoreceptor is continually illuminated.
Thus, in positive phototaxis, the step-down response just described
acts as part of a feedback mechanism, keeping stimulation of the
photoreceptor at a constant level.

It should be noted that this step-down response is a behavior
typical of photophobic responses. The fundamental "programmed"
response of this hypothetical cell is a turn toward its shading-
body side, elicited by a step-down signal. In this case, however,

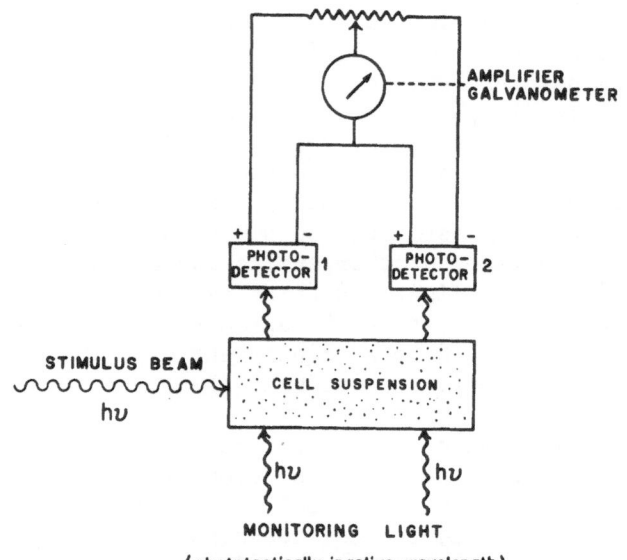

Fig. 2.   Schematic diagram of a population system for measuring
          phototaxis, viewed from above. For the sake of simplicity,
          several essential components have been omitted (e.g. heat
          filters, filters to shield photodetectors from stimulus
          light). The photodetectors are connected in a comparison
          circuit. Asymmetrical distribution of cells as a result
          of phototaxis is registered as a difference in optical
          density measured by the two photodetectors. See text for
          details. (Adapted from Feinleib and Curry, 1967, and
          Colombetti and Lenci, 1983.)

the step-down signal is not a result of cell movement from a lighted
zone into a dark zone of its environment. Instead, the signal is
created <u>internally</u>, by rotation of the cell relative to the light
source. This two-instant model for phototaxis has some intriguing
implications. First, it suggests that phototactic orientation could
result from a series of photophobic responses; that is exactly what
is thought to occur in <u>Euglena</u>. The second and more fundamental im-
plication, true for all two-instant mechanisms, is that the information needed to detect light direction in phototaxis is the same as
that needed for photophobic response: the cell must be able to
detect a temporal change in light intensity.

Methods for Studying Photomovement

     The photophobic and phototactic behavior of microorganisms has
been studied using "population systems" and "individual-cell" systems
(Häder, 1984, and this volume). Population systems are basically

photometric devices that monitor the net movement of large numbers
of cells in response to light (Lindes et al., 1965; Feinleib and
Curry, 1967; Nultsch and Throm, 1975; Ascoli et al., 1978; Pfau et al.,
1983). Methods such as cinematography, streak photomicrography, and
videomicroscopy have been used to record movements of individual
cells (Feinleib and Curry, 1967; Hand and Schmidt, 1975; Boscov and
Feinleib, 1979; Barghigiani et al., 1979; Bessen et al., 1980;
Häder et al., 1981; Nultsch, personal communication).

A schematic drawing of a simple "population system" for measuring
phototaxis is shown in Fig. 2. A monitoring light of a wavelength
ineffective in eliciting phototaxis passes through a chamber containing a suspension of motile cells, and impinges on a pair of
photodetectors connected in a comparison circuit. Initially, the
cells are homogeneously distributed; the same intensity of monitoring
light reaches both detectors and there is no difference in their output. A stimulus beam is then directed into one end of the chamber.
If that stimulus elicits positive phototaxis, the cells swim toward
the light source. As they do so, the optical density in the region
of photodetector-1 becomes higher than the density in the photode-
tector-2 region. This asymmetrical cell distribution is registered
as a difference in output of the two photodetectors.

An automated, computerized version of this kind of device has
been built by Nultsch and his co-investigators (Pfau et al., 1983).
In that apparatus, a continuous culture system is coupled with a
population system for recording phototaxis. Population systems are
convenient because they can record rapidly the net response of
millions of cells. An inherent problem of such systems, however, is
that they cannot differentiate between a change in swimming speed
and a change in straightness of swimming path. The system of Pfau
et al. (1983) has been designed to solve that problem by including
an automated assay of motility; the number of cells passing through
a very small field per unit time is monitored photometrically.

In order to interpret the photobehavior of microorganisms unambiguously, one must monitor the movements of individual cells
(Colombetti and Lenci, 1983). The most common individual-cell method
in current use is videomicrography. (See example in Fig. 4, below).
Tracing hundreds of paths of individual cells projected on a video
monitor is quite accurate, but tedious. Recently, semi-automated
and automated video techniques have been introduced, including computer-aided analysis (Häder et al., 1981; Häder, this volume;
Lipson and Häder, 1984; Hand, personal communication; Morel-Laurens,
unpublished).

BEHAVIORAL STUDIES IN CHLAMYDOMONAS

The green flagellate Chlamydomonas offers advantages for
studying phototaxis. It displays exceptionally clear-cut phototactic

orientation, as well as a vivid photophobic response. Moreover,
Chlamydomonas lends itself well to genetic analysis and many mutants
are available, modified in pigmentation and flagellar structure.
In the last few years, there has been a blossoming of information
and ideas about photomovement in Chlamydomonas, allowing us to begin
developing models for the underlying sensory-transduction pathways
(Nultsch, 1983).

As in other green flagellates, photomovements in Chlamydomonas
are primarily "blue-light" phenomena, independent of the photosynthetic machinery. The main peak of the phototactic action spectrum
lies between 500 nm and 510 nm, with a secondary maximum at ca. 440
nm (Nultsch et al., 1971; Nultsch, 1983). Red light is ineffective
as a stimulus. (Carotenoids and flavins have been suggested as
candidates for the photoreceptor pigment, but a recent report pro-
poses rhodopsin as the favored candidate; Nultsch, 1983; Foster et
al., 1984; see section on "The Photoreceptor System.") Like
Euglena, Chlamydomonas has a reddish stigma (or "eyespot") which
contains carotenoids. However, this structure differs from the
stigma of Euglena in that it is intraplastidic, lying just under
the chloroplast envelope, and is composed of several layers (see
below).

Chlamydomonas normally swims in a kind of "breast-stroke,"
propelling the cell forward and causing it to rotate about its
longitudinal axis. The flagellar waveforms during normal forward
swimming (and during backward swimming) have been analyzed in detail,
making use of a uniflagellated mutant of Chlamydomonas, uni-1 (Brokaw
et al., 1982; Brokaw and Luck, 1983). Like other flagellates,
Chlamydomonas follows a helical path, with the same side of the cell
directed toward the axis of the helix. Recent photomicrographs by
Kamiya and Witman (1984) show wild-type (WT) C. reinhardtii swimming
in a left-handed helix (viewed from behind), with the stigma directed
toward the outside of the helix.

The phenomenology of photomovement in Chlamydomonas is well
known. A dramatic stop response (photophobic response) can be observed if cells kept in a background of red light are exposed to a
6 µs high-intensity flash of blue or white light. Analyzed in slow
motion, the stop response reveals a complex pattern (Boscov and
Feinleib, 1979). The cell stops within 50 ms after a flash, remains
stationary for ca. 100 ms, and then swims backward by about one cell
length. Next, the cell may stop again, or move laterally or backward. In any event, within 1 s after the flash, the cell resumes
normal forward swimming, usually in a new direction. We think that
this photophobic behavior is a response to the step-up in intensity
at the initiation of the flash, because a similar response often
occurs at the onset of continuous illumination, whereas it is rarely
observed when the light is turned off (Feinleib and Curry, 1971; see
also Smyth and Berg, 1982).

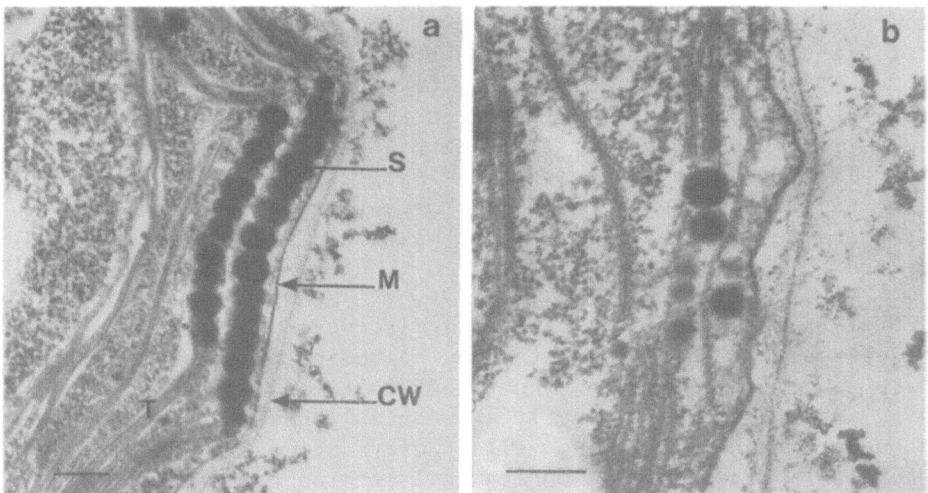

Fig. 3. Stigma region of Chlamydomonas in longitudinal section.
         S = stigma granule, M = electron-dense layer consisting of
         chloroplast envelope and plasma membrane, CW = cell wall,
         T = thylakoid. Fixation with glutaldehyde, post-fixation
         with OsO₄. Calibration bars: 0.2 µm. (a) WT. Note
         electron density and hexagonal packing of stigma granules;
         (b) Mutant ey 627. Note variability in size and spacing
         of stigma granules, and irregular spacing between membrane
         layers.

     During phototactic orientation, Chlamydomonas steers quite pre-
cisely toward or away from the light source. As the light intensity
is raised, the swimming paths become more closely aligned with the
light beam, both in positive and in negative phototaxis (Feinleib
and Curry, 1971).

     The identity of the photoreceptor and the mechanism of detecting
light direction have yet to be established unequivocally for
Chlamydomonas. Recently, however, data from several lines of evidence have converged, pointing to a plausible site and mechanism for
the early events of photosensory perception.

Photomovement in "Eyeless" Mutants

     Behavioral studies of a so-called "eyeless" mutant of C.
reinhardtii (ey 627; Hartshorne, 1953; Smyth et al., 1975) reveal
that the stigma is essential for normal phototactic orientation, but
not for the photophobic stop response (Morel-Laurens and Feinleib,
1983). Figure 3a is an electron micrograph of the stigma region of
a WT cell, in longitudinal section. The stigma is 1.0-1.3 µm long,

and consists of two to four evenly spaced rows (viewed in section)
of osmiophilis granules. Most of the granules are uniform in size;
smaller granules may be found at the ends of a row. Each row is
subtended by a thylakoid. Overlying the whole structure are the
chloroplast envelope and the plasma membrane. These membranes are
more electron dense in the stigma region than they are elsewhere
in the cell. In three dimensions, the stigma forms a curved plate,
parallel to the surface of the cell (Morel-Laurens and Bird, in
press; see also Lembi and Lang, 1965; Walne and Arnott, 1967;
Foster and Smyth, 1980).

Morel-Laurens and Bird (in press) discovered that the "degree
of eyelessness" in the ey mutant could be manipulated by controlling
the rate of cell division in synchronous cultures. In a rapidly
dividing mutant culture, in early exponential phase, stigmata are
almost undetectable. In those cells where a stigma is visible, the
structure is rudimentary; a longitudinal section through the stigma
region of such a cell is shown in Fig. 3b. One can see only a few
scattered granules, variable in size and spacing. The spacing between membrane layers also appears to be irregular.

Surprisingly, the "eyeless" mutant can produce a complete stigma
under the right conditions. In a stationary-phase culture, in which
cell division is very slow, many mutant cells show a stigma with
an ultrastructure similar to that of the WT.

Photomovement behavior in these ey mutants was studied using the
videomicroscope system shown in Fig. 4 (Morel-Laurens and Feinleib,
1983). A high-resolution video camera equipped with an IR-sensitive
Newvicon tube is mounted over a microscope. The algal suspension
is placed on the microscope stage in a clear plastic chamber and the
cell movements are monitored from below with a dim red light. The
image is viewed on a video monitor and is simultaneously taped. The
stimulus source is an Ar ion laser with continuous output in the
blue-green, at 488 nm. The stimulus beam enters from one side of
the chamber, permitting us to record oriented movement (phototaxis)
toward or away from the light source. Videotapes are examined by
single frames, and the path of each cell is traced.

We found that "eyeless" mutants taken from a culture in early
exponential phase, when stigmata were either absent or rudimentary,
were capable of phototaxis, but their orientation was severely impaired. Typical tracings of WT and mutant swimming paths in red
light are shown in Fig. 5, a and b, respectively. Cell position
is shown every 1/3 s for 2 s. In both cases, the swimming paths
are fairly straight and there is no evidence of a "preferred direction." Figures 5c and d show swimming paths for WT and mutant,
respectively, during exposure to the continuous blue-green stimulus
at high intensity: 100 mW·cm^{-2}. Tracing was begun 20 s after onset
of stimulation. The WT cells swim almost parallel with the stimulus

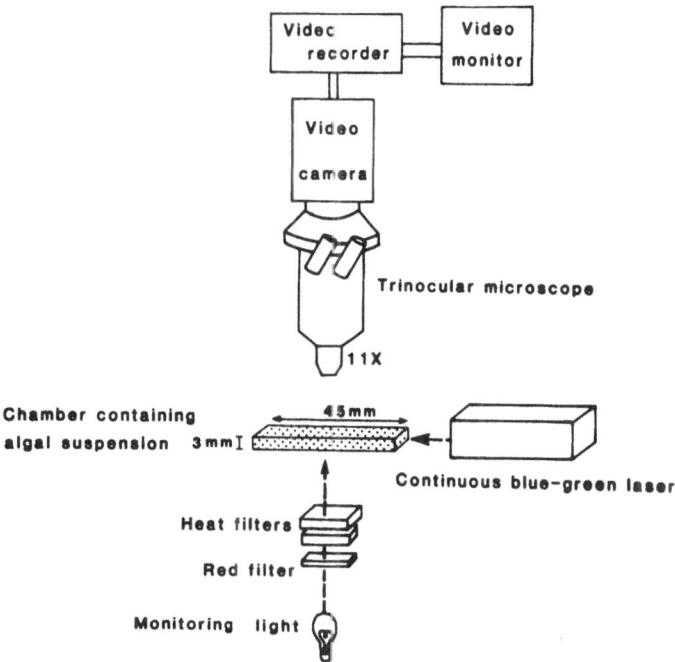

Fig. 4.   Schematic diagram of videomicroscope system used to record
photomovement of individual cells of WT and ey mutant of
Chlamydomonas.

beam, heading away from the light source. By contrast, the swimming
paths of the mutant are highly erratic, with frequent changes in
direction. Observation of these swimming paths does not suffice for
detecting a preferred direction in the mutant.

Straightness of path and preferred direction were analyzed using
vectorial methods described by Batschelet (1965, 1972, 1981) and by
Burr (1979). To determine the swimming direction of each cell, we
measured the polar coordinates of successive positions in the swimming
path, thereby obtaining a vector for each segment of the path. Since,
for the purpose of this analysis, we were interested in the direction
rather than the rate of travel, each segment of the path was assigned
an arbitrary length of 1: this is called a unit vector. Thus, all
segments of the path were weighted equally, regardless of the distance travelled. The individual unit vectors were then averaged to
obtain a mean vector for the path of that cell; i.e. the average
direction taken by the cell. The length of the mean vector, r, provides an index of straightness of path; it is, in essence, a measure
of the clustering of the positions of the cell about its mean direction of travel. The maximum value, r = 1, means the cell has a
perfectly straight path. In this experiment, the median r-value for
the swimming paths of the WT cells was 0.94; the median r-value for
the mutant cells was 0.36.

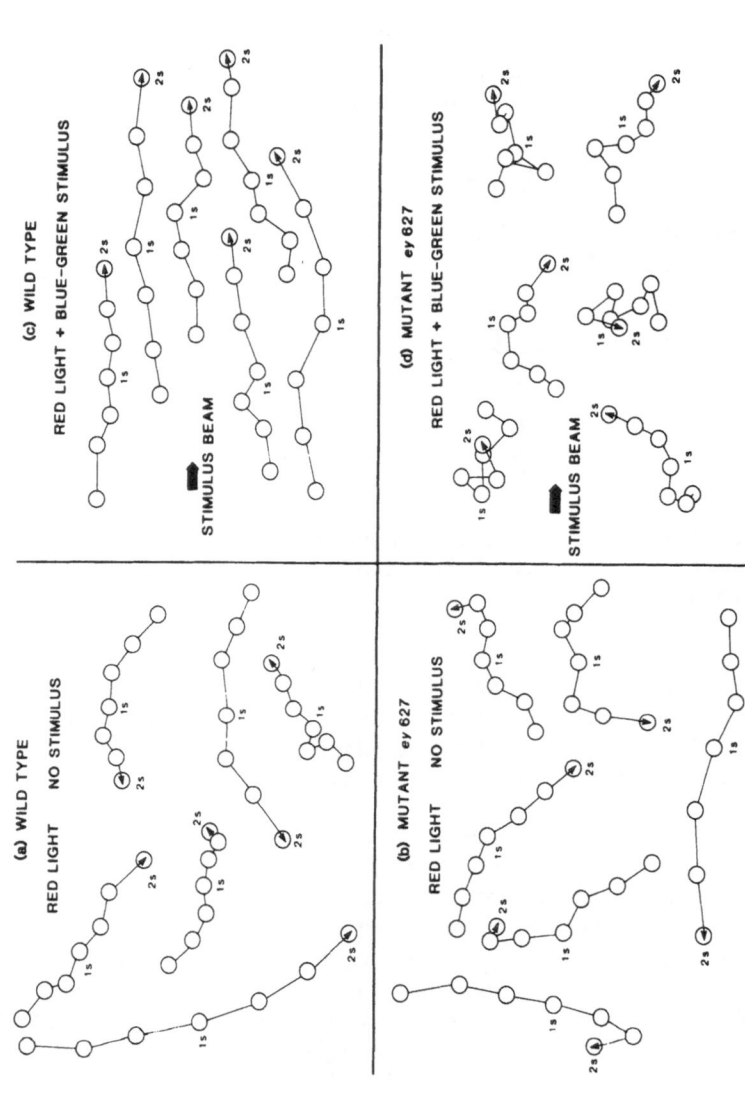

Fig. 5.   Tracings of swimming paths of Chlamydomonas.   Cell positions are shown every 1/3 s for 2 s.   Arrows indicate direction of movement.   (a) WT cells in red light only; (b) ey mutant cells in red light only; (c) WT cells from 20 to 22 s after onset of unilateral stimulation with blue-green light; (d) mutant cells 20 5o 22 s after onset of unilateral stimulation with blue-green light (after Morel and Feinleib, 1983).

A similar type of analysis was applied to cell populations. The mean vector for a sample of cells, i.e. the preferred direction for those cells, was calculated by averaging the mean vectors for the individual swimming paths. Data were analyzed for statistical significance using the Rayleigh test or the V test (Batschelet, 1972; Zar, 1974; Durand and Greenwood, 1958; Burr, 1979).

The relative numbers of cells heading in various directions can readily be illustrated by plotting their mean vectors on a polar-wedge histogram. Figure 6a shows the percentage of WT cells swimming in various directions in red light. No preferred direction is evident. The mutant cells show a similar distribution in red light (not illustrated here).

Figure 6b shows the distribution of WT cells 20-24 s after the onset of stimulation with blue-green light. The laser beam enters at 0°. The cells display a strong negative phototaxis, with 90% swimming within an angle of ±30° of the stimulus-beam axis. The arrow at the center of the histogram represents the mean vector for this sample of cells (exactly 180° in this case), and the length of that arrow is an index of what has been called the "strength of phototaxis."[*]

Figure 6c is a comparable histogram for a sample of ey mutant cells, from a culture in early exponential phase, 20-24 s after onset of stimulation. The ey mutant is also negatively phototactic, but orientation is much weaker than in the WT. The "strength of phototaxis" is 0.39. (Circular distribution statistics were applied to data from experiments of this type; the orientation of the mutant was found to be significantly non-random; Morel-Laurens and Feinleib, 1983.)

By contrast, Figure 6d shows the response of mutant cells taken from a stationary-phase culture, in which many of the cells had a stigma similar to that of WT (Morel-Laurens and Feinleib, unpublished). A majority of these cells show a clear negative phototaxis, swimming within an angle of ±30° of the stimulus-beam axis, like the WT cells.[**] This histogram suggests that, when the stigma of the mutant is almost normal, the orientation process is almost normal.

---

[*]The "strength of phototaxis" is defined by Burr (1979) as the length of that component of the mean vector which lies along the predicted direction of orientation under the V test.
[**]A small number of the cells in this sample show positive phototaxis. We observed a similar phenomenon in other samples of the mutant cells from stationary-phase cultures. We do not yet know why these cells respond somewhat differently than the WT at the same stimulus intensity.

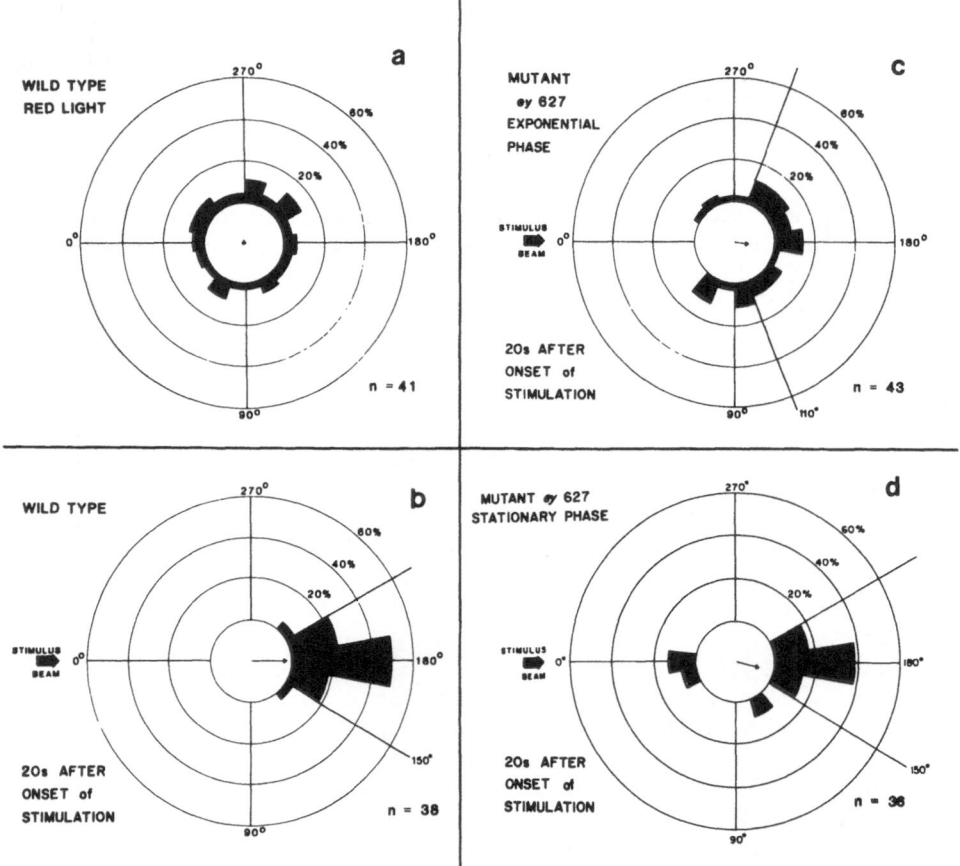

Fig. 6.   Polar-wedge histograms illustrating distribution of swimming
          directions in Chlamydomonas cells. Percentages of cells
          with mean vectors in various directions are plotted on a
          circle at 20° intervals. The arrow represents the mean
          vector for the cell sample; its length is proportional to
          r, with r = 1 set equal to the radius of the inner circle.
          (a) WT in red light; (b) WT during illumination with blue-
          green stimulus, entering at 0°; (c) ey mutant from culture
          in early exponential phase, during illumination with blue-
          green stimulus; (d) ey mutant from culture in stationary
          phase, during illumination with blue-green stimulus. (a-c
          after Morel and Feinleib, 1983).

     Despite the observed differences in phototaxis, all cells
tested, mutant and WT, showed a distinct stop response upon

introduction of the stimulus, indicating that the stigma is not
necessary for this photophobic behavior (Morel-Laurens and Feinleib,
1983). Stigmata of Chlamydomonas have recently been isolated and
identification of pigments is in progress (Morel-Laurens and Martin,
unpublished).

We now know that the stigma is involved in phototactic orientation, but how does that orientation work? Behavioral studies by
Boscov and Feinleib (1979) are consistent with a two-instant orientation mechanism, but the mechanism may be quite different from that
of Euglena.

The Photoreceptor System

The region of plasma membrane peripheral to the stigma has been
proposed as the photoreceptor site for photomovement in Chlamydomonas
(Walne and Arnott, 1967; Foster and Smyth, 1980). A freeze-fracture
electron-micrograph study of the plasma membrane (and the outer
chloroplast membrane) in the region of the stigma lends credence to
this hypothesis: intramembranous particles in this region occur at a
higher density and show a different size distribution than in other
areas of the cell (Melkonian and Robenek, 1980). It has been suggested that some of these particles may represent photoreceptor molecules. To date, no pigment has been isolated from that region; nor
has one been identified by microspectrophotometry. However, Foster
et al. (1984) recently reported evidence that the photoreceptor is a
rhodopsin. When retinal analogs were incorporated into a "blind"
mutant of Chlamydomonas (defective in its photomovement), normal
photobehavior was restored. The peak of the phototactic action
spectrum depended on which retinal analog was added. Only 11-cis
and trans-retinal, the chromophore of bovine rhodopsin, yielded an
action-spectrum peak corresponding to that of WT Chlamydomonas.

An intriguing alternative to the "periodic-shading" mechanism
described above for Euglena has been proposed for phototactic orientation in Chlamydomonas and other algae with layered eyespots (including many Chlorophyceae, and some Dinophyceae and Crysophyceae;
Foster and Smyth, 1980). The stigma of Chlamydomonas is presumed to
act as an interference reflector or quarter-wave stack, strongly
reflecting light onto a photoreceptor region peripheral to it. As
shown schematically in Fig. 7, a quarter-wave stack consists of
alternating thin layers of high and low refractive index. If the
optical pathlength between adjacent layers is equal to $\frac{1}{4}$ λ of a light
incident normal to the surface of the stack, reflection of the light
will be strong due to constructive interference. The reflected waves
will reinforce one another (see explanation, Fig. 7 caption). Moreover, there will be constructive interference between the incident
wave and the reflected waves, e.g. at a distance of $\frac{1}{4}$ λ from the
front surface of the stack, creating a maximum average intensity at
that distance.

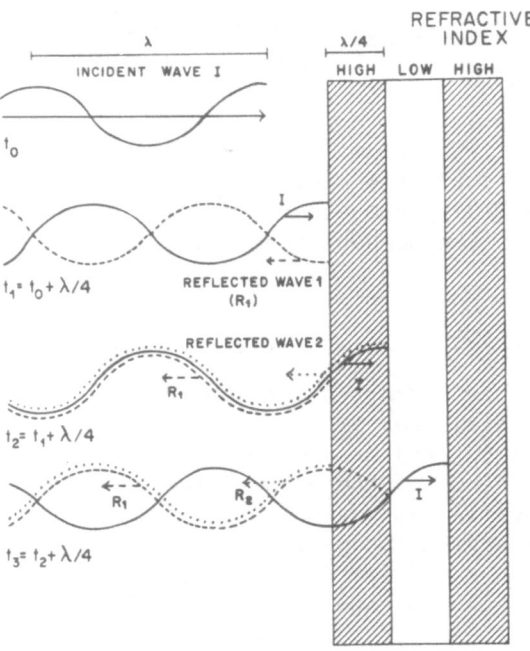

QUARTER-WAVE STACK

Fig. 7.   Schematic illustration of quarter-wave stack. The stack
          consists of alternating layers of high refractive index
          (cross-hatched) and low refractive index. Incident light
          of wavelength λ (solid line) travels to the right, normal
          to the stack surface. The optical pathlength of each layer
          of the stack is $\frac{1}{4} \lambda$. The vertical succession of sketches
          (t_0 - t_3) represents the incident wave and its reflected
          waves at four successive times. The interval between two
          successive sketches equals the time each wave takes to ad-
          vance $\frac{1}{4} \lambda$. At t_1, the incident wave impinges on a surface
          of high refractive index. Part of the light is reflected,
          undergoing a 180° phase inversion. That reflected wave
          (R_1, dashed line) travels to the left. At t_2, the incident
          wave reaches a layer of low refractive index and is re-
          flected without inversion (R_2, dotted line). The two re-
          flected waves are now in phase and will remain in phase
          (t_3), because R_1 has undergone a 180° phase inversion, while
          R_2 has undergone a 180° phase delay (time to move $\frac{1}{4} \lambda$ in and
          $\frac{1}{4} \lambda$ out). The reflected waves thus reinforce each other.
          There will also be constructive interference between the
          incident wave and the reflected waves, e.g. at a distance
          of $\frac{1}{4} \lambda$ in front of the stack, creating a maximum average
          intensity at that distance.

Foster and Smyth measured the distances between layers of a _Chlamydomonas_ stigma in an electron micrograph prepared using a spray-freezing technique (Linder and Staehelin, 1979). Based on these measured distances and on assumptions about the refractive indices of the layers, Foster and Smyth concluded: (i) that the eyespot has the characteristics required of an interference reflector, (ii) that the photoreceptor is located in the plasma membrane, $\frac{1}{4} \lambda$ from the front surface of the reflector, and (iii) that maximal reflectivity for light incident normal to the surface would occur at ca. 480 nm.[*] This wavelength is close to the maximum for the phototactic action spectrum (Nultsch et al., 1971; Nultsch, 1983). With incident illumination, one can indeed observe under the microscope that the eyespot of _Chlamydomonas_ reflects light.

According to Foster and Smyth, this arrangement of quarter-wave stack eyespot plus photoreceptor functions as a highly directional light-sensitive "antenna," analagous to a radar antenna. Like a radar detector, the algal antenna is tilted with respect to the axis of rotation. If the axis of the cell's rotation (the axis of its swimming path) is aligned perfectly with a light beam, the antenna receives an almost constant signal. But if the cell deviates from that path, as it would do frequently because of noise, the antenna receives an "error signal." This signal fluctuates periodically as the antenna changes its direction with respect to the light source during cell rotation, yielding information about the direction of the deviation (and possibly about its magnitude). The cell responds by making course corrections to minimize fluctuations in the signal (much as in the periodic-shading mechanism described above for _Euglena_). To establish whether this interference-reflector version of a two-instant mechanism operates in _Chlamydomonas_, precise optical measurements should be made on intact antennae--no easy task.

Behavioral Studies on Sensory Transduction: The Role of Calcium

The rapid photoresponses of _Chlamydomonas_ require rapid transmission of information from the photoreceptor (presumably associated with the stigma), to the flagella, several μm away. There is now excellent evidence that Ca is involved in the sensory-transduction pathway leading to flagellar reorientation, both in photophobic and in phototactic behavior of _Chlamydomonas_. (The role of Ca in controlling ciliary reversal in the "avoiding reaction" of _Paramecium_ has been thoroughly documented; see Machemer, this volume. Calcium is also involved in control of flagellar reorientation in _Euglena_; Doughty and Diehn, 1982.)

---

[*]These calculations of Foster and Smyth were based on a simplified "ideal" quarter-wave stack model, assuming, e.g. that the layers are perfectly parallel and almost infinite in length.

During the backward-swimming phase of a photophobic response in
Chlamydomonas, the flagella undergo a transient change from beating
in the normal "forward mode" (breast-stroke) to beating in a "reverse
mode." The flagella extend forward, nearly parallel to each other,
and propagate relatively symmetrical planar waves, undulating from
base to tip (Schmidt and Eckert, 1976; Brokaw and Luck, 1983). In
general, the flagella of Chlamydomonas have been found to beat in the
forward mode at low external Ca concentrations and in the reverse
mode at high Ca concentrations. However, the Ca concentration at
which reversal of swimming mode occurs depends on the experimental
system used.

Lowering the concentration of free Ca in the medium can inhibit
photostimulated flagellar reversal in living cells (Schmidt and
Eckert, 1976). In a control solution containing 10^{-3} M Ca,
Chlamydomonas (wall-less mutant CW 92) displayed a typical photophobic response to a flash of white light, including a period of
flagellar beating in the reverse mode. At 10^{-5} M free Ca, backward
swimming was retarded. At 10^{-6} M Ca, some cells swam forward, some
swam backward, and some oscillated back and forth with no net movement. Below 10^{-6} M Ca, backward swimming was completely inhibited;
cells continued to swim forward despite the flash stimulus. This
inhibition could be reversed simply by restoring cells to the control solution. However, photostimulated backward swimming was inhibited irreversibly by D-600, known to block voltage-sensitive Ca
channels. In these intact cells, changing external Ca levels did
not affect the mode of swimming unless the cells were also stimulated
by light.

By contrast, an isolated flagellar apparatus could be induced
to change its mode of swimming in the absence of photostimulation,
merely by controlling the Ca concentration in the solution (Hyams
and Borisy, 1978). "Behavioral studies" were conducted on flagellar
apparatus (including basal bodies) isolated from a wall-less mutant
of Chlamydomonas and reactivated in a solution containing ATP. At
free Ca concentrations at or below 10^{-6} M, the isolated apparatus
swam forward, using a normal "breast-stroke." At concentrations
above 10^{-6} M, the apparatus swam backward, with the flagella beating
in the "reverse mode." The swimming direction could repeatedly be
reversed by changing the Ca concentration appropriately. Flagellar
apparatus demembranated by treatment with detergent showed similar
behavior as a function of external Ca level, except that the response
curve was shifted by one log unit toward lower Ca concentrations.
Hyams and Borisy concluded that Ca exerts an effect on photomovement
by acting at the level of the flagella.

This elegant study was carried one step further by Bessen et
al. (1980), who examined the "behavior" of isolated axonemes of
Chlamydomonas. They were able to induce single axonemes obtained
from C. reinhardtii to beat in either mode characteristic of living

cells, by manipulating the free Ca level in the reactivation solution. At a relatively high concentration of free Ca (10^{-4} M) multiple-flash dark-field micrographs revealed individual axonemes swimming along almost straight paths, undulating in the "symmetrical" waveform typical of reverse swimming. At free Ca concentrations at or below 10^{-6} M, the axonemes changed over to the "asymmetrical" waveform characteristic of forward swimming, i.e. each axoneme executed one-half of a breast-stroke! (In most cases, axonemes in low-Ca solution became attached by one end to the slide or cover slip, but the rest of the shaft was free to bend and to pivot about the point of attachment.)

Because their axonemal preparations were free of basal bodies and associated structures, Bessen et al. were able to conclude that Ca binds directly to an axonemal component to modify the flagellar waveform. A candidate for the Ca binding site is calmodulin, which has been identified in the axonemal fraction of Chlamydomonas flagella, as well as in the "membrane-plus-matrix" fraction and in the cell body (Gitelman and Witman, 1980).

A simple scheme has been proposed for flagellar reversal in the photophobic response of Chlamydomonas, based on the model introduced by Naitoh and Eckert (1974) for ciliary reversal in Paramecium (Hyams and Borisy, 1978; Bessen et al., 1980). Photostimulation leads to membrane depolarization, which in turn results in Ca influx, presumably via opening of voltage-sensitive Ca channels. Calcium binds to an axonemal site (possibly in the central tubule-central sheath complex; Witman et al., 1978), and effects a transient reversal in flagellar beating. At the end of a photophobic response, Ca is pumped out, the resting membrane potential is restored, and the flagella resume their normal "forward mode" of beating. (Photoinduced potential changes have been recorded in the green biflagellate Haematococcus; Litvin et al., 1978; Ascoli and Ristori, 1980.)

Calcium also plays a major role in determining the strength and direction (sign) of phototaxis in Chlamydomonas (Nultsch, 1979, 1983, and personal communication; Morel-Laurens, personal communication). It is important to note that Chlamydomonas always swims in the "forward mode" (breast-stroke) during phototactic orientation, whether phototaxis is positive or negative.

At very low external Ca concentrations, phototaxis is strongly inhibited, whereas motility is only partially impaired. Using an automated population system coupled to a homocontinuous culture apparatus (Pfau et al., 1983), Nultsch grew Chlamydomonas in media to which various concentrations of Ca had been added, and measured both phototaxis and motility photometrically (personal communication). If Mg was substituted for Ca in the culture medium, reducing free Ca to ca. 6×10^{-7} M, the cells grew well, but their phototaxis was decreased to less than 10% of its maximal value (recorded at 10^{-4} M

Ca). In that same low-Ca medium, motility was decreased to a mean
of only 50% of its control value, but was found to be highly variable. Microcinematography revealed many non-motile cells, which
tended to form clumps.* Those cells which remained motile varied
widely in swimming speed, from 10% to 90% of the control value.
Upon addition of Ca (to a final concentration of ca. 10^{-5} or 10^{-4} M),
normal motility was restored.

The dependence of phototaxis and motility on free Ca concentration has recently been investigated in short-term experiments, using
a video-microscope system (Morel-Laurens, personal communication).
Cells were cultured in medium containing 10^{-3} M Ca and were transferred to a Ca test solution just prior to phototactic assay. The
level of free Ca in the test solution was controlled using EGTA and
the final free Ca concentration was calculated. At the control level
of 10^{-3} M free Ca, phototaxis was strongly negative and motility was
normal. At very low free Ca (ca. 5×10^{-9} M), phototaxis disappeared
completely; video recordings of swimming paths revealed no evidence
of oriented response to light.

Retention of good motility at that very low Ca level required
careful handling of the cells during transfer to the test solution
(e.g. spinning them out of the growth medium at slow speed). Under
these carefully controlled conditions, the swimming rate of cells in
a solution containing ca. 5×10^{-9} M Ca remained about the same as
that of control cells for at least half an hour. These data demonstrate that phototaxis can be inhibited completely while motility
remains almost normal.

The differential effects of Ca on phototaxis and motility in
Chlamydomonas have been interpreted as indicating that Ca probably.
acts at two sites in the sensory-transduction pathway for phototaxis
(Nultsch, 1983, and personal communication). Nultsch has proposed a
working hypothesis for the transduction pathway, which proceeds
approximately as follows. Unilateral irradiation excites the photoreceptor molecules in a membrane (plasma membrane or outer chloroplast envelope) overlying the stigma, causing a conformational change
in the photoreceptor protein. This change leads to a transient increase in Ca permeability of the plasma membrane in that region.
The resulting Ca influx causes a local membrane depolarization. This
"receptor potential" may then spread electrotonically to the region
of the flagellum on the stigma side of the cell. Voltage-sensitive
Ca channels now open in the membrane of that flagellum, causing a
regenerative Ca influx. The Ca activates an axonemal component (as
in the model for photophobic response described above), effecting a

---

*Similar abnormalities in motility at low Ca had been described by
 Morel-Laurens (personal communication).

change in the beat pattern of the flagellum on the irradiated side
of the cell. As a result, the cell turns toward (or away from) the
light. Nultsch pointed out a major unknown in the pathway: What
determines whether phototaxis is positive or negative.

Recent work indicates that the "switch" from positive to negative phototaxis in Chlamydomonas may also be under Ca control.
Nultsch has measured phototaxis of a cell population as a function
of Ca added to the medium, from ca. 6×10^{-7} M to 10^{-3} M (personal
communication). The resulting dose-response curve bears a striking
resemblance to published curves for net response vs. light intensity
(e.g. Feinleib and Curry, 1971). At low Ca concentrations, net
phototaxis is positive. Below 10^{-5} M Ca, response is weak (less
than 20% of its maximal value), but it then rises sharply to its
maximum at 10^{-4} M Ca. As Ca concentration is raised beyond that
level, increasing numbers of cells become negatively phototactic.
As a result, net response decreases and finally become negative at
5×10^{-3} M Ca.

Morel-Laurens (personal communication) is able to "mimic" the
effects of increasing light intensity on phototaxis by increasing the
free Ca concentration. However, she reports that the Ca concentration at which the positive-to-negative "switch" occurs is strongly
influenced by other factors: growth phase of the culture, stimulus
intensity, and time of observation following the onset of stimulation.

Presumably, when a Chlamydomonas cell is illuminated from one
side, Ca exerts a differential effect on the two flagella, leading to
orientation. It is tempting to imagine that activity of one flagellum is altered under conditions leading to positive phototaxis,
whereas the other flagellum is affected in negative phototaxis. There
is now compelling evidence that the two flagella are differentially
affected, both by light and by Ca.

In the uni-mutant of Chlamydomonas (mentioned earlier), photostimulation leads to a transient decrease in beat frequency (Smyth
and Berg, 1982). This observation is particularly interesting because the uni-mutant always retains the same flagellum: the so-
called trans flagellum, located on the side of the cell opposite
from the stigma (Fig. 8). Because the cell has only one flagellum,
normal beating in the "forward mode" causes it to spin in place
instead of propelling it forward in the helical path characteristic
of WT cells. In the far-red/infrared monitoring light, the flagellum was found to beat with a frequency of 50 - 80 Hz. A 2 s pulse of
blue light elicited a transient decrease in flagellar beat frequency
(by 10-20% in the cells for which response was analyzed in detail).

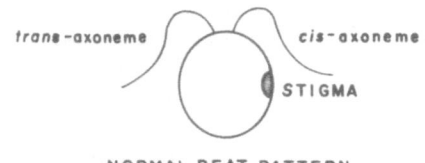

NORMAL BEAT PATTERN

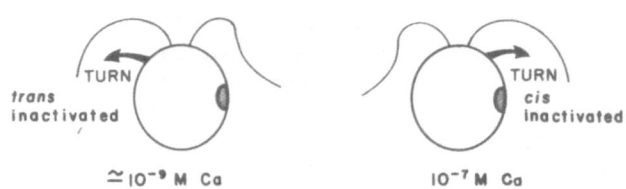

Fig. 8. Schematic illustration of differential effects of Ca on
axonemal beating in demembranated cell models of Chlamydomonas (based on data of Kamiya and Witman, 1984). Top,
for reference: normal forward mode of flagellar (or
axonemal) beating, as in intact cells. Bottom left:
inactivation of the trans axoneme at very low free Ca
(ca. 10^{-9} M); the cis axoneme continues to beat normally.
This pattern of beating would turn the cell away from the
side where the stigma is located. Bottom right: inactivation of the cis axoneme at slightly higher free Ca
(10^{-7} M); the trans axoneme continues to beat normally.
This pattern of beating would turn the cell toward its
stigma side. In each case, the inactivated axoneme is
held in a rigid arc.

Most of the cells that responded did so upon the step-up in
light intensity at the beginning of the pulse. As the authors suggest, if a normal cell responded in that way to a step-up in light
intensity, i.e. if its trans flagellum became inactive relative to
its cis flagellum, the cell would turn away from its stigma side,
initiating negative phototaxis (as in Fig. 8, bottom left). According to Smyth and Berg, a decrease in flagellar beat frequency in
response to a flash probably accounts for the decrease in swimming
rate (ca. 10%) observed by Boscov and Feinleib (1979) in cells
turning in response to a flash. This explanation seems highly
plausible.

Calcium control of differential flagellar activity has been
elegantly demonstrated using detergent-extracted demembranated
"models" of Chlamydomonas (Kamiya and Witman, 1984). These cell
models are not photoresponsive, but their swimming "behavior" can be
modified by manipulating the free Ca level in the reactivation

solution. The sketch in Fig. 8 (based on the data of Kamiya and Witman) shows that, at very low free Ca (ca. 10^{-9} M) the trans axoneme was inactivated, while the cis axoneme continued to beat normally. This differential activity would turn the cell away from the side where the stigma is located. At free Ca concentrations of 10^{-6} or 10^{-7} M, the opposite effect was observed: the cis axoneme was inactivated, while the trans axoneme continued to beat normally. This pattern of beating would turn the cell toward the stigma side. Dark-field photomicrography revealed that the inactivated axoneme was held in a rigid arc.

At 10^{-8} M free Ca, most cell models swam in a helical path, much like that of the WT, with the stigma directed toward the outside of the helix. Curiously, when cell models were introduced into a low free-Ca solution (ca. 10^{-9} M), many of them initially followed this kind of helical path. Others swam in circles, with the stigma facing outward. With time, more and more of the cell models swam in circles, and these circles became progressively smaller, suggesting that the trans axoneme was becoming progressively inactivated in the low-Ca solution. Conversely, when a cell model was first placed in a high free-Ca solution (10^{-6} M or 10^{-7} M), it swam in a helical path, with the stigma directed toward the axis of the helix. Again, helical swimming gave way to swimming in small circles, as the cis axoneme became progressively inactivated. (Inactivation of the trans flagellum at very low Ca was also observed in intact cells.)

If we assume that the photoreceptor for phototaxis lies just peripheral to the stigma, we might begin to draw up a simple model, reasoning on the basis of the work of Kamiya and Witman. We know that, in a solution of low Ca concentration, the cis flagellum remains active, turning the cell away from the stigma side (Fig. 8, bottom left). We might, therefore, predict that a stimulus light which brings about a lowering of internal Ca would elicit negative phototaxis. Conversely, we know that, in solutions containing high Ca, the trans flagellum remains active, turning the cell toward the stigma side (Fig. 8, bottom right). Thus, we might predict the occurrence of positive phototaxis at high internal Ca. However, phototaxis experiments to date might lead us to predict just the opposite role for Ca! Phototaxis was positive in solutions of low Ca concentration and negative at high Ca (Nultsch, personal communication; Morel-Laurens, personal communication).

This discrepancy of results between phototaxis experiments and flagellar-activity experiments would best be resolved if the two kinds of study on Ca effects were conducted under closely comparable conditions: using the same culture medium, the same algal strain, and the same range of Ca concentrations. In practice, comparable studies will be difficult to conduct. A major problem is that internal Ca concentration cannot readily be manipulated in an intact cell--but only an intact cell will respond phototactically.

An alternative approach is to modify the model predicted by
flagellar activity studies to fit the data from phototaxis studies,
and then to test that modified scheme. One way of altering the
model in an attempt to fit the data is to postulate a latency of
one-half the rotation period before the cell turns in response to a
step-up (or step-down) stimulus. In elaborating this scheme, one
must take into account such factors as the rotation rate of the
cell, the latency and duration of any observed decrease in flagellar
beat frequency (Smyth and Berg, 1982), and the exact position of the
stigma relative to the plane of the flagella (Nultsch, personal
communication).

An ideal behavioral experiment to do next would be to immobilize
an intact Chlamydomonas cell by holding it by its "tail" end, direct
a microbeam of blue light at the stigma, and observe what the flagella do and when they do it. Techniques for mastering this trick
are currently being developed (Foster, personal communication;
Nultsch, personal communication; Witman, personal communication).

REFERENCES

Ascoli, C., Barbi, M., Frediani, C., and Mure, A., 1978, Measurement
    of Euglena motion parameters by laser light scattering tech-
    niques, Biophys. J., 24:585-599.
Ascoli, C., and Ristori, T., 1980, Proc. Natl. Congr. Cybern.
    Biophys., Pisa.
Barghigiani, C., Colombetti, G., Franchini, B., and Lenci, F.,
    1979, Photobehavior of Euglena gracilis: Action spectrum for
    the step-down photophobic response of individual cells, Photo-
    chem. Photobiol., 29:1015-1019.
Batschelet, E., 1965, Statistical Methods for Analysis of Problems
    in Animal Orientation and Navigation, AIBS Monograph,
    Washington, DC.
Batschelet, E., 1972, Statistical methods in the analysis of problems
    in animal orientation and certain biological rhythms, in: "NASA
    Symposium on Animal Orientation and Navigation," S. R. Galles,
    K. Schmidt-Koenig, G. J. Jacob and R. F. Belleville, eds.,
    NASA, Washington, pp. 61-91.
Batschelet, E., 1981, "Circular Statistics in Biology," Academic
    Press, London.
Bessen, M., Fay, R. B., and Witman, G. B., 1980, Calcium control of
    waveform in isolated flagellar axonemes of Chlamydomonas, J.
    Cell Biol., 86:446-455.
Boscov, J. S., and Feinleib, M. E., 1979, Phototactic response of
    Chlamydomonas to flashes of light. II. Response of individual
    cells, Photochem. Photobiol., 30:499-505.
Brokaw, C. J., and Luck, D. J. L., 1983, Bending patterns of Chlamy-
    domonas flagella. I. Wild-type bending patterns, Cell
    Motility, 3:131-150.

Brokaw, D. J., Luck, D. J. L., and Huang, B., 1982, Analysis of the movement of Chlamydomonas flagella: The function of the radial-spoke system is revealed by comparison of wild-type and mutant flagella, J. Cell Biol., 92:722-733.

Buder, J., 1915, Zur Kenntnis des Thiospirillum jenense und seine Reaktionen auf Lichtreize, Jahrb. wiss. Botan., 56:529-584.

Buder, J., 1917, Zur Kenntnis der phototaktischen Richtungsbewegungen. Jahrb. wiss. Botan., 58:105-220.

Burr, A. H., 1979, Analysis of phototaxis in nematodes using directional statistics, J. Compar. Physiol., 134:85-93.

Colombetti, G., and Lenci, F., 1983, Photoreception and photomovements in microorganisms, in: "The Biology of Photoreception," D. J. Cosens and D. Vince-Prue, eds., Symp. Soc. Exp. Biol. XXXVI, pp. 399-422.

Diehn, B., 1973, Phototaxis and sensory transduction in Euglena, Science, 181:1009-1015.

Diehn, B., 1979, Photic responses and sensory transduction in protists, in: "Handbook of Sensory Physiology," VII/6A, H. Autrum, ed., Springer Verlag, Berlin, pp. 23-68.

Diehn, B., Feinleib, M. E., Haupt, W., Hildebrand, E., Lenci, F., and Nultsch, W., 1977, Terminology of behavioral responses of motile microorganisms, Photochem. Photobiol., 26:599-560.

Doughty, M. J., and Diehn, B., 1982, Photosensory transduction in the flagellated alga, Euglena gracilis. III. Induction of Ca^{2+}-dependent responses by monovalent cation ionophores, Biochim. Biophys. Acta, 682:32-43.

Durand, D., and Greenwood, T. A., 1958, Modifications of the Rayleigh test for uniformity in analysis of two-dimensional orientation data, J. Geol., 66:229-238.

Feinleib, M. E., 1975, Phototactic response of Chlamydomonas to flashes of light. I. Response of cell populations, Photochem. Photobiol., 21:351-354.

Feinleib, M. E., 1980, Photomotile responses in flagellates, in: "Photoreception and Sensory Transduction in Aneural Organisms," F. Lenci and G. Colombetti, eds., Plenum Publishing, London, pp. 45-68.

Feinleib, M. E. H., and Curry, G. M., 1967, Methods for measuring phototaxis of cell populations and individual cells, Physiol. Plant., 20:1083-1095.

Feinleib, M. E. H., and Curry, G. M., 1971, The relationship between stimulus intensity and oriented phototactic response (topotaxis) in Chlamydomonas, Physiol. Plant., 25:346-352.

Foster, K. W., Saranak, J., Patel, N., Zarrilli, G., Okabet, M., Kline, T., and Nakanishi, K., 1984, A unicellular eukaryotic model for visual studies: The bovine-like rhodopsin system of Chlamydomonas reinhardtii, Photochem. Photobiol., 39:10S.

Foster, K. W., and Smyth, R. D., 1980, Light antennas in phototactic algae, Microbiol. Rev., 44:572-630.

Gitelman, W. E., and Witman, G. B., 1980, Purification of calmodulin from Chlamydomonas: Calmodulin occurs in cell bodies and flagella, J. Cell Biol., 87:764-770.

Häder, D.-P., 1984, Wie orientieren sich Cyanobakterien im Licht, *Biologie in unserer Zeit*, 14:78-83.

Häder, D.-P., Colombetti, G., Lenci, F., and Quaglia, M., 1981, Phototaxis in the flagellates *Euglena gracilis* and *Ochromonas danica*, *Arch. Microbiol.*, 130:78-82.

Häder, D.-P., and Poff, K. L., 1979a, Light-induced accumulations of *Dictyostelium discoideum* amoebae, *Photochem. Photobiol.*, 29:1157-1162.

Häder, D.-P., and Poff, K. L., 1979b, Photodispersal from light traps by amoebas of *Dictyostelium discoideum*, *Exp. Mycol.*, 3:121-131.

Halldal, P., 1959, Factors affecting light response in phototactic algae, *Physiol. Plant.*, 12:742-752.

Hand, W. G., and Schmidt, J., 1975, Phototactic orientation by the marine dinoflagellate *Gyrodinium dorsum* Kofoid. II. Flagellar activity and overall response mechanism, *J. Protozool.*, 22:494-498.

Hartshorne, J. N., 1953, The function of the eyespot in *Chlamydomonas*, *New Phytol.*, 52:292-297.

Hildebrand, E., 1983, Halobacteria: The role of retinal-protein complexes, *in*: "The Biology of Photoreception," D. J. Cosens and D. Vince-Prue, eds., Symp. Soc. Exp. Biol. XXXVI, pp. 207-222.

Hyams, J. S., and Borisy, G. G., 1978, Isolated flagellar apparatus of *Chlamydomonas*: Characterization of forward swimming and alteration of waveform and reversal of motion by calcium ions *in vitro*, *J. Cell Sci.*, 33:235-253.

Kamiya, R., and Witman, G. B., 1984, Submicromolar levels of calcium control the balance of beating between the two flagella in demembranated models of *Chlamydomonas*, *J. Cell Biol.*, 98:97-107.

Lembi, C. A., and Lang, N. J., 1965, Electron microscopy of *Carteria* and *Chlamydomonas*, *Am. J. Botany*, 52:464-477.

Lenci, F., and Colombetti, G., 1978, Photobehavior of microorganisms: A biophysical approach, *Annu. Rev. Biophys. Bioeng.*, 7:341-361.

Linder, J. C., and Staehelin, L. A., 1979, A novel model for fluid secretion by the trypanosomatid contractile vacuole apparatus, *J. Cell Biol.*, 83:371-382.

Lindes, D., Diehn, B., and Tollin, G., 1965, Phototaxigraph: Recording instrument for determination of rate of response of phototactic microorganisms to light of controlled intensity and wavelength, *Rev. Sci. Instr.*, 36:1721-1725.

Lipson, E. D., and Häder, D.-P., 1984, Video data acquisition for movement responses in individual organisms, *Photochem. Photobiol.*, 39:437-441.

Litvin, F. F., Sineshchekov, O. A., and Sineshchekov, V. A., 1978, Photoreceptor electric potential in the phototaxis of the alga *Haematococcus pluvialis*, *Nature*, 271:476-478.

Melkonian, M., and Robenek, H., 1980, Eyespot membranes of *Chlamydomonas reinhardtii*: A freeze-fracture study, *J. Ultrastruct. Res.*, 72:90-102.

Morel-Laurens, N. M. L., and Bird, D. J., 1984, Effect of cell division on the stigma of wild-type and an "eyeless" mutant of Chlamydomonas, J. Ultrastruct. Res., in press.

Morel-Laurens, N. M. L., and Feinleib, M. E., 1983, Photomovement in an "eyeless" mutant of Chlamydomonas, Photochem. Photobiol., 37:189-194.

Naitoh, Y., and Eckert, R., 1974, The control of ciliary activity in protozoa, in: "Cilia and Flagella," M. A. Sleigh, ed., Academic Press, New York and London, pp. 305-352.

Nultsch, W., 1975, Phototaxis and photokinesis, in: "Primitive Sensory and Communication Systems: The Taxes and Tropisms of Micro-organisms and Cells," M. J. Carlisle, ed., Academic Press, London, New York, San Francisco, pp. 29-90.

Nultsch, W., 1979, Effect of external factors on phototaxis of Chlamydomonas reinhardtii. III. Cations, Arch. Microbiol., 123:93-99.

Nultsch, W., 1983, The photocontrol of movement in Chlamydomonas, in: "The Biology of Photoreception," D. J. Cosens and D. Vince-Prue, eds., Symp. Soc. Exp. Biol. XXXVI, pp. 521-539.

Nultsch, W., and Häder, D.-P., 1979, Photomovement of motile micro-organisms, Photochem. Photobiol., 29:423-437.

Nultsch, W., and Throm, G., 1975, Effect of external factors on phototaxis of Chlamydomonas reinhardtii, I. Light, Arch. Mikrobiol., 80:351-369.

Nultsch, W., Throm, G., and I. V. Rimscha, 1971, Untersuchungen an Chlamydomonas reinhardtii Dangeard in homokontinuierlicher Kultur, Arch. Mikrobiol., 80:351-369.

Pfau, J., Nultsch, W., and Rüffer, U., 1983, A fully automated and computerized system for simultaneous measurements of motility and phototaxis in Chlamydomonas, Arch. Microbiol., 135:259-264.

Pfennig, N., 1968, Thiospirillum jenense (Thiorhodaceae) Lokomotion und phototaktisches Verhalten, Inst. für den wissentschaftlichen Film, Göttingen, E678.

Poff, K. L., and Loomis, Jr., W., 1973, Control of phototactic migration in Dictyostelium discoideum, Exp. Cell Res., 82:236-240.

Schmidt, J. A., and Eckert, R., 1976, Calcium couples flagellar reversal to photostimulation in Chlamydomonas reinhardtii, Nature, 262:713-715.

Smyth, R. D., and Berg, H. C., 1982, Change in flagellar beat frequency of Chlamydomonas in response to light, Cell Motility Suppl., 1:211-215.

Smyth, R. D., Martinek, G. W., and Ebersold, W. T., 1975, Linkage of six genes in Chlamydomonas reinhardtii and the construction of linkage test strains, J. Bacteriol., 124:1615-1617.

Song, P.-S., Tapley, Jr., K. J., and Berlin, J. D., 1983, The photoreceptor in Stentor coeruleus, in: "The Biology of Photoreception," D. J. Cosens and D. Vince-Prue, eds., Symp. Soc. Exp. Biol. XXXVI, pp. 503-520.

Walne, P. L., and Arnott, H. J., 1967, The comparative ultrastructure
    and possible function of eyespots: _Euglena granulata_ and
    Chlamydomonas eugametos, _Planta_, 77:325-353.
Witman, G. B., Plummer, J., and Sander, G., 1978, _Chlamydomonas_
    flagellar mutants lacking radial spokes and central tubules.
    Structure, composition, and function of specific axonemal
    components, _J. Cell Biol._, 76:729-747.
Zar, J. H., 1974, "Biostatistical Analysis," Prentice-Hall, Engle-
    wood Cliffs.

PHOTOSENSING IN CYANOBACTERIA

Wilhelm Nultsch

Botanisches Institut
Fachbereich Biologie der Universität Marburg
Karl v. Frisch-Str.
D-3550 Marburg 1, W. Germany

INTRODUCTION

It now seems to be well established that the antecedents of the cyanophyceae (=blue-green algae), recently often called cyanobacteria because of their procaryotic cell organization, arose on earth about 2.3 billion years ago. Up to this time, the atmosphere was reducing, i.e. free from molecular oxygen. Since photosynthesis of bacteria is anoxygenic, oxygen sensitive processes such as dinitrogen fixation could be carried through at any place on earth. The cyanobacteria, however, invented the oxygenic photosynthesis, in which water serves as electron donor for CO_2-reduction instead of the sulfur compounds or organic materials used by the photosynthetic bacteria. Since the second product of water splitting is molecular oxygen which is released to the atmosphere, the cyanobacteria are mainly responsible for the generation of an oxidizing atmosphere in the early history of life.

The development of an oxygen-rich atmosphere, however, created two serious problems for the cyanobacteria themselves as well as for other organisms. On the one hand, the fixation of molecular nitrogen (N_2) necessary for protein synthesis is inhibited by oxygen because of the high oxygen sensitivity of nitrogenase, the main enzyme of N_2 fixation. Therefore, N_2-fixing organisms are capable of N_2 fixation only under anaerobic conditions, unless they solved the problem by compartmentalization. In some cyanobacteria, such as Nostocaceae, this is achieved by the development of special cells, so-called heterocysts, which contain only the anoxygenic photosystem I (PS I) besides the N_2-fixing machinery, whereas the vegetative cells containing the complete photosynthetic apparatus (PS I and II) do not fix dinitrogen.

On the other hand, energy-rich oxygen species such as singlet molecular oxygen ($^1O_2$), superoxide (O_2^-) and the hydroxyl radical ($\cdot OH$), can be formed in side reactions of oxygenic photosynthesis (Halliwell, 1981). As some of these oxygen species are extremely aggressive and cytotoxic, cells capable of oxygenic photosynthesis would be photodamaged or even photokilled when they are exposed to light. Since this is unavoidable to perform a light driven process, the development of mechanisms which protect the cells from photo-damage was necessary. This photoprotection has been achieved in different ways: $^1O_2$ is quenched by carotenoids arranged closely to the reaction centers in the thylakoid membranes, and O_2^- is made innoxious by the enzyme superoxide dismutase. At high fluence rates, however, these protective mechanisms may be insufficient for shadow plants which are adapted to low fluence rates. Therefore, some motile forms of the cyanobacteria have developed photosensing systems which enable them to search out areas of light conditions in the biotope which are favorable for photosynthesis but do not cause photodamage.

Two different types of photosensory motor responses were developed during evolution in cyanobacteria as in many other organisms:
  (1) Photophobic responses which are caused by sudden changes in the photon fluence rates, and
  (2) Phototaxis which denotes the orientation of movement with respect to the light direction.
In addition, light can also affect the speed of movement. This phenomenon, called photokinesis, will not be dealt with here, since in this case light serves as energy source of movement which can hardly be regarded as a sensory transduction process.

PHOTO-PHOBIC RESPONSES

In cyanobacteria, the photo-phobic response is simply a stop of movement which is normally followed by a resumption of movement in the opposite direction, irrespective of whether it is caused by a decrease (step-down) or an increase (step-up) of the fluence rate. Thus, photophobic reactivity can be measured by counting the numbers of reacting and non-reacting trichomes under the microscope, in video recordings or with microcinematography. Since this method is rather time consuming, a population method has been devised by Nultsch (1962a, 1968) in which square light fields are projected onto a homogeneous preparation of cyanobacteria in agar. The resulting accumulations in the light "trap" as a result of step-down responses can easily be measured by densitometric scanning. The same method can also be used for measuring step-up responses. In this case, the organisms leave the light field, so that it becomes empty (Häder and Nultsch, 1973).

Most of the work in studying photo-phobic responses in cyano-bacteria has been done with <u>Phormidium uncinatum</u>. This organism

is highly sensitive to light. Its absolute threshold (zero threshold)
for the step-down photophobic response is about 0.1 lx (Drews, 1959;
Nultsch, 1962a), whereas its discrimination threshold is about 0.05
(Nultsch and Häder, 1970), i.e. the same order of magnitude as in
purple bacteria. Therefore, the Phormidium trichomes are able to
perceive even small differences of light intensity. If a photographic negative is projected onto a homogenous preparation, the
trichomes accumulate in the different areas according to their
brightness, forming a photographic "positive" this way, as recently
shown by Häder (1984) who repeated Buder's famous purple bacteria
experiment.

Action spectra studies with the step-down response of Ph.
uncinatum (Nultsch, 1962a) have shown that in general, the photo-
phobically active radiation is absorbed by the photosynthetic pigments, mainly by the phycobiliproteins (Fig. 1). Red light absorbed
by chlorophyll a is also active, whereas the effectiveness of blue
light is far out of proportion to its absorption by the Soret band.
The same is true for Ph. ambiguum (Nultsch, 1962b), with the

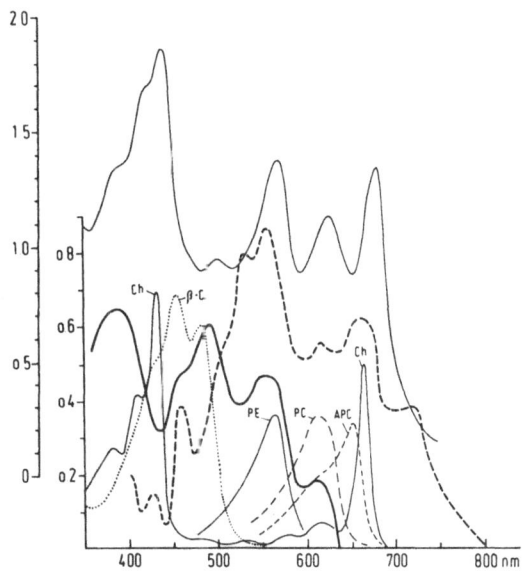

Fig. 1. Action spectra of the step-down photo-phobic response
          (heavy dashed line) and phototaxis (heavy solid line) of
          Ph. uncinatum. For comparison the in vivo absorption
          spectrum (fine solid line) and the absorption spectra of
          solutions of chlorophyll a (Ch), β-carotene (β-C), phyco-
          erhthrin (PE), phycocyanin (PC) and allo-phycocyanin (APC)
          are shown. Abscissa: wavelength in nm; ordinates: ab-
          sorbance and photobehavior in relative units. Modified
          after Nultsch (1962 a,b).

exception that in this form which lacks C-phycoerythrin the main
maximum is found at about 625 nm, the absorption maximum of C-
phycocyanin. As shown by Nultsch and Richter (1963), the photophobic action spectrum of Ph. uncinatum is essentially consistent
with the photosynthetic one. Studies with photosynthetic inhibitors,
uncouplers, and artificial redox system (Nultsch, 1967, 1968;
Nultsch and Jeeji Bai, 1966) lead to the conclusion that photo-phobic
responses are caused by a sudden change in the steady state of the
photosynthetic electron transport. Since redox-systems with a mid-
point potential near the one of the PS II acceptor are mostly effective, this point of the electron transport chain seems to be especially important.

The roles of PS I and II in the photo-phobic step-down response
of Ph. uncinatum were studied in detail by Häder and Nultsch (1973)
and Nultsch and Häder (1974). Further studies with uncouplers and
inhibitors resulted in the so-called electron pool-hypothesis,
according to which photo-phobic responses are regulated by an
electron pool near the PS II acceptor, probably plastoquinone (Häder,
1975, 1976). This confirms the results of the redox experiments
(Nultsch, 1968).

The photosynthetic electron transport from the inside to the
outside of the thylakoid membranes is coupled with a countertrans-
port of protons into the thylakoids. As a consequence, the cytoplasm
becomes negative with respect to the external medium, i.e. hyperpolarized. This electric potential can be measured with microelectrodes
(Häder, 1978 a,b). The coupling of the photo-phobic response with
this electric potential could be shown by a comparison of the action
spectra of the light induced potential changes and of the photophobic response (Fig. 2). After the addition of the triphenyl-
methyl-phosphonium cation which makes membranes permeable and this

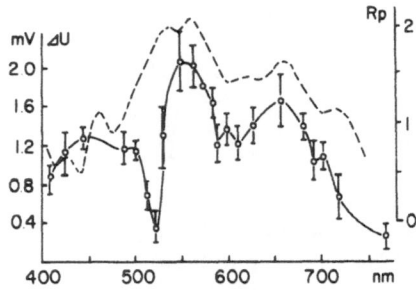

Fig. 2.   Action spectra of internally measured light-induced electric
          potential changes (circles, solid line) and photo-phobic
          responses (dashed line) of Ph. uncinatum. Abscissa:
          wavelengths in nm; ordinate: electric potential changes in
          mV (left) and photo-phobic reaction values (R_p, right).
          After Häder (1978b).

way abolishes the electric potential, the photo-phobic reactivity
ceases (Häder, 1979).

Very low fluence rates and their changes cannot be expected
to cause very strong potential changes. Thus, an amplification of
the signal is necessary. Since the removal of cations, especially
Ca^{2+}, and the addition of the Ca^{2+} ionophore calcimycin drastically
inhibit the step-down photo-phobic response of Ph. uncinatum (Häder,
1982; Häder and Poff, 1982), it has been concluded that the membrane
depolarization causes a transient opening of voltage dependent Ca^{2+}
channels and, hence, a Ca^{2+} influx. As shown in Fig. 3, the darkening
of Ph. uncinatum trichomes causes, in fact, a $^{45}Ca^{2+}$ uptake. Based
on these results, Häder (1984) proposed the following model of the
photo-phobic reaction chain (Fig. 4). In a moving Phormidium trichome
the front tip is negative with respect to the rear part. Once the
front tip is shaded or darkened, the proton gradient breaks down,
causing a depolarization of the membrane, and, hence, an opening of
calcium channels. The resulting Ca^{2+} influx into the front cells
amplifies the signal and makes the front tip positive with respect
to the rear, this way causing a reversal of movement, i.e. a phot-
phobic response.

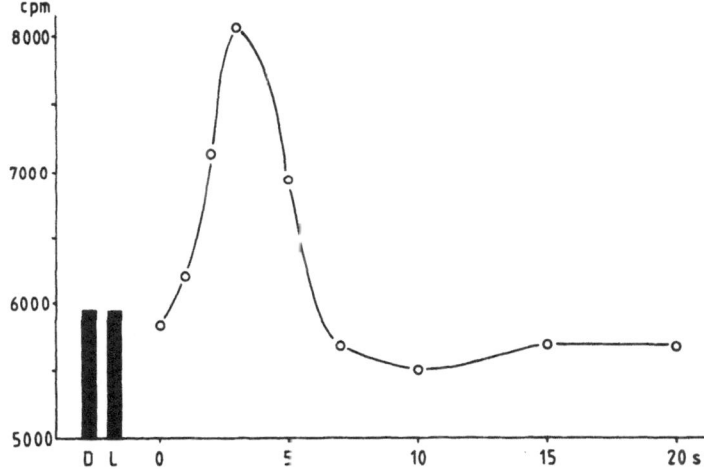

Fig. 3. ^{45}Ca uptake by Ph. uncinatum under various light treatments.
         Ordinate: cpm per $1.49 \pm 0.08 \times 10^7$ cells; D, dark control;
         L, light control; abscissa: time in s after transfer from
         light (40 min) to darkness with subsequent ^{45}Ca uptake
         during a 5 s dark period. After Häder and Poff (1982).

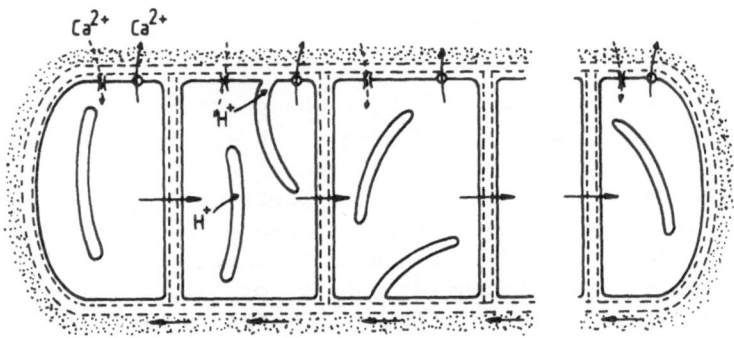

movement

Fig. 4.   Model of the photo-phobic reaction chain of Ph. uncinatum.
By building up and breaking down of the proton gradient
produced by light, small electric potential changes are
caused which, in turn, trigger voltage dependent Ca^{2+}
channels. Ca^{2+} is pumped out by active transport. After
Häder (1984).

Recently, Häder (1984) observed even step-up photophobic responses in Ph. uncinatum at very high fluence rates whose mechanism
is still unknown. Thus, the step-down response enables the trichomes
to stay in light conditions favorable for photosynthesis, whereas the
step-up response prevents the organisms from moving into direct sunlight and, hence, protects them from photodamage.

PHOTOTAXIS

Positive and negative phototaxis, i.e. movement towards the
light source or away from it, can be brought about in two different
ways: either by an active steering or by a change of the autonomous
rhythm of reversal on exposure to unilateral light.

The second type is realized in the Phormidium species investigated so far (Drews, 1959; Nultsch, 1961), and probably in all motile
cyanobacteria which rotate about their long axis during movement,
e.g. in Oscillatoriaceae. These organisms display an alternating
forward and backward movement without preferring any direction so
that they spread evenly over an agar plate. Even on the onset of
unilateral illumination, they do not change their movement direction
actively. However, in individuals which are in a more or less parallel position to the light beam, movement toward the light source is
prolonged, while movement away is shortened, resulting in positive
phototaxis, as indicated by the shift of the spreading area towards

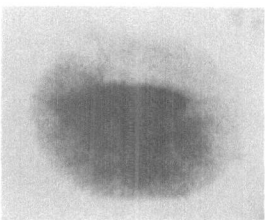

Fig. 5. Spreading of <u>Ph</u>. <u>autumnale</u> over an agar plate under uni-
        lateral illumination at low fluence rates (positive photo-
        taxis). Modified after Nultsch (1961).

the light source (Fig. 5). With negative reactions, the situaticn
is the opposite. In trichomes which are oriented perpendicularly
to the light beam, no change of the autonomous rhythm is observed.
However, since movement of cyanobacteria filaments is never absolutely straight, with time all individuals of a population come into
a position more or less parallel to the direction of light so that
the population approaches the light source.

    The lack of a steering mechanism in these organisms is apparently
due to the rotation of the trichomes which makes the measurement of
a light gradient by a one instant mechanism difficult though not
impossible. Detection of the light direction by a two instant mechanism is also not possible, since the photoreceptor pigments are
more or less evenly distributed in the periphery of the cells, and
a special apparatus for light modulation, e.g. a stigma, does nct
exist.

    In the Nostocaceae studied so far, such as <u>Anabaena variabilis</u>
and <u>Cylindrospermum licheniforme</u>, and probably in all gliding cyanobacteria which do not rotate during movement, both positive and
negative phototaxis are due to a true steering mechanism. Once a
trichome is irradiated from the side, it bends toward the light
source at low but away from it at high fluence rates. This is true
also for U-formed trichomes which turn their bent "tip" until the
axis of symmetry of the U-like figure is oriented parallel to the
light direction (Fig. 6). As a result a population transferred to
the center of an agar plate moves towards the light source at low
and away from it at high fluence rates (Fig. 7). Only at the
transition point from positive to negative, or when phototaxis is
inhibited, the population spreads more or less circularly.

    Phototaxis can also be evaluated by measuring the orientation
of single trichomes, using microscope observation (Nultsch et al.,
1979), video recording or microcinematography. These methods,
however, are rather time consuming, and the standard deviation is

Fig. 6. A. variabilis trichomes are oriented parallel to the light
         direction. Arrows point to bent "tips." Many heterocysts
         are visible. Magnification: about 33 x.

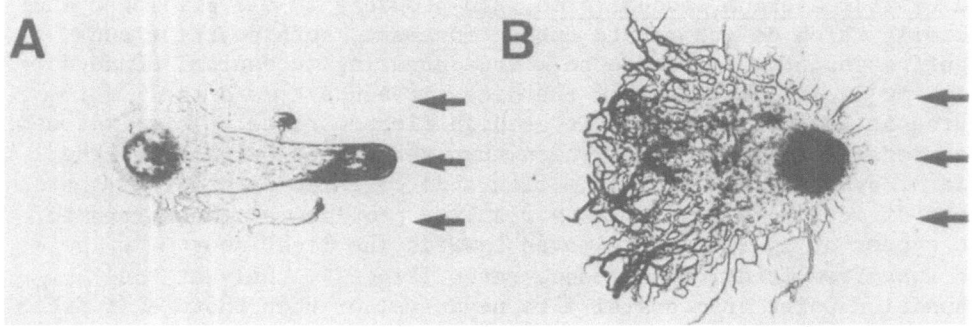

Fig. 7. Phototactic movement of an A. variabilis population toward
         the light source (positive, A) and away from it (negative,
         B).

relatively high since the trichomes often have an irregular form
so that the angle formed with the light direction can hardly be
measured. Therefore, population methods are preferred in which
the movement tracks towards the light source and away from it are
measured, and the distances per time unit in either direction are
calculated (Nultsch et al., 1979). An improved version of this
method has been used recently (Schuchart and Nultsch, 1984). As
shown in Fig. 8, the distribution of the organisms is measured by
scanning the whole spreading area with a densitometer. The records
are evaluated by planimetry, and the distances weighted by multiplication with a distance factor. The differences of the values
of the positive and negative spreading area are used as relative
measures of phototactic movement.

The action spectra of positive phototaxis measured with Ph.
uncinatum and A. variabilis show striking differences. In Ph.
uncinatum near UV and visible light up to 640 nm are active (Fig. 1).
Although the strong effectiveness of radiation between 450 and 510
nm points to yellow pigments, and the peak in the UV favors flavins
being the photoreceptors, the third maximum around 560 nm and the
shoulder at 620 nm clearly indicate that the phycobiliproteins C-
phycoerythrin and C-phycocyanin are involved in photoperception.

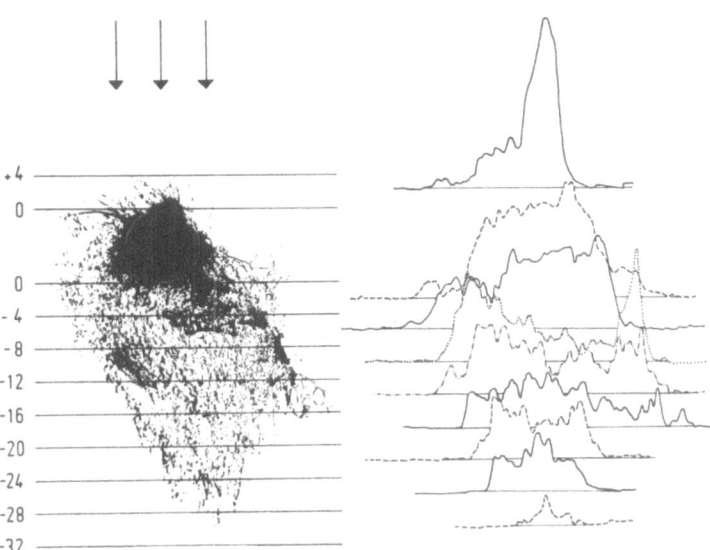

Fig. 8.   Photogram of A. variabilis trichomes spreading over an
           agar surface away from the light source (A). The corre-
           sponding recordings of 4 mm wide scannings are shown in
           B.   After Schuchart and Nultsch (1984).

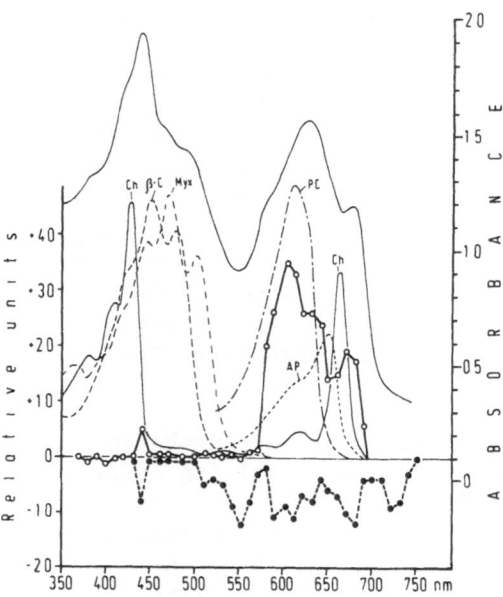

Fig. 9. Action spectra of positive (open circles, heavy solid line)
and negative phototaxis (filled circles, heavy dashed line)
of A. variabilis. For comparison the in vivo absorption
spectrum (fine solid line) and the absorption spectra of
solutions of chlorophyll a (Ch), β-carotene (β-C), myxoxan-
thophyll (Myx), C-phycocyanin (PC) and allophycocyanin (AP)
are shown. Abscissa: wavelength in nm; ordinates:
phototaxis in relative units (left) and absorbance (right).
Modified after Nultsch et al. (1979).

Chlorophyll a is not active as a photoreceptor in phototaxis of
Ph. uncinatum, since the action spectrum shows a minimum around the
Soret band, and red light absorbed by chlorophyll a is entirely
ineffective.

   In A. variabilis, which lacks C-phycoerythrin, the main maximum
of the phototactic action spectrum coincides with the absorption
maximum of C-phycocyanin (Fig. 9). Contrary to Ph. uncinatum, however, even light absorbed by chlorophyll a is active, indicated by
the red maximum around 675 nm and the small but distinct peak at
440 nm. The shoulder between 620 and 650 nm is probably due to
allo-phycocyanin absorption. Thus, in A. variabilis, the phototactically effective radiation is absorbed by the photosynthetic
pigments with the exception of carotenoids. This is true also for
negative phototaxis (Fig. 9). In addition, however, the range
between 500 and 560 which is inactive at fluence rate levels of
positive phototaxis causes significant negative phototaxis. Moreover, red light above 700 nm causes negative phototaxis at any

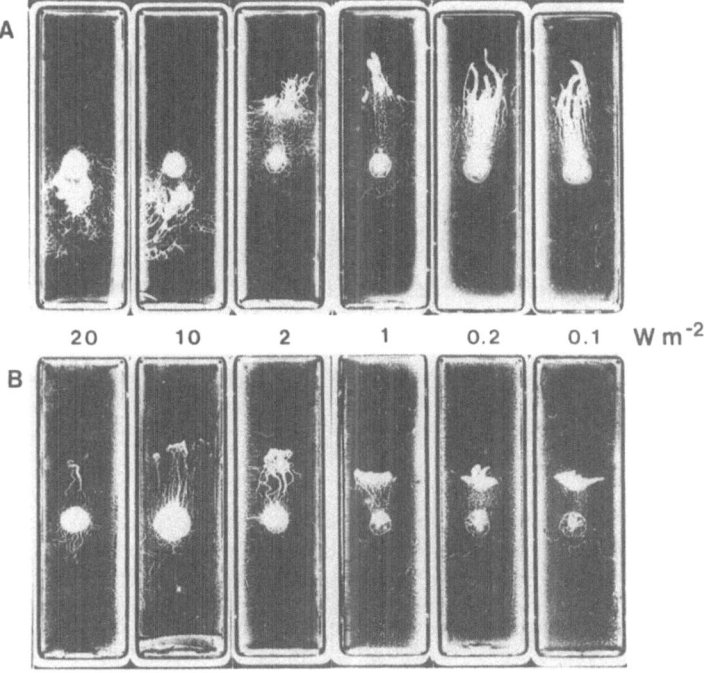

Fig. 10.   Photograms of phototactic reactions of A. variabilis in
           white light in the absence (A) and in the presence (B)
           of 1 mM NaN$_3$.

fluence rate. The photoreceptor pigment absorbing the light in
these ranges of wavelengths is still unknown. The so-called P750
found in Anacystic nidulans can be ruled out, since it does not
occur in A. variabilis (Nultsch et al., 1983a).

     As A. variabilis does not rotate about its long axis (Nultsch
and Wenderoth, 1983), it appears to make use of a one-instant mechanism to detect the spatial light absorption gradient between the
irradiated and the shaded side of the trichome. This gradient is
due to the absorption of the phototactically active light by the
photosynthetic pigments in the thylakoids which are arranged peripherally inside the cell. Since the gradient between the irradiated
and the shaded flank exists independently of the fluence rate, it is
suitable to detect the light direction, but gives no information
whether the reaction sign should be positive or negative.

     Recently it has been suggested that the phototactic reaction
sign is regulated by a hypothetical sign reversal generator which in
turn is controlled by an active oxygen species, probably singlet
oxygen, which is produced in photosynthesis (Nultsch et al., 1983a;

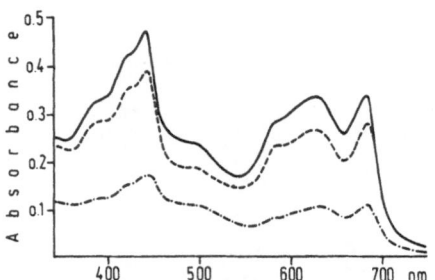

Fig. 11. in vivo absorption of A. variabilis before irradiation
(solid line) and after 5 d irradiation with 13.5 Wm^{-2}
white light in the absence (dashed-dotted line) and in
the presence (dashed line) of 1 mM NaN$_3$. After Nultsch
et al. (1983a).

Schuchart and Nultsch, 1984). As has been shown, sodium azide, a
potent quencher of $^1O_2$, reverses the phototactic reaction sign at
high fluence rates from positive to negative (Fig. 10).

In this connection it must be mentioned that high fluence
rates at which negative phototaxis is observed cause photobleaching
in A. variabilis. This photobleaching effect is largely prevented
by 1 mM NaN$_3$ (Fig. 11). Microspectrophotometric measurements with
heterocysts which contain only PS I and with vegetative cells which
contain both PS I and II have shown (Nultsch et al., 1983b) that
the phycobiliproteins which transfer the absorbed energy to PS II
are much more sensitive to high fluence rates than chlorophyll a
(Fig. 12). Carotenoids which protect chlorophyll a from photooxidation are least sensitive to irradiation. As a consequence, their

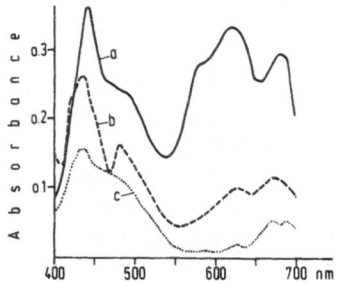

Fig. 12. Microspectrophotometrically measured absorption spectra
of vegetative cells of A. variabilis before (a) and after
irradiation with high fluence rates for 28 h (b) and
72 h (c). Abscissa: wavelength in nm; ordinate: absorbance. After Nultsch et al. (1983b).

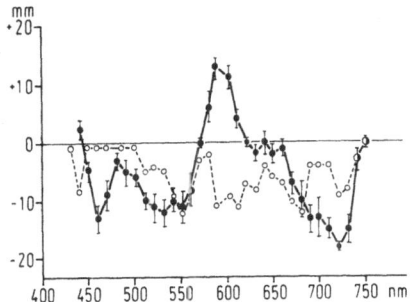

Fig. 13. Spectral sensitivity curves of phototaxis of <u>A. variabilis</u>
measured at a photon fluence rate of 1.12×10^{-8} mol cm^{-2}
s^{-1} in the absence (open circles, dashed line) and in the
presence (closed circles, solid line) of 1 mM NaN$_3$.
After Nultsch et al. (1983a).

relative amount compared with the other pigments increases during
photobleaching so that the color of the trichome becomes yellowish-
brown.

These phenomena can be interpreted from an ecological point of
view: high fluence rates which cause photodamage reverse the phototactic reaction sign to negative so that the trichomes withdraw
from the surface into the soil. If they are prevented from doing
this, the cells respond with photobleaching, i.e. they decrease the
absorption of the photosynthetically active radiation, but they
keep the level of photoprotective carotenoids relatively high. If
the photoprotection is increased by the externally added N$_3^-$ ion,
photobleaching is prevented and, hence, the phototactic reaction
sign becomes positive. The dominant role of phycobiliproteins is
documented by a comparison of the phototactic action spectra measured
at high fluence rates in the absence and the presence of 1 mM NaN$_3$
(Nultsch et al., 1983a). As shown in Fig. 13, the sign of the
phototactic response changes at about 570 nm.

It must be mentioned, however, that azide is not an absolutely
specific quencher of $^1O_2$, since it also quenches triplet states of
photosensitizers (Hasty et al., 1972; Winter et al., 1981). Therefore, experiments with other quenchers have been carried out.

As shown by Schuchart and Nultsch (1984), at 5 mM DABCO [= 1,4-
diazabicyclo(2.2.2)octane], the phototactic response at 13.5 Wm^{-2}
which is negative in the control becomes slightly positive. Furan
derivatives are not suitable substances to quench $^1O_2$ <u>in vivo</u> because of their low solubility in water and/or because of their
cytotoxity. Carotenoids, however, such as the water soluble
crocetin and some solubilized carotenoids (Nultsch and Schuchart,

Table 1. Effect of Carotenoids on the Phototactic
          Transition Point in A. variabilis.

Carotenoid	Transition Point Wm^{-2}	Double bonds
Control	7	-
Crocetin	40	7
C_{30}-ester	66	9
β-carotene	54	11
Canthaxanthin	70	11

unpublished) shift the transition point to higher fluence rates by
about one order of magnitude (Table 1).

Low pH values which diminish the quantum yield of $^1O_2$ production
shift the transition point of phototaxis to higher fluence rates
(Table 2), alkaline pH values to lower fluence rates (Schuchart and
Nultsch, 1984). Gassing with N_2 and argon also shifts the transition
point to higher fluence rates. Since the same effect is achieved by
O_2 gassing, the shift of the inversion point and, hence, the lower
sensitivity to high irradiances, is the result of the $^1O_2$ removal by
the gas stream rather than of the decreased oxygen tension. All
these results support the hypothesis that singlet molecular oxygen
is involved in the control of the phototactic reaction sign.

The iodide ion is a quencher of $^1O_2$ as well as of singlet excited and triplet states of photosensitizers. It inhibits negative

Table 2. Effect of pH Value on the Phototactic
          Transition Point of A. variabilis.

pH	Transition Point Wm^{-2}
5.5	> 120
6.0	15
7.0	5.7
8.0	4.5

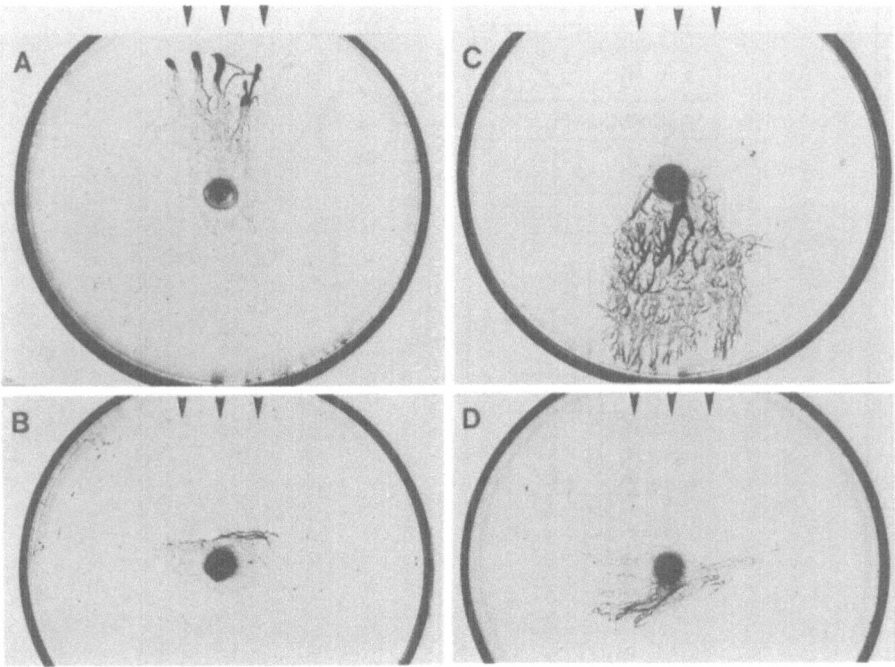

Fig. 14.   Photograms of spreading areas of A. variabilis caused
           by unilateral irradiation (arrows) at 1.35 Wm^{-2} (A, B)
           and at 13.5 Wm^{-2} (C, D), A and C in the absence, B and D
           in the presence of 20 mM KI. After Schuchart and
           Nultsch (1984).

phototaxis at 20 mM, a concentration that impairs motility only
slightly (Schuchart and Nultsch, 1984). However, it inhibits positive phototaxis also (Fig. 14), suggesting that iodide quenches
the excited states of the photoreceptor pigments rather than $^1O_2$.
Azide does not inhibit positive phototaxis. This finding favors
the concept that N_3^- acts by $^1O_2$ quenching.

    Based on these results, the following model (Fig. 15) of the
phototactic reaction chain of A. variabilis has been proposed. The
photoreceptor pigments, probably chlorophyll a and the phycobiliproteins, are located in the thylakoid membranes which are arranged
in the periphery parallel to the cell surface. Light absorption
by these pigments, and, in addition, by the carotenoids, produces a
light gradient between the irradiated and the shaded side of the
cell. This gradient is perceived as directional information
(front-rear, left-right, and vice versa) and is transmitted to the
signal processor. Since at low fluence rates $^1O_2$ production is
low or even zero, the signal is positive, and the input from the

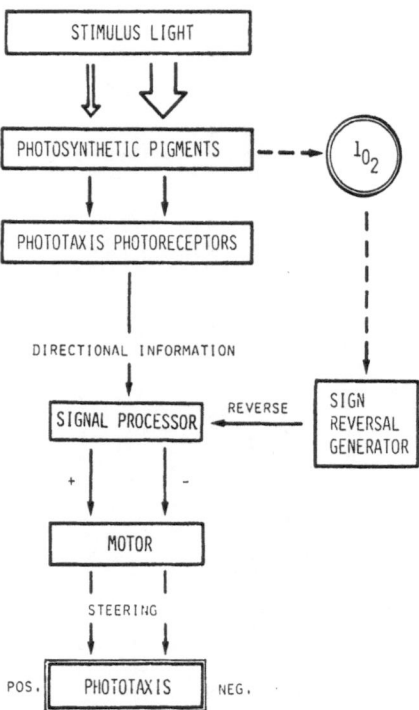

Fig. 15.   Model of the phototactic reaction chain of A. variabilis
           Further explanation in the text.

processor to the motor apparatus is "forward" in trichomes parallel
to the light direction or "bend towards the light" in trichomes
perpendicular to the light direction. Whereas in the first case
trichomes go ahead, in the latter case the speed is accelerated at
the shaded flank and/or decelerated at the irradiated flank. At
high fluence rates, however, $^{1}O_2$ is produced and activates the sign
reversal generator which reverses the signal to negative, whereas
the directional information remains unchanged. The signal processor
gives the opposite input to the motor apparatus, and the trichomes
move either backwards or bend away from the light source.

    Thus, both photobehavioral responses, phototaxis and photo-
phobic responses, enable the cyanobacteria to search for light
conditions favorable for photosynthesis, but to avoid areas of high
light intensities which may cause photodamage.

REFERENCES

Drews, G., 1959, Beitrage zur Kenntnis der phototaktischen reaktionen
    der cyanophyceen, Arch. Protistenk., 104:389.

Häder, D.-P., 1975, The effect of inhibitors on the electron flow
    triggering photo-phobic reactions in cyanophyceae, Arch.
    Microbiol., 103:169.

Häder, D.-P., 1976, Further evidence for the electron pool hypothe-
    sis. The effect of KCN and DSPD on the photo-phobic reaction
    in the filamentous blue-green alga Phormidium uncinatum,
    Arch. Microbiol., 110:301.

Häder, D.-P., 1978a, Evidence of electrical potential changes in
    photophobically reacting blue-green algae, Arch. Microbiol.,
    118:115.

Häder, D.-P., 1978b, Extracellular and intracellular determination
    of light-induced potential changes during photophobic reactions
    in blue-green algae, Arch. Microbiol., 119:75.

Häder, D.-P., 1979, Effect of inhibitors and uncouplers on light-
    induced potential changes triggering photophobic responses,
    Arch. Microbiol., 120:57.

Häder, D.-P., 1982, Gated ion fluxes involved in photo-phobic re-
    sponses of the blue-green alga, Phormidium uncinatum, Arch.
    Microbiol., 131:77.

Häder, D.-P., 1984, Wie orientieren sich Cyanobakterken im Licht,
    Biologie in unserer Zeit., 14:78.

Häder, D.-P., and Nultsch, W., 1973, Negative photo-phototactic
    reactions in Phormidium uncinatum, Photochem. Photobiol., 18:
    311.

Häder, D.-P., and Poff, K. L., 1982, Dependence of the photophobic
    response of the blue-green alga, Phormidium uncinatum, on
    cations, Arch. Microbiol., 132:345.

Halliwell, B., 1981, Chloroplast Metabolism: The Structure and
    Function of Chloroplasts in Green Leaf Cells, Oxford University
    Press, Oxford.

Halliwell, B., and Gutteridge, J. M. C., 1984, Oxygen toxicity,
    radicals, transition metals and disease, Biochem. J., 219:1.

Hasty, N., Merkel, P. B., Radlick, P. and Kearns, D. R., 1972,
    Role of azide in singlet oxygen reactions: Reaction of azide
    with singlet oxygen, Tetrahedron Lett., 1:49.

Nultsch, W., 1961, Der Einfluß des Lichtes auf die Bewegung der
    Cyanophyceen. I. Mitt. Phototopotaxis von Phormidium au-
    tumnale, Planta, 56:632.

Nultsch, W., 1962a, Der Einfluß des Lichtes auf die Bewegung der
    Cyanophyceen. III. Mitt. Photophobotaxis von Phormidium
    uncinatum, Planta, 58:647.

Nultsch, W., 1962b, Phototaktische Aktionsspektren von Cyanophyceen,
    Ber. Deutsch. Bot. Ges., 75:443.

Nultsch, W., 1967, Untersuchungen über den Einfluß von Entkopplern
    auf die Bewegungsaktivität und das phototaktische Reaktions-
    verhalten blaugrüner Algen, Z. Pflanzenphysiol., 56:1.

Nultsch, W., 1968, Einfluß von Redox-Systemen auf die Bewegungsak-
    tivität und das phototaktische Reaktionsverhalten von
    Phormidium uncinatum, Arch. Mikrobiol., 63:295.

Nultsch, W., and Richter, G., 1963, Aktionsspektrum des photosyn-
    thetischen $^{14}CO_2$-Einbaus von Phormidium uncinatum, Arch.
    Mikrobiol., 47:207.

Nultsch, W., and Jeeji-Bai, 1966, Untersuchungen über den Einfluß
    von Photosynthese-Hemmstoffen auf das phototaktische und
    photokinetische Reaktionsverhalten blaugrüner Algen, Z.
    Pflanzenphysiol., 54:84.

Nultsch, W., and Häder, D.-P., 1970, Bestimmung der photopho-
    botaktischen Unterschiedsschwelle bei Phormidium uncinatum,
    Ber. Deutsch. Bot. Ges., 83:185.

Nultsch, W., and Häder, D.-P., 1974, Über die Rolle der beiden
    Photosysteme in der Photo-phobotaxis von Phormidium uncinatum,
    Ber. Deutsch. Bot. Ges., 87:83.

Nultsch, W., Schuchart, H., and Höhl, M., 1979, Investigations on
    the phototactic orientation of Anabaena variabilis, Arch.
    Microbiol., 122:85.

Nultsch, W., Schuchart, H., and Koenig, F., 1983a, Effects of
    sodium azide on phototaxis of the blue-green alga Anabaena
    variabilis and consequences to the two-photoreceptor systems-
    hypothesis, Arch. Microbiol., 134:33.

Nultsch, W., Benedetti, P. A., and Gualtieri, p., 1983b, Micro-
    spectrophotometric investigations of photobleaching in
    Anabaena variabilis cells and heterocysts and its prevention
    by sodium azide, Z. Pflanzenphysiol., 111:327.

Nultsch, W., and Wenderoth, K., 1983, Partial irradiation experiments
    with Anabaena variabilis, Z. Pflanzenphysiol., 111:1.

Schuchart, H., and Nultsch, W., 1984, Possible role of singlet
    molecular oxygen in the control of the phototactic reaction
    sign of Anabaena variabilis, J. Photochem., 25:317.

Winter, G., Shioyama, H., and Steiner, U., 1981, Electron transfer
    quenching of dye triplets by NO_2^- and N_3^-. A spin-orbit coupling
    effect on the radical yield, Chem. Phys. Lett., 81:547.

SENSORY TRANSDUCTION IN EUGLENA

Bodo Diehn

EURIMPEX Inc.
4100 E. Howe Road
Bath, MI 48808

INTRODUCTION

In previous chapters, generalized schemes of the steps in sensory transduction have been presented. They typically consist of boxes labeled "Reception," "Integration," etc., to summarize the processes by which a stimulus alters the activity of the organism's motor apparatus. In our laboratory, we have studied in detail the control of behavior by external stimuli in the unicellular flagellate Euglena gracilis, and I would like to demonstrate here that for this organism, we can replace the black box by a molecular description of transduction processes. In order not to make Euglena appear too well understood, I will also discuss some observations on the interaction of different sensory systems that we can at present only describe in behavioral rather than molecular terms.

The gross morphology of Euglena is shown in Fig. 1. Evidence is mounting that photoreception as well as the transduction processes described in this contribution occur in the paraflagellar body (PFB) of the cell. Since I will not be discussing the nature of the photoreceptor pigment, the reader is referred to Diehn and Klint (1970), Mikolajczyk and Diehn (1976) and Barghigiani et al. (1979) for an introduction to studies on that subject. What I will discuss in some detail now is the sequence of molecular events by which a light stimulus, once perceived, is converted to an alteration in the cell's swimming behavior.

THE PHOTOSENSORY SYSTEM

We have chosen to investigate the "step-down" photophobic response" of Euglena because that response is the behavioral basis of

165

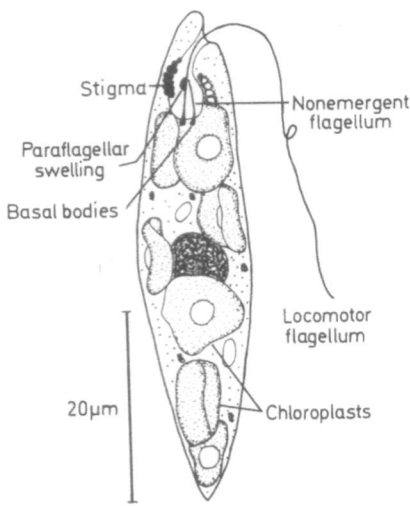

Fig. 1. Gross morphology of <u>Euglena</u> gracilis (from Diehn, 1976).

photoaccumulation as well as phototaxis in this cell. It occurs if
a light stimulus of moderate intensity is suddenly removed (for a
review of terminology and a general discussion of the behavior of
microorganisms, see Diehn, 1976). In order to facilitate discussion
of the step-down response, it helps to first look at the cell's
normal swimming behavior in the absence of stimulation (Fig. 2).

The organism, rotating as it swims forward, moves in straight
path segments that are interrupted by spontaneous changes in direction that typically occur every 3 to 6 s. That is, the frequency
of directional change (FDC) in the absence of stimulation is approximately 10 - 20.

When a step-down stimulus is imposed, the cell's swimming behavior changes to that shown in Fig. 3. Within less than 0.2 s,
the cell goes into a "tumbling" mode that is caused by sideways reorientation of the flagellum. This period of "continuous" flagellar
reorientation" (CFR) lasts for a time that we measure as part of our
response assay. It is followed by a succession of "periodic flagellar reorientation" (PFR) events, i.e. periodic tumbling events, after
which normal behavior is reestablished. The duration of time during
which PFR events occur is also measured. We define the overall response time, Tr, as beginning with CFR and ending with the last PFR.

An independent assay of the step-down response is provided by
measurements of the cells' rate of accumulation in a "light trap,"
since this phenomenon is mediated by step-down responses which the
cells execute upon encountering a light/to/dark border. For this
assay we are using an automated instrument that we have named

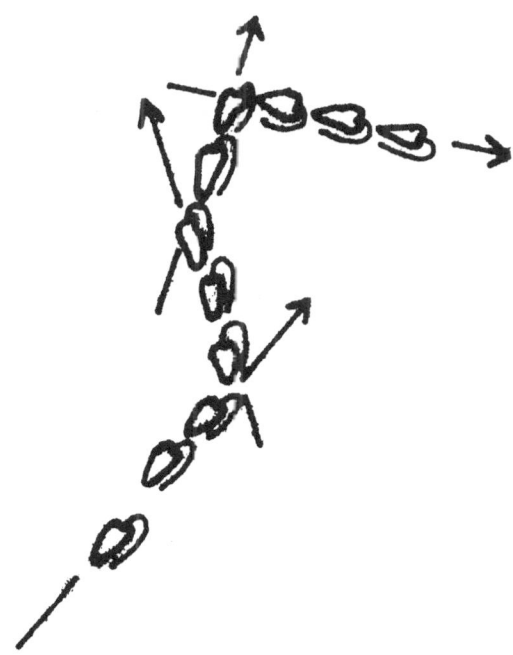

Fig. 2.   Swimming behavior of E. gracilis in the absence of
          stimulation.

"phototaxigraph" (Lindes et al., 1965). For further experimental
details, the reader is referred to Doughty and Diehn (1979).

THE MOLECULAR BASIS OF PHOTOSENSORY TRANSDUCTION

1) Calcium Controls Flagellar Activity

     Our approach to studying the control of flagellar activity in
our organism started with the recognition that in all 9 + 2 axonemes,
such as found also in Euglena flagella, the motility is based on a
Mg-ATP utilizing mechanochemical cycle that is regulated by calcium
ions. Thus, when we started the work described here (Doughty and
Diehn, 1979), our initial studies were focused on the possible involvement of calcium.

     We reasoned that if calcium played a role in altering flagellar
activity, then changing the concentration of the ion within the
intraflagellar space should result in an observable effect. Simply
changing the extracellular calcium concentration by altering the
composition of the medium, had, of course, no significant effect on

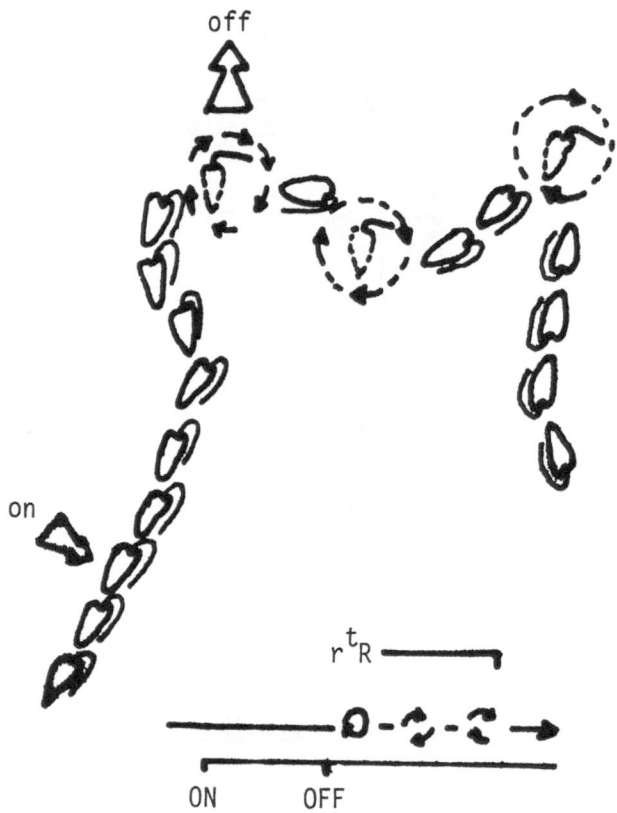

on

ON   OFF

Fig. 3. The effect of a step-down photic stimulus on the movement
         of Euglena.

cellular activity (more on this later, however), since one would not
expect the cell to be freely permeable to a control ion. However,
when calcium was added together with the ionophore A 23187 which is
known to cause membrane permeability for this ion, the cell's behavior changed dramatically (Fig. 4): Increasing the calcium level
causes both a striking increase in the frequency of directional
change (frequency of "tumbling"), and in the duration of the individual tumbles. In fact, at the higher calcium levels, the behavior
of the cells is indistinguishable from that of cells experiencing
a photophobic response, except that the behavior persists rather
than exhibiting the adaptation phenomenon that terminates the CFR
in the photic response. That this effect is caused only by calcium
is shown in Table 1.

     Does the calcium come from the external medium, or from an in-
tracellular compartment? I mentioned earlier that varying the

Table 1. A23187 Induced Klinokinesis and Cell Tumbling
          Behavior: Specificity of Action.

Cation	Concentration M	F.D.C. −A23187	F.D.C. +A23187	Notes
None[*]	−	4.2 ± 3.3	13.5 ± 2.7	
Mg	1.25×10^{-5}	6.6 ± 2.1	7.4 ± 2.7	
Co	2×10^{-5}	9.6 ± 4.1	12.3 ± 3.9	
Ba	1.25×10^{-5}	7.2 ± 2.9	11.5 ± 3.9	
Ni	2.5×10^{-5}	9.7 ± 3.5	10.8 ± 3.3	
Zn	5×10^{-5}	4.2 ± 3.4	21.7 ± 4.0	
Ca	10^{-5}	3.4 ± 3.0	44.8 ± 6.5	**

*Basal buffer contains zero free Ca ion concentration
 effected by addition of 1 mM EGTA, solution buffered to
 pH 6.9 with Hepes/Pipes-KOH (5 mM).
**Cells executing tumbling responses of mean duration of
 2.1 ± 0.5 s.

external calcium concentration does not affect cell behavior in the
absence of stimulation. However, there is an effect on the light-
induced response: Tumbling duration is significantly enhanced by
increased external calcium. This effect is much smaller for other

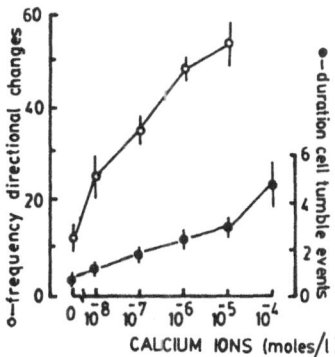

Fig. 4. Ionophore A23187-induced change in cell swimming path
         (frequency of directional changes in changes/min) and in-
         duction of cell tumbling behavior (duration of cell tumble
         events in s) as a function of extracellular calcium ion
         concentration (Redrawn from Doughty and Diehn, 1979).

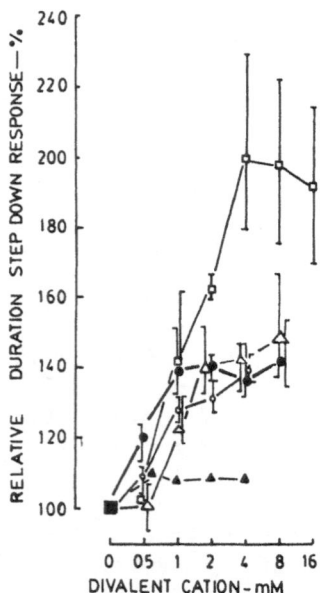

Fig. 5.   Effect of different extracellular concentrations of dif-
          ferent divalent cations on the duration of the step-down
          photophobic responses: ● - CoCl₂, Δ - BaCl₂, O - CaCl₂,
          ▲ - MgCl₂. (Redrawn from Doughty and Diehn, 1979).

divalent cations (Fig. 5). One can tentatively conclude that the
calcium which triggers flagellar reorientation is indeed derived from
the external medium.

To clinch the argument, we added lanthanum ion, a known calcium
channel blocker, to the medium. The resulting inhibition of the
photophobic response shows that the calcium which controls this response is derived from the extracellular environment (Fig. 6).

Calcium is thus at least one of the chemical signals. Is it the
final signal? If that were so, we should be able to show an effect
of this ion on the flagellar axoneme in vitro. This is exactly what
was observed when we studied the effect on the ATPase activity of
isolated, de-menbranated Euglena axonemes (Fig. 7): The activity is
depressed by 2 μM of Ca. This is consistent with the requirement
that during a phobic response, the 180 degree bend at the lower end
of the flagellum must be partially relaxed such as to redirect the
thrust sideways.

2) What Controls the Calcium Influx?

The next question is, of course, whether calcium is also the
first and only signal, i.e. whether Ca influx into the intraflagellar

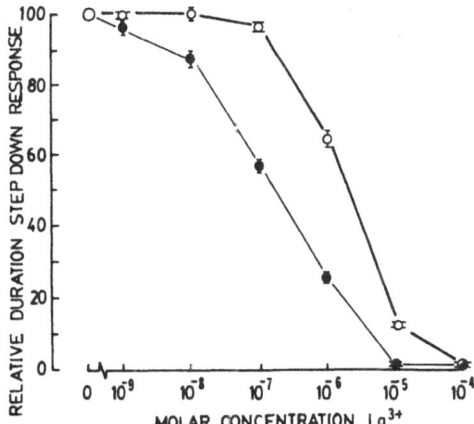

Fig. 6. Effect of lanthanum ions on the duration of the step-down
         photophobic response: ● - initial cell tumbling response
         to light removal; O - total duration of photophobic re-
         sponses (redrawn from Doughty and Diehn, 1982).

space is directly controlled by light. This is not a probably situation on purely operational grounds. The cell's sensory system needs
control points as well as points at which the sensory input from the
various stimuli are integrated, and a one-step mechanism does not
easily provide these. In fact, the dose-response curve of the photophobic reaction is sigmoidal at the temperature at which the organisms are cultured (Fig. 8), indicating the presence of a regulatory
mechanism.

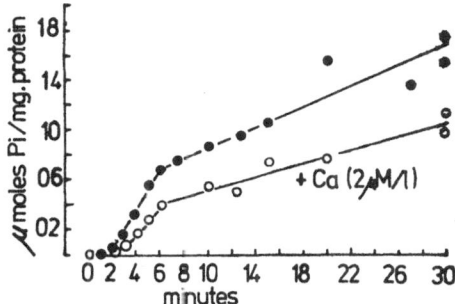

Fig. 7. Effect of calcium ions on MgATPase activity of isolated,
         demembranated flagellar axonemes of Euglena.

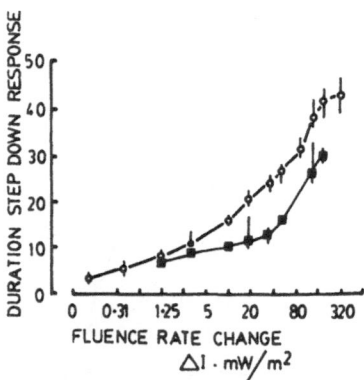

Fig. 8. Light-induced step-down photophobic response: duration
of cell tumbling responses (t_R in s) as a function of incident fluence rate change for cells at 19°C (O) and 23°C
(●). (Redrawn from Doughty et al., 1980).

However, let's assume for argument's sake that light directly
controls Ca ion influx, i.e. that the membrane exhibits light-dependent active electrogenesis, since the movement of calcium as the
only ion would result in a voltage gradient across the membrane.
In that case, the photophobic response should be affected by substances that are known to alter voltage-dependent ion fluxes. From
the results of a series of such experiments, it is evident that the
sensory membrane is not excitable and that, therefore, the role of
light is not simply to trigger activation of membrane calcium ion
conductance directly.

Having studied the effects of divalent cations, and having
found that of them, only calcium is involved in the photosensory
transduction chain, it made sense to us to turn to the monovalent
cations in the search for such components of the chain as might
control calcium permeability. As Fig. 9 shows, increasing the external sodium ion concentration greatly enhances the duration of the
photophobic response to a constant stimulus. There is a smaller
effect of potassium ion, and a negligible one of ammonium. If one
assumes that it is an influx of sodium which opens the calcium gate,
then one would expect an effect on the photophobic response of
ouabain, a substance that is known to block a transmembrane pump
that is responsible for simultaneously moving sodium ion out of
and potassium ion into cells. As seen in Fig. 10, ouabain enhances
the duration of tumbling. This enhancement through inhibition of
Na efflux suggests that it is an influx of Na which opens the Ca
gate. What we appear to have, then, is a light-activated Na(out)/K
(in) pump that ceases to work upon darkening, at which point passive
sodium influx becomes prevalent. The enhancement of the response by
external K might well be due to the facilitation of internal K

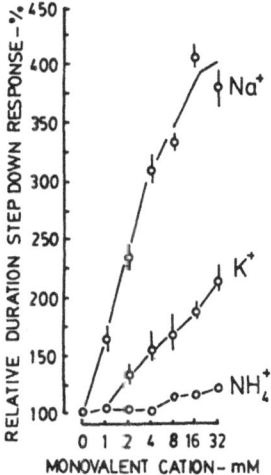

Fig. 9. Effect of different extracellular concentrations of dif-
         erent monovalent cations on the duration of the step-down
         responses (data from Doughty et al., 1980).

buildup during illumination, which would aid flux reversal upon
pump inactivation.

      At this point, the suggestion presents itself that one should
be able to simulate photic stimulation by artificially inducing an

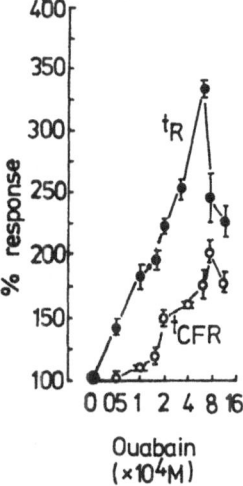

Fig. 10. Blue light induced cell tumbling: Effect of ouabain
          (after Doughty et al., 1980).

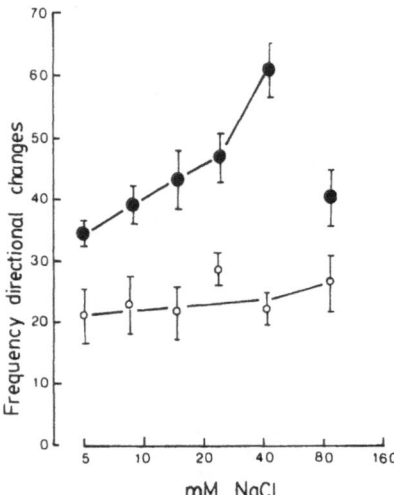

Fig. 11.   Gramicidin-induced change in cell swimming behavior (fre-
           quency of directional changes in changes/min) as a function
           of extracellular NaCl concentration.   O - control cells
           without ionophore; ● - cells with gramicidin (from Doughty
           and Diehn, 1982).

influx of sodium ions, just as we had done with Ca and A 23187 at a
later point in the transduction sequence. Gramicidin is a substance
that makes membranes permeable to sodium. To our initial surprise,
its addition to the medium did NOT cause continuous tumbling. We
then realized that sodium ions should not be expected to move into
the cell against the voltage gradient its translocation would cause.
When we added the membrane-permeant anion tetraphenylborate together
with sodium in the presence of gramicidin, we did see the expected
induction of tumbling (Fig. 11). As one would predict, this effect
required external calcium ion and was inhibited by lanthanum.

        Left to demonstrate was that if the membrane were made freely
permeable to potassium ion, externally applied potassium should reduce spontaneous tumbling. We did this experiment with the potassium
ionophore valinomycin, and the results were as predicted (Doughty
and Diehn, 1982).

A POSSIBLE PHOTOSENSORY TRANSDUCTION SEQUENCE

        From the above experimental results, one can construct a probable
sequence of the molecular events that occur upon step-down photic
stimulation. As the starting point, we consider an illuminated cell
in which the putative sensory Na/K pump is operating. The transduction sequence would be as follows:

Light Intensity Decrease

↓

Na/K Pump inhibited transiently

↓

Internal [Na] increases / [K] decreases

↓

Ca Gates opened

↓

Ca alters Flagellar ATPase

↓

Flagellar Reorientation

This scheme is of necessity incomplete. It does not explain
why the sodium ion influx upon step-down stimulation is only transient, nor does it propose how light would control the Na/K ATPase.
Nevertheless, it does provide a basic coherent description of the
major processes in the photosensory transduction pathway of Euglena.

Lest one think that there is little work left to do with this
cell, I must point out that not only is there another photic response,
the "step-up photophobic response" which has received very little experimental attention, but that Euglena can also respond behaviorally
to mechanical stimulation (Mikolajczyk and Diehn, 1979), and to chemical stimuli (Colombetti and Diehn, 1978; Miller and Diehn, 1978).
In the remainder of this chapter, I will discuss some of our observations on the interactions of photic and chemical stimuli in the
cell's transduction system.

CHEMOSENSORY RESPONSES TOWARD OXYGEN

In our laboratory, we discovered that if a Euglena suspension
is confined to a layer 0.1 mm or less in depth and excluded from
contact with air, expanding ring patterns made up of cells swimming
closely together will form spontaneously (Colombetti and Diehn, 1978).

The reason for the higher cell density in the ring is that cells
can enter this area freely, but experience (chemo-)phobic responses
at the ring boundaries. These responses "reflect" the cells back
into the pattern if their otherwise random movement happens to direct
them out of the ring.

It appears that Euglena moves outward in the expanding ring-
shaped pattern because it seeks an optimal environmental oxygen concentration. In the interior of the region bounded by the ring, the
oxygen has been used up by the cells' respiration (this type of experiment is done under yellow illumination which does not support
photosynthesis; white or red light will inhibit the formation of
ring patterns and destroys existing rings). Outside the ring, the

oxygen concentration is high enough to cause a step-up chemophobic response. A detailed description of the phenomenon is given by Colombetti and Diehn (1978).

INTERACTIONS OF PHOTIC AND CHEMICAL STIMULI

It is fairly easy to simultaneously apply light and chemical stimulation to a Euglena population. One simply adds photic stimuli to a cell suspension in which a chemoaccumulation ring pattern has developed, taking care that the stimulus does not alter the oxygen concentration near the cells by stimulating photosynthesis or respiration.

Applying a blue light stimulus, of such intensity as will induce a step-down photophobic response upon removal, has a dramatic effect upon the chemophobic ring pattern: the cells immediately begin to disperse, apparently because they no longer react to the (unchanged) oxygen gradient with chemophobic responses. As a consequence, the ring pattern disappears within approximately 10 s. Upon continuing illumination, the ring forms again, in the same position as before, within another 30 s or so. If the light stimulus is removed at this time, the cells execute normal-appearing step-down photophobic responses. Simultaneously, the ring pattern boundaries sharpen perceptibly.

The transient dispersal of the chemoaccumulation pattern indicates that the low-intensity photic stimulus (which, as will be recalled from what I said before, does not in itself cause a photomovement response, but simply prepares the cell for one pending an intensity step-down) causes a temporary erasure of chemosensory transduction.

Let's speculate on what this may mean. In this type of experiment, we are using light to probe the chemotransduction mechanism.

The inhibition of chemophobic responses by low-intensity blue light (which light we have postulated to lower intraflagellar calcium concentration through activation of the Na/K pump) shows that the ultimate sensory signal in the chemophobic response is calcium as well. One can also argue that the new steady-state concentration gradient of sodium ion across the membrane has been established after 30-40 s, since at this time modulation of the calcium ion concentration by the chemoreceptor system's action appears to resume (probably also via the Na/K gradient; a change in this gradient should not erase chemotransduction if the chemosensory pathway could control Ca influx independently). This reasoning is further supported by the observation that the chemophobic responses are enhanced upon the increase in internal Na triggered by the photic step-down stimulus.

The above suggestions are, of course, no more than informed guesses. In the spirit of this book, they are presented to inspire the reader to devise experiments for testing them. In this vein, let me state again my working hypotheses for the mechanism of chemosensory transduction, and then ask you to take with me another leap of the imagination based on a single experiment.

Chemophobic responses are mediated by the same Na/K/Ca cascade that we have shown to be involved in the step-down photophobic response. We have demonstrated that the chemoreceptor molecule is cytochrome c oxidase (Miller and Diehn, 1978). There is little reason to doubt that both the step-up and step-down chemophobic responses are utilizing the Na/K/Ca portion of the transduction chain; upon blue light illumination, the cells freely cross the previous step-down barrier, toward the inside of the ring, as well as the previous step-up barrier toward the outside (this observation is unequivocal, since one can perfectly well see the ring collapse inward only upon illumination with photosynthetic light, see Colombetti and Diehn, 1978).

This tentative conclusion about a joint pathway for both step-up and step-down chemophobic responses leads one to speculate that the step-up photophobic response might also be controlled by the same ionic mechanism that we have demonstrated for the step-down response. As a simple test for this, one might determine whether inducing a step-up photophobic response in cells in a chemophobic ring pattern will have the same effect as does induction of the step-down photophobic response. We did this experiment a while ago, before we had unraveled the ionic mechanism of the photoresponse (Diehn, 1979). A problem one encounters in this experiment is that to induce a step-up response, one needs to apply a high-intensity light stimulus, which regardless of wavelength will stimulate sufficient photosynthetic oxygen evolution to destroy the ring pattern. We solved this problem by working with etiolated (bleached by growing in the dark) cultures of Euglena that were incapable of photosynthesis. The results were as expected: upon applying the step-up stimulus, the boundaries of the chemophobic pattern sharpened, while upon subsequent removal of the high-intensity light, the pattern dispersed transiently.

The above observation does argue in favor of a unified ionic mechanism for all phobic responses. Until the photoreceptor for the step-up response is identified, it will be difficult to imagine how turning OFF high-intensity light would cause the same activation of the putative Na/K pump that we have shown to take place upon turning ON low-intensity light.

REFERENCES

Barghigiani, C., Colombetti, G., Franschini, B., and Lenci, F., 1979,
    Photobehavior of Euglena gracilis: Action spectrum for the
    step-down photophobic response of individual cells, Photochem.
    Photobiol., 29:1015-1019.
Colombetti, G., and Diehn, B., 1978, Chemosensory responses toward
    oxygen in Euglena gracilis, J. Protozool., 25:211-217.
Diehn, B., 1976, Photic responses and sensory transduction in pro-
    tists, in: "Handbook of Sensory Physiology," H. Autrum, ed.,
    Vol. VII/6A, pp. 13-68.
Diehn, B., 1979, The interactions of photic and chemical stimulus/
    response systems in Euglena gracilis, Acta Protozool., 18:7-16.
Diehn, B., and Kint, B., 1970, The flavin nature of the photoreceptor
    pigment for phototaxis in Euglena, Physiol. Chem. Phys., 2:483-
    488.
Doughty, M. J., and Diehn, B., 1979, Photosensory transduction in the
    flagellated alga, Euglena gracilis. I. Action of divalent
    cations, calcium antagonists and calcium ionophore on motility
    and photobehavior, Biochim. Biophys. Acta, 588:48-68.
Doughty, M. J., and Diehn, B., 1982, Photosensory transduction in the
    flagellated alga, Euglena gracilis. III. Induction of calcium
    ion dependent responses by monovalent cation ionophores,
    Biochim. Biophys. Acta, 682:32-43.
Doughty, M. J., Grieser, R., and Diehn, B., 1980, Photosensory trans-
    duction in the flagellated alga, Euglena gracilis. II. Evidence
    that blue light effects alteration in Na/K permeability of the
    photoreceptor membrane, Biochim. Biophys. Acta, 602:10-23.
Lindes, D., Diehn, B., and Tollin, G., 1965, Phototaxigraph: Record-
    ing instrument for determination of rate of response of photo-
    tactic microorganisms to light of controlled intensity and wave-
    length, Rev. Sci. Instr., 36:1721-1724.
Mikolajczyk, E., and Diehn, B., 1976, The effect of potassium iodide
    on photophobic responses in Euglena: Evidence for two photore-
    ceptor pigments, Photochem. Photobiol., 22:269-271.
Mikolajczyk, E., and Diehn, B., 1979, Mechanosensory responses and
    mechanoreception in Euglena gracilis, Acta Protozool., 18:591-
    602.
Miller, S., and Diehn, B., 1978, Cytochrome c oxidase as the receptor
    molecule for chemoaccumulation (chemotaxis) of Euglena toward
    oxygen, Science, 200:548-549.

MECHANORESPONSES IN PROTOZOA

Hans Machemer

Arbeitsgruppe Rezeptoren
Abteilung für Biologie
Ruhr-Universität
D-4630 Bochum, W. Germany

INTRODUCTION

Unicellular organisms commonly respond to mechanical stimuli impinging upon them. The resulting behavior is often conspicuous, such as cellular contraction or alteration in locomotion. Jennings (1899) detected that the large heterotrichous ciliates Spirostomum and Stentor have a highly mechanosensitive anterior cell end and a much less sensitive posterior end. Cells that were locally disturbed contracted immediately and then swam backward for a while. Unlocalized mechanical stimuli, such as jarring the culture dish, lead to similar responses. Later, it was detected in Paramecium that stimulation of the anterior cell end induced backward swimming, while posterior stimulation led the cell to accelerate the forward swimming mode (Jennings, 1906). Although the "avoiding reaction," which is triggered by an anterior stimulus, became the hallmark of protozoan motor behavior, the topological differentiation in ciliate mechanosensitivity fell into oblivion, only to be rediscovered 60 years later.

This lecture deals with mechanoresponses in ciliates only. In fact, insights into the regulating mechanisms of protozoan sensory-motor coupling have been limited to those few species for which electrophysiological and other recording techniques were established. I shall use the term "mechanoresponses" in a way as to include the behavioral events. Such a definition is useful for comparisons with bacterial cellular behavior.

Motor responses in ciliated cells result from alterations in motility of the cilia. Some basic facts of structure and function of cilia, and of the control of ciliary functions will be briefly

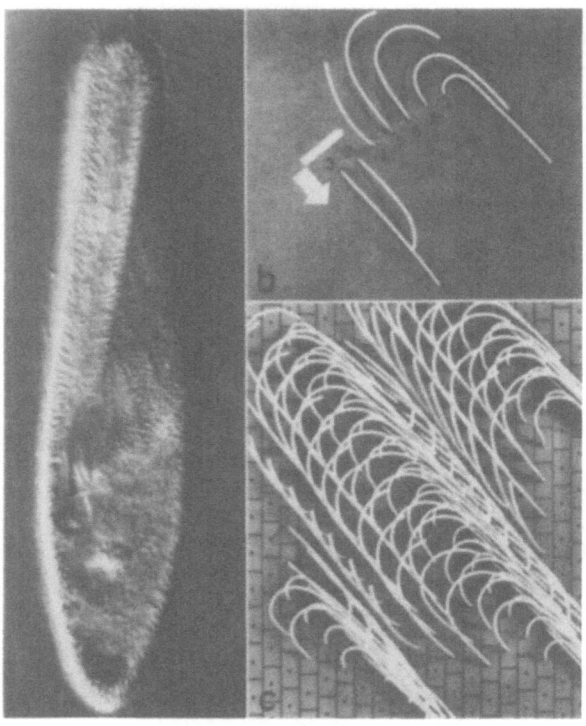

Fig. 1.   (a) Free swimming Paramceium (anterior end up), while the
membrane is at rest.   (b) Model of ciliary cycle (1-3,
power stroke; 4-7, recovery stroke).   Large arrow, direction
of power stroke; small arrow, direction of metachronal wave.
(c) Same model with neighboring cilia added (compare with
metachronism in oral groove of cell).

considered.   The main section of the chapter will be devoted to the
mechanosensory organization of the ciliate cell and to the question
of sensory-motor coupling.

MOTOR ORGANIZATION

     A typical ciliated organism is a free swimming cell powered by
a large number of cilia.   For example, the cell surface of Paramecium
is covered by a "fur" of 3000 to 6000 cilia depending on the species.
Although each cilium acts as an autonomous oscillator, all cilia beat
largely in a coordinated fashion.   The homogeneity of ciliary motor
functions is due, in part, to the manner in which the cilia are
anchored in the cell cortex.   Often the cilia occur in longitudinal
rows, and their basal bodies are regularly associated with root

Fig. 2.   A Stylonychia cell (anterior = left) with ventral and
          marginal cirri (= compound cilia) resting on solid substrate.
          Anterior active membranelles (compound cilia) extend to cy-
          tosome. Grell (1973) after Machemer.

fibers (kinetodesmata) and microtubules. The ciliary base and the
joining structures for anchoring are highly asymmetric so that each
cilium has an identifiable anterior, posterior, left and right face
(Fig. 3B)

     Cilia are commonly active in the unexcited state of the cell.
A "resting" Paramceium swims forward in a lefthanded helix because,
in the absence of a stimulus, the cilia beat toward the posterior and
right. Similarities in the rate and direction of ciliary beating and
in the hydrodynamic interactions lead to phase shifts in the bending
cycles of neighboring cilia. In these "metachronal" wave systems
cilia are fully synchronized along lines parallel to the direction of
the power stroke. The phase shifts between the cilia are maximal at
right angles to the line of synchronization. In the ciliate protozoa
the metachronal waves thus travel at right angles and toward the left
of the power stroke orientation (Fig. 1). It should be noted that
metachronism seen in the ciliated cell reflects, but does not guide,
ciliary activity (Machemer, 1974a).

     Not only can ciliation of the cell surface be limited to one or
a few bands of cilia, but also cilia may form aggregates of compound
motor organelles. The hypotrich ciliates are extremes in ciliary
specialization. The dorso-ventrally depressed Stylonychia, for example, "walks" on 18 ventral compound cilia called cirri (Fig. 2).
Suspended nutrients are swept toward the cytosome by means of sail-
like compound cilia called membranelles. Although all intraciliary
spaces are electrically coupled to the cytosol, the membranelles
are active while the cirri remain quiescent in the resting state of
the cell (Deitmer et al., 1984). Thus, a hypotrich cell can either
"sit" in place, walk or even swim. Locomotion occurs at different
speeds in both the forward or backward directions (Machemer, 1965),
depending on the state of cellular excitation. The diversity of
the motor responses in ciliates presumably evolved in conjunction

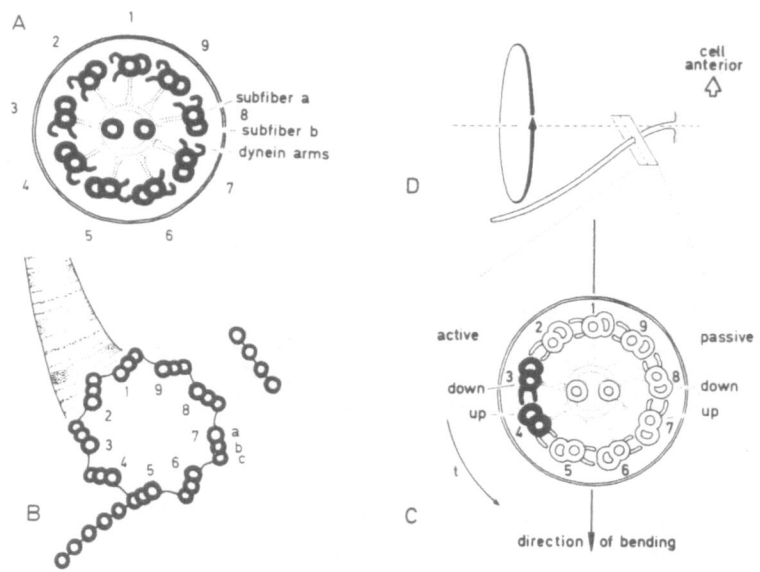

Fig. 3.   (A) Ciliary cross-section as viewed from tip to base.
Dynein arms project in direction of numbering of doublets
(1 = anteriormost doublet).   (B) Cross-section of Paramecium
basal body; skewed triplets each composed of doublet (a,b)
with incomplete tubule (c) added.   Striated kinetodesmal
root attached to triplets 1-3; microtubular bands originate
next to triplets 8,9 (tangential) and triplet 5 (radial
orientation).   (C) Active sliding between one pair of
doublets causes bending of cilium (straight arrow) and
passive sliding between antipode pair.   Observed counterclockwise gyration of cilium (D) implies a counterclockwise
transfer of regions of active sliding with time (t).   A,C,
D, Machemer (1977); B, modified after Pitelka (1968).

with an elaborate system of sensing and processing of various stimulus
inputs.

THE CILIUM, A ROTARY SLIDING MACHINE

     Behavioral responses of a ciliate cell are typically caused by
modifications in ciliary activity. The ciliary machine is, therefore, an indispensible coupling link between stimulus and response.
Cilia, unlike muscle, are not separate tissue for contraction but are
organelles integrated into the ciliate cell and therefore covered by
the common surface membrane. A ring of nine peripheral doublets of
microtubules with two single microtubules in the center and an array

of crosslinking elements represent the structural framework of a
cilium called the axoneme (Fig. 3A). The central microtubules end
at the level of the cell surface while the peripheral doublets extend
into the ciliary base. At this base the addition of a third microtubule to each pair leads to the well-known triplet configuration
or kinetosome (Fig. 3B).

Firm basal anchoring of the doublets together with active sliding
between neighboring doublets gives rise to axonemal bending (Afzelius,
1959). Because the sliding forces exerted by the swinging dynein arms
are unidirectional (Sale and Satir, 1977), a particular bending orientation of the cilium correlates with active sliding of a particular
pair of doublets (Machemer, 1977). The transfer of sites of active
sliding between a pair of doublets to the neighboring pair (n/n+1 →
n+1/n+2) is counterclockwise, as viewed from the ciliary tip toward
the base (Fig. 3C). Consequently, sliding activation along a closed
ring of microtubules induces the cilium to gyrate in the counterclockwise direction (Fig. 3D). Polarization in time and in space of
the ciliary beating cycle lead to what has been termed the "power
stroke" and "recovery stroke" of the cilium. Mechanical considerations of the ciliary motor suggest that regulation of ciliary frequency and beating direction occur at the sites of generation of
mechanical shear in the axoneme.

BIOELECTRIC CONTROL OF CILIA

When the membrane of Paramecium is partially disrupted by treatment with detergent (Triton X-100), the quiescent ciliary "models"
can be reactivated to resume cyclic beating in the presence of millimolar concentrations of ATP and Mg. Addition of Ca ions, rising
within a range of 10^{-8} and 10^{-4} M, induces the power stroke of the
reactivated ciliary models to reorient in an increasingly counterclockwise direction and to modify the beating frequency in a complex
manner (Naitoh and Kaneko, 1972). In living cells the intraciliary
concentrations of cations are under control of the surface membrane.
The membrane covering the cilia is believed to be virtually imperme-
able to Ca in the resting state. Membrane depolarization, however,
activates voltage-sensitive Ca channels so that ionic Ca can enter
the cytosol through these channels and an action potential arises
across the membrane. Because the number of open channels rises with
rising depolarization, a graded relationship exists between the
membrane potential and the direction of the ciliary power stroke.

In Paramecium, depolarization resets the orientation of the
power stroke from "posterior-right" to "anterior-right" in reference
to the cell axis (Machemer, 1974b). This re-programming of the
ciliary beat cycle causes the cell to swim backward and is tradition-
ally termed "ciliary reversal." In hypotrich ciliates, such as
Stylonychia, depolarization induces the previously inactive cirri to

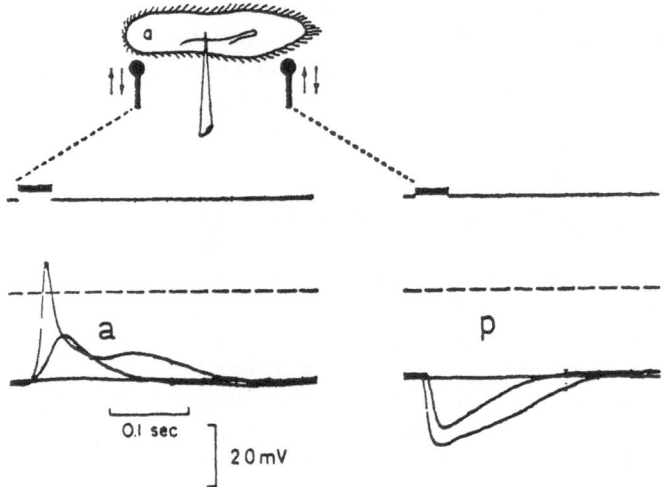

Fig. 4. Graded bipolar receptor responses in <u>Paramecium</u> to anterior
         (a) and posterior (p) mechanical pulses of three different
         intensities. Naitoh and Eckert (1969).

beat anteriorly. These and any other types of ciliary motor response
following membrane depolarization are labeled by the generalized term
"Depolarization-induced Ciliary Activation" (DCA) (Machemer and de
Peyer, 1982).

     The cilia also respond to membrane hyperpolarization, although
this voltage signal does not presumably modify ciliary membrane conductances. Following hyperpolarization, the power stroke is reori-
ented in the clockwise direction and the ciliary beat frequency increases. In <u>Paramecium</u>, this leads to a raised swimming speed,
while forward locomotion starts from rest in <u>Stylonychia</u>. Naitoh
(1983) calls this type of motor response "ciliary augmentation." A
more generalized determination is "Hyperpolarization-induced Ciliary
Activation" (HCA).

     Various types of stimuli can affect the membrane potential of the
ciliate cell to shift to more negative or more positive levels. These
alterations in membrane potential are rapidly transmitted to the
cilia, which respond by changing their rate and direction of beating
in a graded manner. The mechanisms by which the shifts in membrane
potential are recoded for intracellular signals, which eventually
control the axonemal function, have not yet been fully elucidated.

TWO MECHANORECEPTOR RESPONSES

     An old observation appeared in a new light when <u>Paramecium</u> was
mechanically stimulated during electrophysiological recording of the

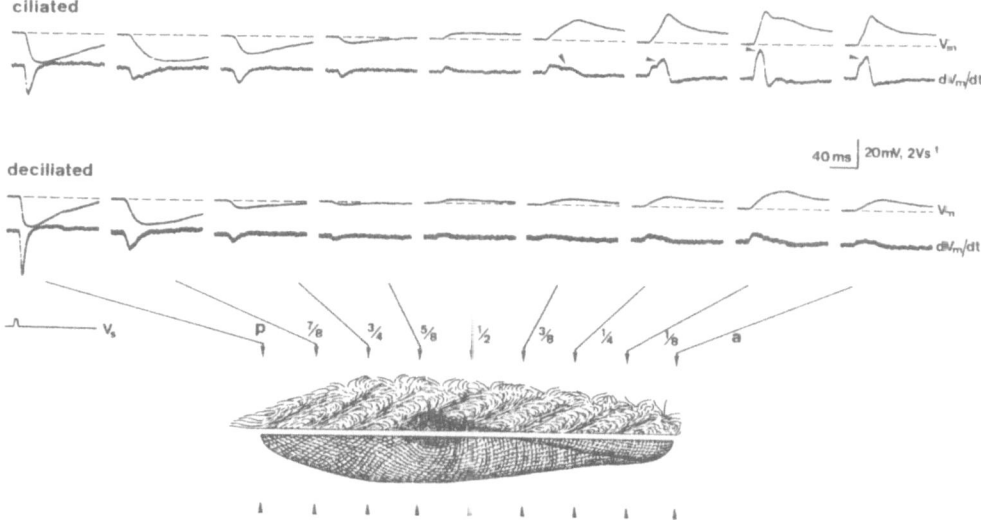

Fig. 5. Topographical modification of mechanosensitivity of the
Paramecium surface membrane. Local pulses (arrows) elicit
receptor potentials in ciliated and deciliated cells, but
the action potential is missing following removal of the
cilia [compare traces of membrane potential (Vm) and time
derivative of potential (dVm/dt)]. Arrowheads in time
derivative signify regenerative upstroke of depolarization.
Vs, stimulus driving pulse. Ogura and Machemer (1980).

membrane potential (Fig. 4). Local anterior stimulation led to a
depolarization and posterior stimulation to a hyperpolarization.
Such transient shifts in membrane potential following mechanical
stress to the cell are called mechanoreceptor potentials. Membrane
depolarization, caused by mechanostimulation, can trigger the activation of voltage-sensitive channels to produce a stimulus-graded
action potential in excitable cells such as Paramecium. The triggering receptor component appears as the "foot" of the rising slope
of the action potential (Figs. 4 and 5).

The time course of a hyperpolarizing receptor potential is not
masked by secondary membrane processes (Fig. 4). It is seen that
receptor potentials of either polarity are graded with the intensity
of the stimulus. The depolarizing receptor potential may be separated from the regenerative membrane process using voltage conditioning techniques (Fig. 6). An injected inward current preceding the
anterior mechanical stimulus shifts the receptor potential to more
negative levels, so that it remains below the threshold of an action
potential. The receptor potential decay is determined by passive
membrane properties, i.e. the time constant of the membrane.

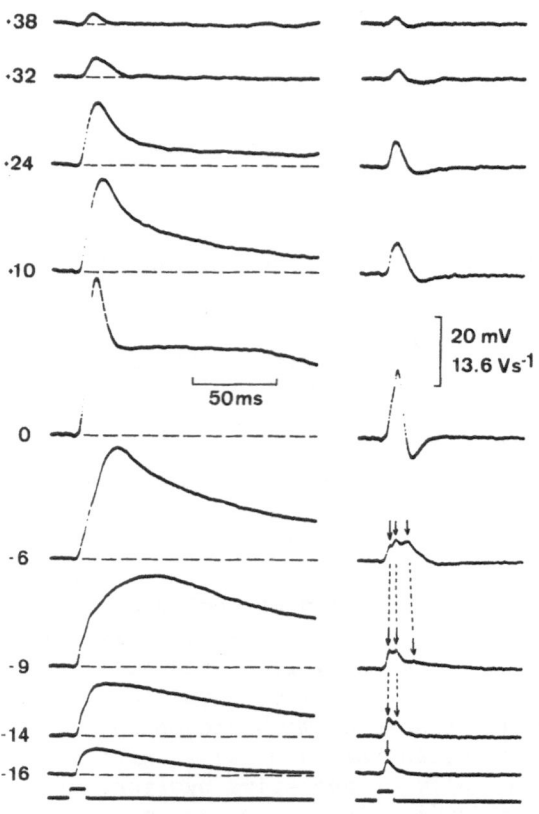

Fig. 6.   Conditioning of the membrane (numbers at left = shift in
           mV from resting potential) unmasks the depolarizing re-
           ceptor potential (see trace at -16 mV); no reversal of the
           depolarizing receptor potential was detectable with posi-
           tive displacements of the membrane of up to +38 mV. Right
           column with potential time derivatives illustrates the
           suppression of regenerative components of the depolarization
           (dashed arrows). Bottom traces, stimulus driving pulse.
           Data from Stylonychia. De Peyer and Machemer (1978a).

     Hyperpolarizing as well as depolarizing receptor potentials are
summed when two mechanical stimuli occur at small time intervals
(Fig. 7) or simultaneously at two sensitive sites of the surface
membrane (Fig. 8). The summation experiments did not reveal a detectable refractoriness of the receptor responses. Mechanical stimuli of reduced velocity lead to low rates of rise of the membrane
potential (Fig. 9), suggesting that temporal summation may be involved.

     Membrane potential conditioning experiments have indicated that
the hyperpolarizing receptor potential can reverse polarity. In an

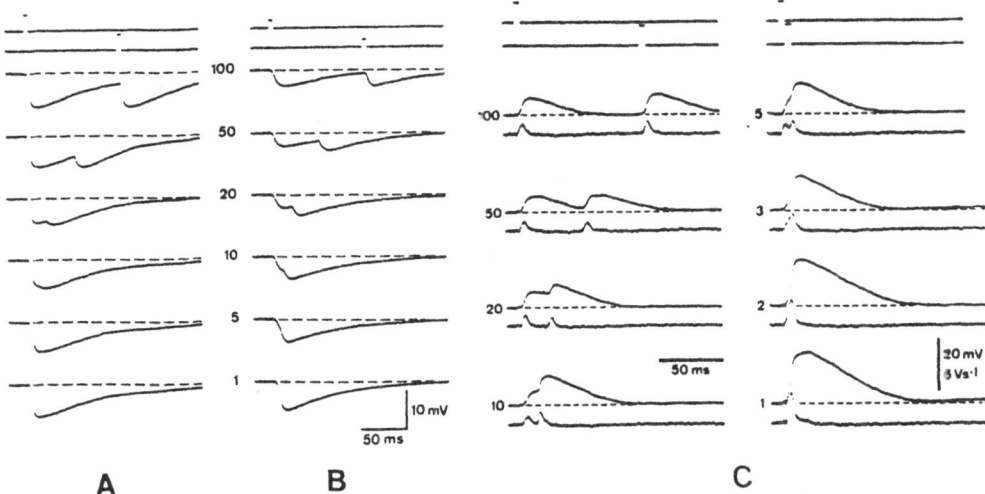

Fig. 7. Summation of receptor potentials in Stylonychia following
         mechanical stimuli (samples: upper traces) separated in
         time (numbers near traces: ms). (A) No summation occurs
         with supramaximal hyperpolarizing stimuli because responses
         are at the reversal potential (= K equilibrium potential).
         (B) Typical summation using submaximal stimulation. (C)
         Summation of depolarizing receptor potentials from con-
         ditioned potential. A regenerative depolarization arises,
         after the summed potential has passed the threshold of the
         action potential (see time derivatives below voltage
         traces). De Peyer and Machemer (1978a).

experimental solution of 4 mM KCl and 1 mM $CaCl_2$ the reversal potential in Stylonchia was 10 mV negative of the resting potential.
The reversal potential limits the receptor potential amplitude,
and is thereby an ultimate ceiling of summation (Fig. 7A).

     Similar conditioning of the membrane voltage prior to anterior
mechanostimulation did not reveal a reversal in polarity of the
receptor potential (Fig. 6). This suggests that the reversal potential of this response occurs at voltages far more positive than
the resting potential.

     A first survey of the mechanoreceptor potentials in ciliates
has shown that depolarizing and hyperpolarizing responses have some
properties in common such as gradedness, absence of refractoriness
and summation. The ionic bases are obviously different, as indicated
by their polarity and different reversal potentials. These properties of ciliate mechanosensation are comparable to those of other
receptor systems such as the postsynaptic membrane. In the ciliate
cell, as well as in the synapse, receptor potentials are suited for
signal processing at the membrane level.

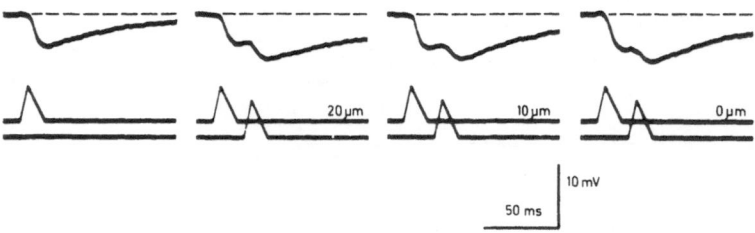

Fig. 8. Summation following two local stimuli (second to fourth
column) being separated in space by 20, 10, and 0 μm
(bottom traces). The summed potentials remain unaltered.
Data from Stylonychia. De Peyer and Machemer (1978a).

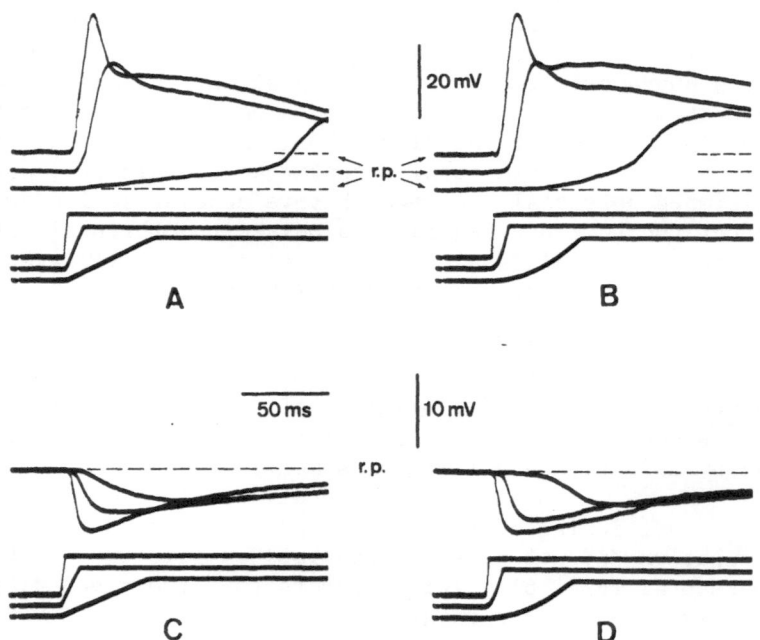

Fig. 9. Depolarizing (A, B) and hyperpolarizing membrane responses
(C, D) in Stylonychia following series of stimuli with increasingly reduced rates of rise. A, C, linear, B, D,
exponential stimuli. It is seen that the early rising
slopes of the receptor responses (upper traces) are coupled to the velocity of the stimulus (lower traces).
r.p., resting potential. De Peyer and Machemer (1978a).

MECHANORECEPTOR CHANNEL DISTRIBUTION

Local application of calibrated stimulus pulses using fine-tipped piezo-driven probes revealed that the depolarizing sensitivity of the membrane decreased, and hyperpolarizing sensitivity increased with shifts from the anterior toward the posterior end of the cell (Fig. 5). As would be expected, there exists an intermediate section of the cell where mechanostimulation does not elicit a receptor potential. This very restricted area is located at a latitude near the cytostome.

A graded modification in mechanosensitivity was also noted between the dorsal and the ventral side of the cell ("ventral" being defined by the location of the cytostome). At the same latitude of the cell body, depolarizing receptor responses were comparatively larger upon ventral stimulation and hyperpolarizing responses were larger upon dorsal stimulation (Fig. 10). Recent observations in Paramecium suggest the existence of circular gradients in mechanosensitivity between the right and, via dorsal and left, the ventral cell surface (Machemer-Röhnisch and Machemer, 1984).

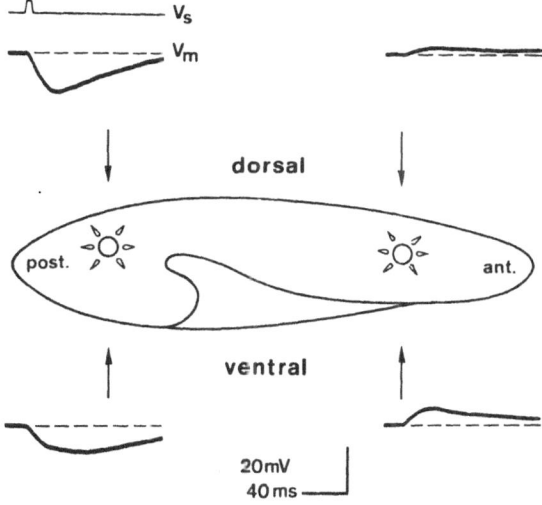

Fig. 10. Dorso-ventral differentiation of mechanosensitivity in Paramecium; stimuli (Vs) applied at the latitudes of the pulsating vacuoles. Dorsal stimulation led to larger hyperpolarizing responses, and to smaller depolarizing responses, as compared to ventral stimulation. Vm, membrane potential. Ogura and Machemer (1980).

While the complex topological organization of ciliate mechanosensitivity assists a motor response which is oriented with respect
to the stimulus source, the structural and physiological bases of
this system are not clear. It has been assumed for quite a long
period (Kafka, 1914), and reiterated until recently, that cilia are
the natural candidates for transducing mechanical input because
they project beyond the cell surface in the manner of "feelers."
This hypothesis was disproved after tests of physiologically deciliated paramecia revealed the persistence of mechanosensitivity
(Fig. 5). Cells deprived of their cilia failed, however, to generate
action potentials following depolarization (Ogura and Takahashi,
1976; Dunlap, 1977). The deciliation experiments gave the first
direct evidence of topographical separation in the ciliate membrane
of channels for stimulus sensation and membrane excitation.

If mechanoreceptor channels are limited to the soma membrane
of the cell, as suggested from the deciliation experiments, how do
they distribute in the surface membrane to generate the observed
scheme of depolarizing and hyperpolarizing mechanosensitivity? There
are two possible answers: (1) Depolarizing receptor channels may be
limited to the anterior part and hyperpolarizing channels to the
posterior part of the cell, or (2) either type of channels extends
over the larger portion of the cell surface in the manner of gradients; the two gradients have opposite directions and they overlap.

The latter hypothesis was confirmed using local stimulation
techniques in Paramecium under voltage clamp. When the membrane
potential was held at the reversal potential of the hyperpolarizing
response, inward receptor currents were often seen upon stimulation
of a posterior as well as anterior site of the cell. Such inward
currents correspond to the depolarizing receptor potential of the
unclamped membrane. Conversely outward receptor currents, equivalent
to the hyperpolarizing receptor potential, occurred when the cell was
stimulated anteriorly with the membrane potential clamped to the
anterior reversal potential (Fig. 11).

These observations suggest that the absence of a receptor response in an intermediate cell region is due to simultaneous and
equal activation of depolarizing and hyperpolarizing receptor
channels. This serves to shortcircuit inward and outward receptor
currents and thereby prevents a shift in membrane potential.

Bipolar mechanosensitivity and gradient-type channel distribution may apply to many free swimming ciliates, although the data
are still insufficient for generalizations. In the heterotrich
group of ciliates, such as Stentor and Spirostomum, recent observations are confirming earlier claims that in the first line the
anterior cell end is mechanically sensitive (Wood, 1975). Electrophysiological work in the gymnostome Didinium indicates that a
hyperpolarizing sensitivity is missing while a depolarizing

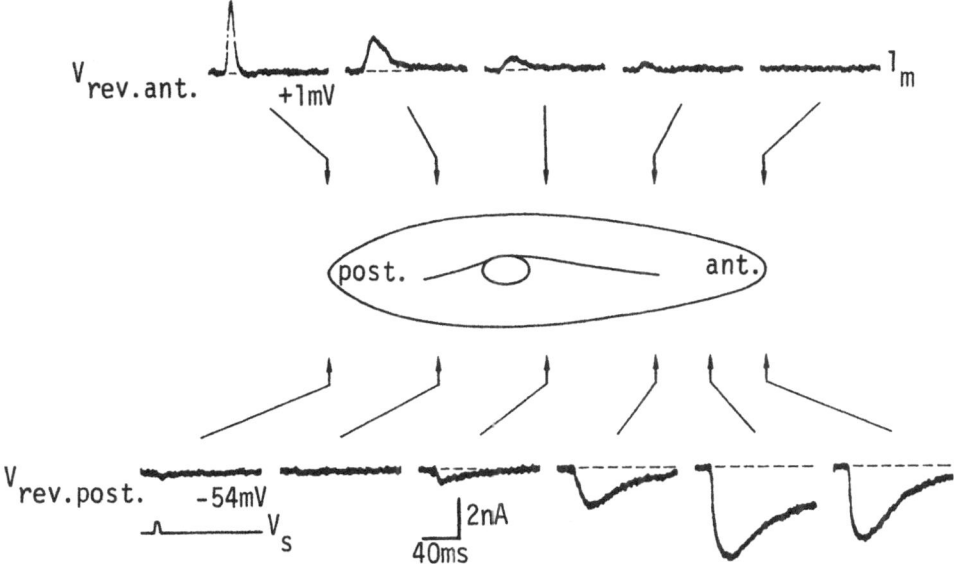

Fig. 11. Suppression of one component of the receptor current
          applying voltage clamp at the reversal potentials of the
          anteriorly (Vrev. ant., +1 mV) and posteriorly elicited
          receptor responses (Vrev. post., -54 mV) in Paramecium.
          Removal of driving force for K-outward current unmasks
          pure Ca receptor current component at Vrev. post., which
          rises anteriorly. At Vrev. ant. the K component in-
          creasingly prevails over the Ca component when the stimulus
          is shifted posteriorly. Ogura and Machemer (1980).

receptor response exists, which is comparable to that found in
Paramecium (Hara et al., 1980).

THE HYPERPOLARIZING RECEPTOR CHANNEL

        Observations of reversal potentials more negative than the
membrane resting potential have suggested that potassium is the
main charge carrier of the hyperpolarizing receptor response (Naitoh
and Eckert, 1973). Local stimuli placed at the very posterior cell
end in Stylonychia and Paramecium produced receptor responses which
reversed sign at membrane potentials which were related to the external K concentration in a strictly Nernstian manner (Fig. 12A).
The reversal potential remained constant upon modification of Ca
in the external fluid (Fig. 12B), suggesting that the posterior receptor channel is a pure K-electrode. When the external K concentration was raised beyond 30 mM the reversal potentials changed
polarity allowing the assessment of intracellular activities of K
(Stylonychia: 33 to 37 mM; Paramecium: 34 mM).

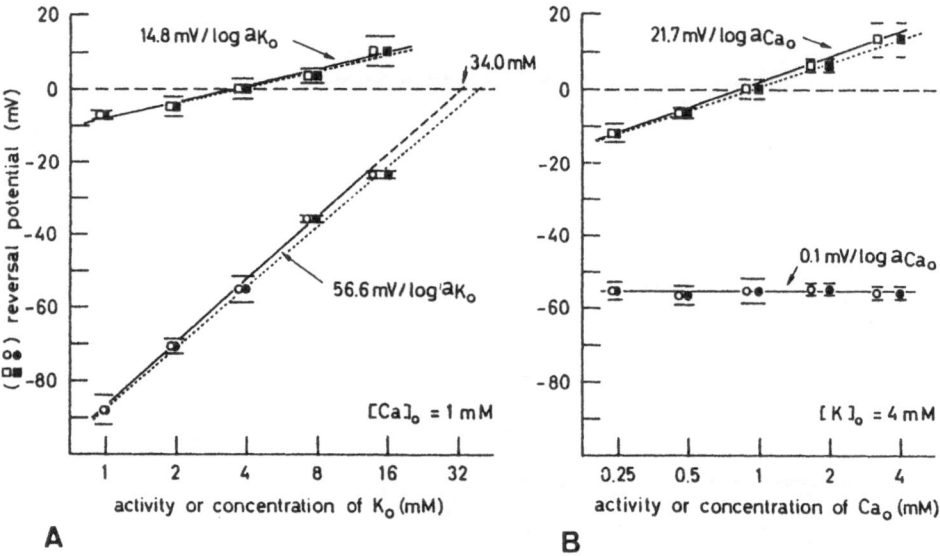

Fig. 12.   Potassium dependence (A) and calcium dependence (B) of
the reversal potentials of anteriorly elicited receptor
current (squares) and posteriorly elicited receptor
current (circles) in Paramecium. Closed symbols, ionic
concentrations; open symbols, activities (a). It is
seen that the anterior receptor current includes Ca and
K components; the posterior receptor current is a pure
K current. Ogura and Machemer (1980).

Because K is abundant in the cytosol, the present evidence of
specificity for potassium of the posterior receptor channel is not
fully conclusive; other monovalent cations may permeate the channel
when electrochemical gradients are produced. External applications
of Na (Naitoh and Eckert, 1973) and of Cs (Deitmer, 1982) had very
little or no effect on the receptor response.

The hyperpolarizing mechanoreceptor channel of Stylonychia was
found to be highly susceptible to the blocking action of TEA
(tetraethylammonium) and 4-AP (4-aminopyridine) and thereby differs
from voltage-dependent and leakage conductances for K (Naitoh and
Eckert, 1973; Deitmer, 1982). This suggests that the mechanically
sensitive channel is unrelated to the K conductance of both the
resting and electrically stimulated membrane.

Tests of the K receptor conductance under voltage clamp revealed the time characteristics of the current underlying a receptor
potential. The current rises in a sigmoidal manner including a latency of about 2 ms, reaches maximum within 10 ms and decays with a
time constant of 7 to 8 ms in Stylonychia (Deitmer, 1981) and

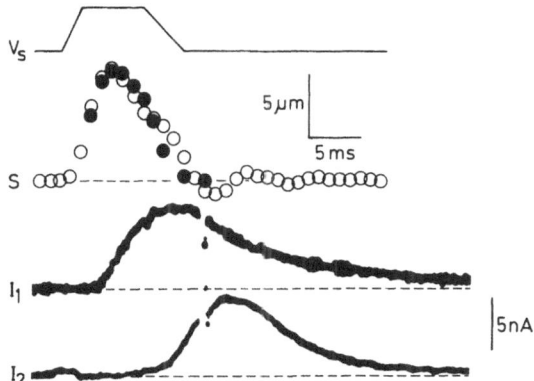

Fig. 13. Characteristics of hyperpolarizing stimulation in Paramecium under voltage clamp. The stimulating probe (S) rose
0.8 ms after start of the electric driving pulse (Vs).
The receptor current following somatic stimulation (I_1)
had a latency of 2 ms, rose within 10 ms to peak, and de-
cayed exponentially (time constant 8 ms) at 18°C. Receptor
currents from ciliary stimulation (I_2) were more delayed
and showed a more irregular time course. Circles in (S)
are cross-sections of probe recorded at 1 ms intervals
(filled circles, time course of probe during cell stimulation). Machemer-Röhnisch and Machemer (1984).

Paramecium (Machemer-Röhnisch and Machemer, 1984) (Fig. 13). The
receptor current rate of rise correlates with the stimulus amplitude
and the response latency correlates with the velocity of the stimulus. Sustained stimulus amplitudes do not contribute to the time
course of the current and thereby of the receptor potential (Fig.
14).

The decay times of the receptor current and the receptor potential differ by a factor of 3 or more from each other (Fig. 15);
the slower decay of the receptor potential follows the time constant
of the passive membrane. Time derivatives of the receptor potential
can well duplicate the receptor current time course (Fig. 5) because
changes in membrane potential parallel those of the K-conductance.

It is interesting to note that the K-mechanoreceptor channel
shows some degree of voltage dependence (Deitmer, 1981). Stimulation
of the depolarized membrane in Stylonychia leads to a degression from
the linear I/V relationship that is not predicted for a strictly unimodal receptor channel (Fig. 16). Because the time periods for rise
and decay of the receptor current also decrease with depolarization,
it is possible that the observed voltage-dependent depression of the
receptor current is due, in part, to an increased rate of current

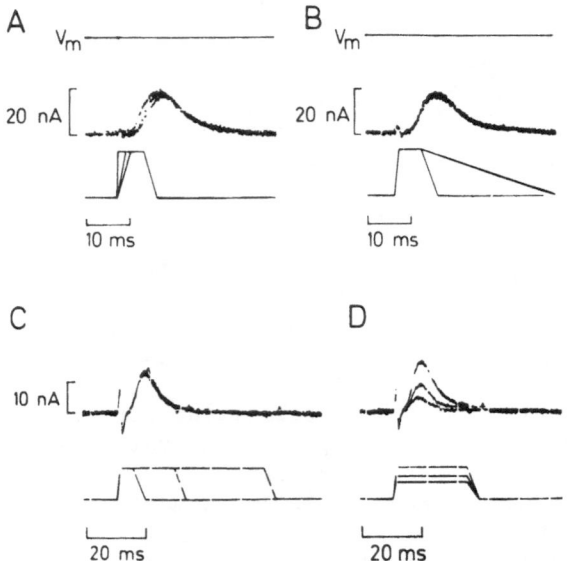

Fig. 14.   Effects of stimulus time characteristics on hyperpolarizing
           receptor current in Stylonychia under voltage clamp (Vm,
           membrane potential).   (A) Decreasing driving pulse rates
           of rise (lower trace) increase current latency (middle
           trace).   The rate of fall of the driving pulse (B) and the
           duration (C) do not affect the current.   (D) The receptor
           current amplitude is a function of the size of the driving
           pulse and thereby of the excursion of the probe.   Deitmer
           (1981).

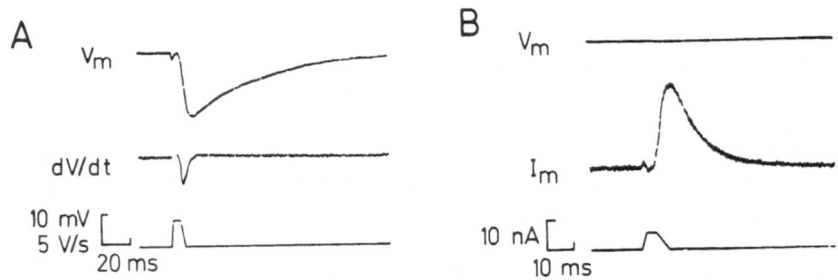

Fig. 15.   Time course of hyperpolarizing receptor potential (A, Vm)
           and of receptor current (Im) under voltage clamp (B, Vm)
           following similar stimulation (lower traces) in Stylonychia.
           Time derivative (dV/dt) illustrates smooth decay of receptor
           potential.   Observe that decay time constant of receptor
           current (B) is a fraction of that of the receptor potential
           (A).   Deitmer (1981).

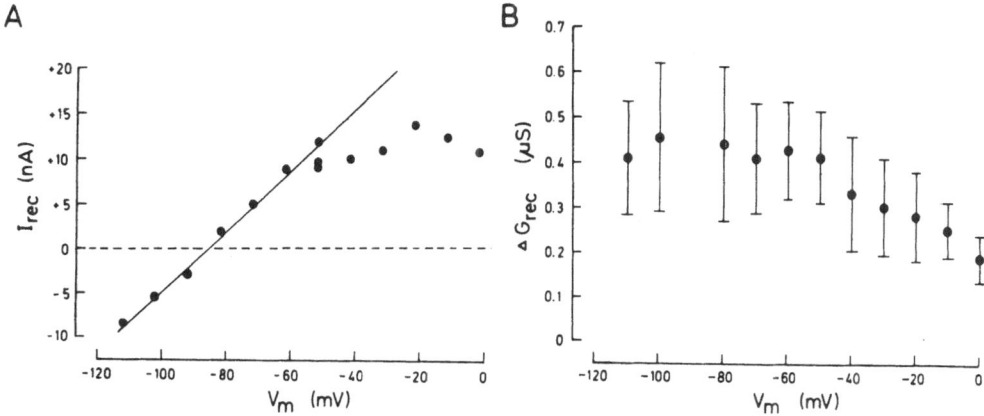

Fig. 16. Current/voltage relation of the posterior mechanoreceptor
in <u>Stylonychia</u>. The receptor current is a linear function
of the hyperpolarized membrane under voltage clamp (A).
With depolarizations from the resting potential (-52 mV)
the current amplitude is smaller than expected for a
unimodal receptor channel (A). Replotting of the data
shows (B) that the mechanoreceptor conductance increase
may be reduced upon depolarization. Deitmer (1981).

inactivation. Alternatively, the current across open channels may
saturate independently of the driving force for K.

THE DEPOLARIZING RECEPTOR CHANNEL

     The potential deflection and the inward current, which are seen
following anterior stimulation of the cell, are caused by more than
one type of activated receptor channel. Nernst plots of the reversal
potentials of the anteriorly elicited mechanoreceptor responses indicate the involvement of Ca and K ions. In <u>Stylonychia</u> the slope of
the reversal potential was about 23 mV per 10-fold change in external
Ca concentration which is below the 29 mV value expected for a pure
Ca electrode (De Peyer and Machemer, 1978a; De Peyer and Deitmer,
1980). In <u>Paramecium</u> this slope was similar (21.7 mV) and corresponded to a 14.8 mV change in reversal potential per decade in K
concentration. Thus, Ca and K ions permeate the anterior membrane
upon mechanical stimulation. The observed slopes of the reversal
potentials found in <u>Paramecium</u> fully correspond to predictions of
the modified Hodgkin-Horowicz equation of the reversal potential

$$V_{rev} = E_{Ca} \frac{\Delta gCa}{\Delta gCa + \Delta gK} + E_K \frac{\Delta gK}{\Delta gCa + \Delta gK} \qquad (1)$$

and its application to modified solutions with one cation species
held constant (Ogura and Machemer, 1980)

$$\frac{\Delta gCa}{\Delta gK} = \frac{\Delta E_K - \Delta V_{revK}}{\Delta V_{revK}} = \frac{\Delta V_{revCa}}{\Delta E_{Ca} - \Delta V_{revCa}} \qquad (2)$$

For cells bathed in varying solutions of CaCl and KCl, either slope
yields a conductance ratio, $\Delta gCa/\Delta gK$, of 2.9, suggesting that no
other ion, such as Cl, is involved in the receptor current.

Replacement of Ca in the bath by Mg, Sr or Ba showed that other
divalent cations can also serve as charge carriers of the inward
receptor current. Mn, which is detrimental to the membrane in the
absence of Ca, failed to permeate in Stylonychia (De Peyer and
Deitmer, 1980), but not so in Paramecium (Satow et al., 1983). Monovalent cations (Na, Li) did not carry the inward receptor current.
These data correspond to observations of poor cation selectivity
in the hair cells of the vertebrate inner ear (Corey and Hudspeth,
1979).

Inward and outward receptor currents have different characteristics and also vary in the species tested (Table 1). The minimal
response latency, i.e. the time interval between the beginning of
membrane deformation and the onset of the current, is around 1 ms.
This value remains well below that of the minimal latency observed
for the outward receptor current. In Paramecium the maximal rate of
rise and typical amplitude of the inward current are only a fraction
of those recorded in Stylonychia.

Table 1. Characteristics of Mechanoreceptor Currents in
          Paramecium (P) and Stylonychia (S) at 17-18°C.
          Data from Deitmer (1981), De Peyer and Deitmer
          (1980), Machemer and Machemer-Röhnisch (1984)
          and Ogura and Machemer (1980)

Current	Latency (ms)	Rise Time (ms)	Decay Time Const. (ms)	Max. Rate of Rise (nA/ms)	Max. Ampl. (nA)
Outward	P:2-2.5				3	17
		8-10	7-8		
	S: ?			9	20
Inward	P: 1[*]	5-18	≥25	0.2	1
	S: ?	4	1.8; 7.2	6	12

[*]Unpublished observations by Machemer

The decay time constants consist of two components in a Ca-containing solution; again, the decay is faster in _Stylonychia_ than in _Paramecium_.

It is so far unknown whether or not the anterior receptor current is using one type of channel for both Ca and K. Presumably there are two separate channels for these ions because externally applied monovalent cations inhibited the depolarizing receptor potential in _Paramecium_ (Satow et al., 1983) and TEA, which blocks the K-receptor current, has little effect on the depolarization following anterior stimulation in _Stylonychia_ (Deitmer, 1982).

Since the response latency of the posterior receptor current exceeds that of the anterior current, it is conceivable that the K-receptor current is Ca-activated in a manner similar to outward rectifier currents in ciliates and tissue cells (Meech, 1978; Eckert and Brehm, 1979; Deitmer, 1983). It could be assumed that Ca released from internal sequestering sites activates the K receptor current of the posterior cell end. This is unlikely, however, because reversal potentials of an inward Mg receptor current had a reduced Nernst slope (21 mV) as compared to a Ca receptor current (23.5 mV) (De Peyer and Deitmer, 1980). The reverse relationship is expected for an anterior receptor current including a Ca-activated K current.

ROLE OF CILIA IN STIMULUS RECEPTION

Since previously deciliated paramecia remain mechanosensitive, the motile cilia appear to play no role in the process of sensation. The presumed separation between the functions of stimulus reception and motor activity is a plausible concept for a ciliated cell, which integrates various sensory input from various sites of the cell surface.

However, the evidence excluding cilia from the reception process is not clear. First, deciliated paramecia show a reduced hyperpolarizing mechanosensitivity as compared to normal cells (Machemer and Machemer-Röhnisch, 1984). Second, modified immobile cilia arise from the posterior ends of various ciliates, i.e. areas which are known to be highly sensitive to mechanostimulation. It is possible, therefore, that parts of the ciliary structure, such as the basal body and/or the axoneme, play a role in mechanosensation.

Mechanical pulses directed at predetermined sites of the bundle of straight and elongated tail cilia in _Paramecium_ led to variations of the receptor current under voltage clamp (Table 2). The receptor currents arose with minimal response time and maximal amplitude upon stimulation of the membrane in the center of the tail. The response time (interval between onset of voltage driving the

Table 2. Receptor current modifications following
          stimulation of predetermined somatic (S) and
          ciliary (C) sites of the tail of <u>Paramecium</u>
          <u>caudatum</u>. It is seen that response times
          were minimal, and amplitudes maximal when
          stimuli were applied to the somatic center
          (SO); ciliary stimulation involved compara-
          tively longer response times and smaller
          response amplitudes. Machemer-Röhnisch and
          Machemer (1984)

Site	Response time (ms)	n	Amplitude (nA)		% of SO	n
SO	3.2 $\pm$0.6	102	5.0	(0.5;16.6)	100	103
S1	3.7 $\pm$1.2	134	2.4	(0; 9.5)	48	155
C1	6.9 $\pm$3.4	150	1.9	(0; 13.0)	38	179
C2	9.3 $\pm$3.0	124	2.0	(0; 14.5)	39	182
C3	10.1 $\pm$2.2	93	2.4	(0; 15.0)	48	163
C4	11.7 $\pm$2.6	38	1.1	(0; 10.8)	22	82

piezo-crystal and onset of receptor current) rose by about 9 ms upon
stimulation at the latitude of the tips of the tail cilia. A stimulus
excursion starting from the lateral base of the tail toward the cilia
was associated with a much longer current response time and smaller
amplitude than the same stimulus directed toward the soma from the
same starting position (compare responses to S1 and C1 in Table 2).

    The additional delays in receptor current generation observed
upon stimulation of the tail cilia suggest that mechanoreceptor
channels are excluded from ciliary membranes. Local mechanical stress
directed at right angles to the ciliary axis is more effective in de-
flecting the cilia (and presumably causes more sliding between axonemal components) than if directed parallel to this axis. With these
assumptions, the data do not indicate that microtubule-sliding and
shearing at the level of the basal body are rate-limiting steps in
mechanotransduction (Machemer-Röhnisch and Machemer, 1984).

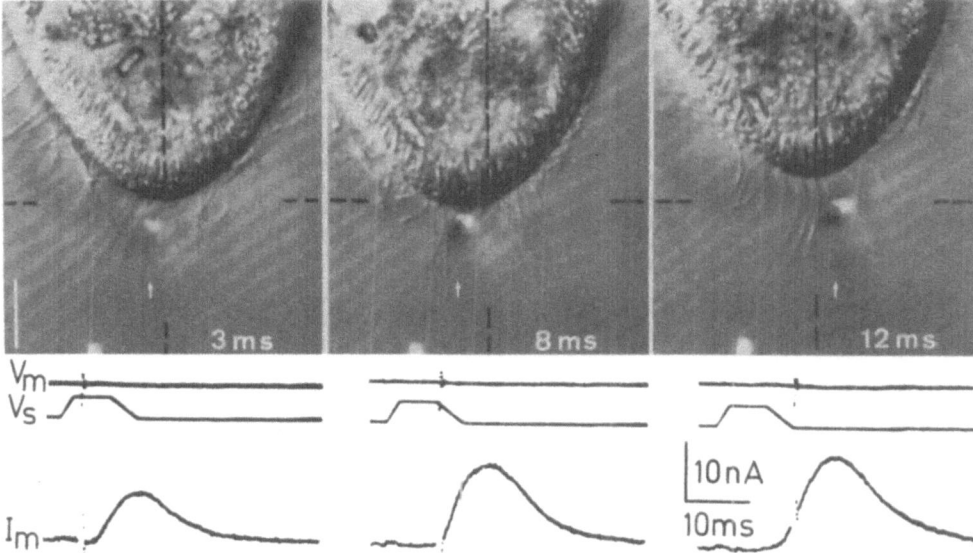

Fig. 17. Electric responses to ciliary mechanostimulation. While
the probe (arrow) traverses from right to left the bundle
of immobile tail cilia of Paramecium, simultaneous photographic and voltage clamp records show that the receptor
current (Im) started to rise 8 ms after onset of the electric driving pulse (Vs); at this time the tail cilia are
strongly deflected and the cell is displaced in stimulus
direction without showing somatic deformation. After 12
ms the receptor current is still rising; the cell swings
backward, while the needle has returned to the starting
position. Machemer-Röhnisch and Machemer (1984).

     Why, then, do immobile cilia arise from highly sensitive areas
of the soma membrane in cells like Paramecium and Stylonychia? Images
of tail cilia being displaced by the stimulus probe suggest that a
local ciliary stimulus can deflect the bundle of tail cilia without
implicating visible deformations of the soma membrane (Fig. 17). At
the same time, the receptor response is reduced as compared to direct
somatic stimulation (Table 2). Only fractions of the mechanical energy appear to be transferred upon the sensitive soma membrane, while
the tail cilia are elastically charged. Consequently, the tail cilia
of Paramecium attenuate and defocus a local mechanical stimulus which
might otherwise damage the surface membrane of the soma.

     The tail cilia may as well serve to enhance a non-local posterior
stimulus, such as pressure exerted by the surrounding fluid upon the
backward swimming cell. This condition was mimicked by directly
stimulating the somatic apex of the cell, which caused a brief forward

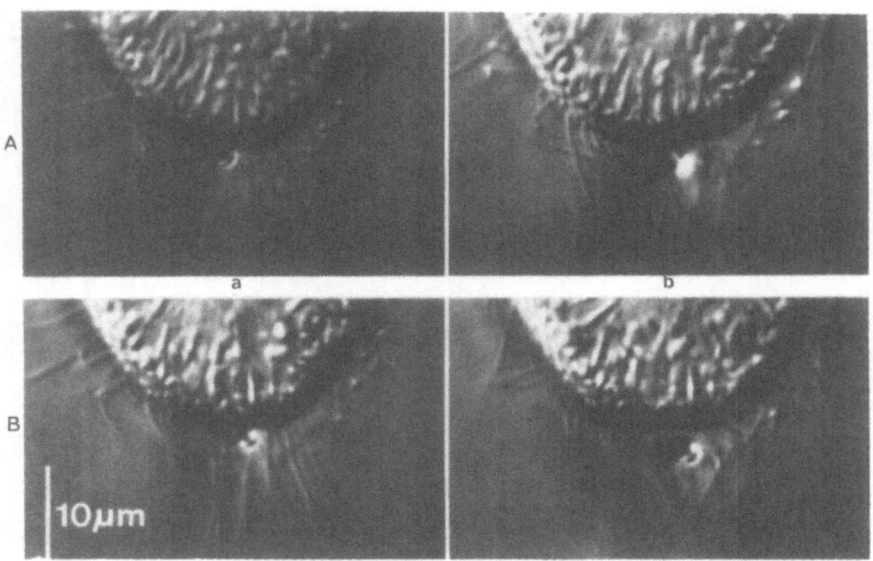

Fig. 18.   Passive deflection of <u>Paramecium</u> tail cilia following pulse
           of hydrodynamic pressure from the rear. Relative movement
           of the cell body was induced by stimuli with low (A) and
           high velocity (B) toward the somatic center (SO, see Table
           2), which caused the cell to swing forward and then back-
           ward. The micrographs correlate with stimulus starting
           time (a, 0 ms) and receptor current peak time (b, 12 - 15
           ms). Machemer and Machemer-Röhnisch (1984).

and then backward swing of the posterior cell end (Fig. 18). Inspec-
tion of the tail cilia shows that they are strongly deflected while
the cell rebounds.

    Receptor currents recorded during this type of stimulation were
larger in ciliated than in deciliated cells. Thus, a general mechanical shock applied to the posterior cell end is most effective in the
presence of tail cilia. Their function is, in this case, a multiple
conduction of the mechanical energy toward ciliary insertion points.
Here, distortions of the local somatic environment are summed to produce a comparatively large receptor response (Machemer and Machemer-
Röhnisch, 1984).

    In summary, the specialized cilia projecting from sensitive areas
of the ciliate soma membrane are not "feelers," i.e. they are not
playing an active role in the sensory transduction process. However,
these cilia can serve as filters, which, due to their physical properties, modify the transfer of mechanical energy to the sensitive
soma membrane.

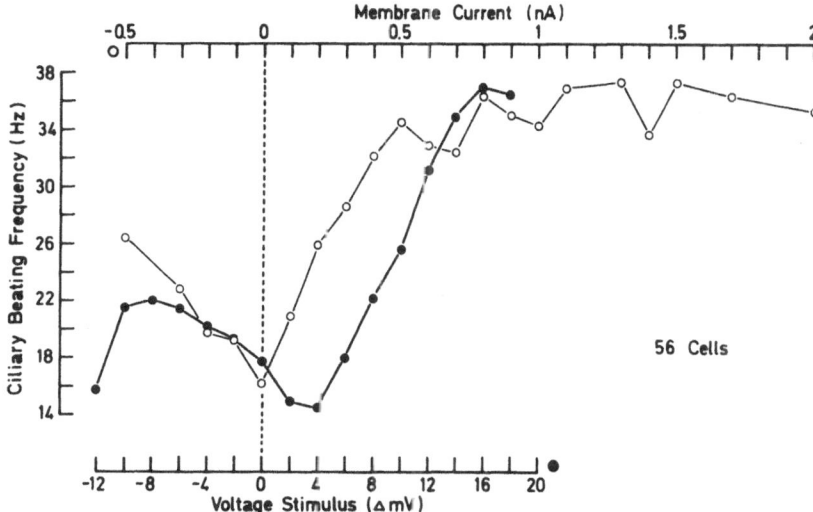

Fig. 19. Ciliary beating frequency correlated with steady-state membrane current (open circles) and voltage shift from the resting potential (closed circles) under voltage clamp applying slow positive-going voltage ramps. The data are averages from 56 cells bathed in solutions with varied concentrations of K (0.4 - 3.2 mM), Ca (0.1 - 5.1 mM) and Mg (0 - 5 mM), each showing the same principal frequency/voltage relationship. Note that the data do neither particularly favor membrane current nor exclude voltage stimuli, as possible signals for controlling the frequency of ciliary beating. Data replotted from Machemer (1976).

INTRACELLULAR PATHWAYS OF THE SIGNAL

It has been shown so far that mechanical stimuli impinged upon the ciliate cell can produce an electric membrane response. The amplitude and polarity of this response depends upon the strength of the stimulus and on the site of cellular application. While the receptor potential is a genuine cellular response, it may not be "understood" _per se_ by intracellular components such as the ciliary axoneme. What kind of transformation of the signal is required for a ciliary motor response to be coupled to the sensory process? Only partial answers to this question can be given to date.

When observed in a voltage-clamped cell, the first cellular response to stimulation is a conductance increase. It is conceivable that the loop of receptor current, set up by the mechanically-induced conductance, includes ciliary membranes. In the case of posterior mechanostimulation, for instance, the outward receptor current entails

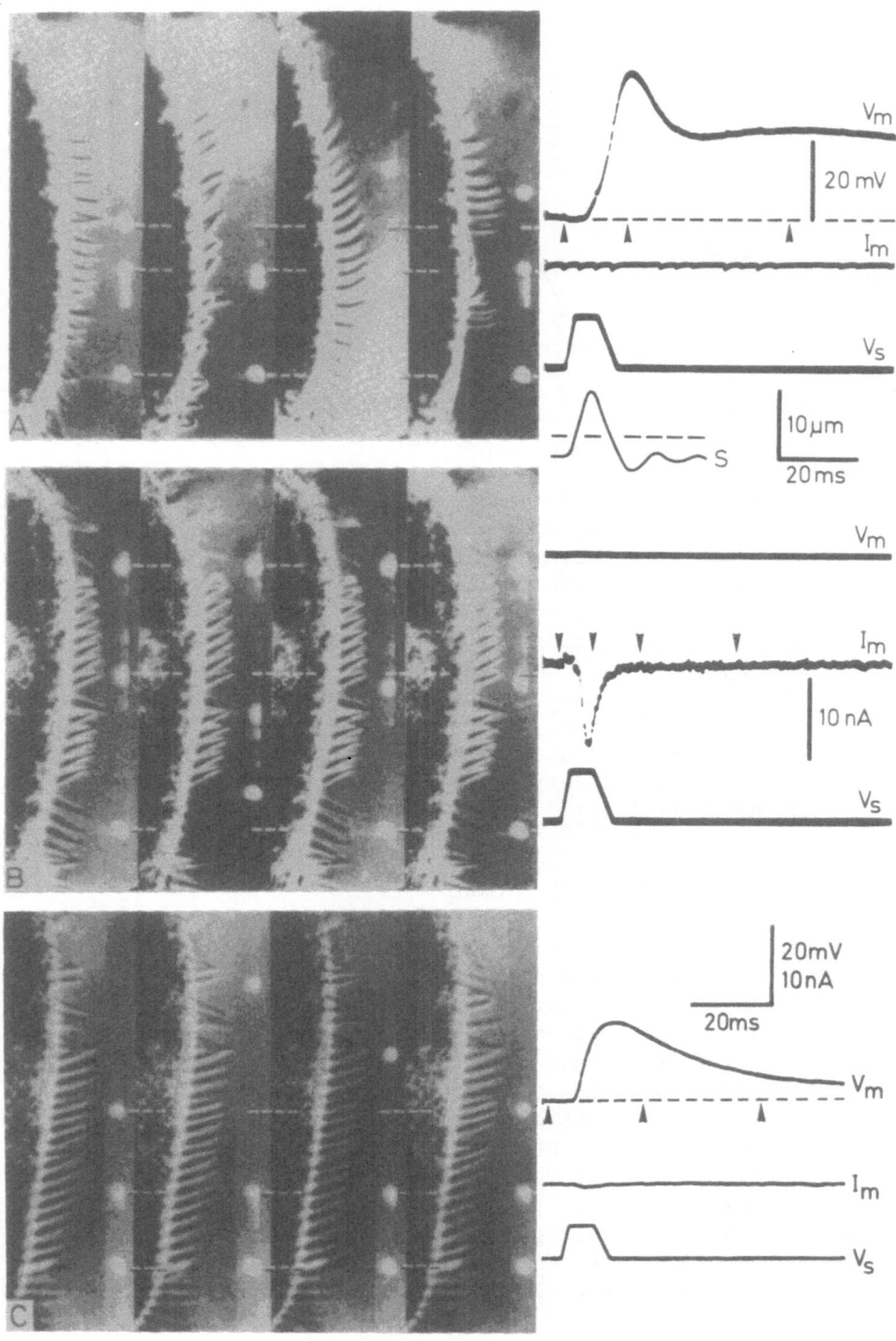

Fig. 20.   Conditions of sensory-motor coupling in <u>Stylonychia</u> to
◄━━━━━━━   depolarizing mechanostimulation. Frames from high-frequency
            film showing cirri (compound cilia, left) were correlated
            in time with mechanical and electric events (arrowheads
            near traces). (A) Electrically driven (Vs) excursion of
            probe (S) toward anterior cell end induces receptor po-
            tential followed by action potential (Vm). The previously
            resting cirri start beating toward anterior cell end (up)
            together with the depolarization. (B) Same cell stimulated
            under voltage clamp does not activate the cirri although
            large inward receptor current (Im) is seen. (C) Anterior
            stimulation after total replacement of Ca by Mg in solution
            gives rise to depolarizing receptor potential, but the
            cirri remain inactive. A, B, Depeyer and Machemer (1978b),
            C, unpublished.

an inward current component which also passes across ciliary membranes, thereby potentially re-distributing charged molecules and
ions of the intraciliary cytosol (Brehm and Eckert, 1978a).

     No direct evidence has been presented for this hypothesis so far.
Indirect evidence, such as correlations between ciliary activation
and membrane current, is equivocal since the frequency of ciliary
beating may be correlated with voltage as well as steady-state current (Fig. 19). Ciliary membranes are virtually non-conducting in
the resting and steady-state of the cell. This conclusion is based
on experiments showing that the input resting resistance, the steady-
state I/V relationship, and membrane potential were not modified
following deciliation with ethanol (Machemer and Ogura, 1979). Only
minor receptor current loops include, therefore, the ciliary membranes
(which represent 50% of the total surface area). Given these electric conditions, phoretic redistributions of intraciliary fluid components are presumably small. Similar arguments apply regarding the
electrophoretic effect upon cilia of depolarizing inward receptor
currents through anterior channels.

     An alternative hypothesis of sensory-motor coupling assumes that
ions passing the cell membrane upon mechanostimulation may be a
signal for ciliary activation. This view seems attractive for cases
of anterior stimulation, which generates an inward Ca receptor current. Experiments combining mechanostimulation with high-speed cinematography in <u>Stylonychia</u> showed that although an anteriorly applied
pulse gave rise to a large Ca receptor current under voltage clamp
conditions, none of the compound cilia were seen to be activated by
this current (Fig. 20A). When the same stimulus was applied with
the membrane potential running free, depolarization and activation
(DCA) of the compound cilia was seen (Fig. 20B). Removal of Ca from
the external fluid and replacement by Mg allowed generation of the
depolarizing receptor potential, but no activation of the cilia was
seen (Fig. 20C).

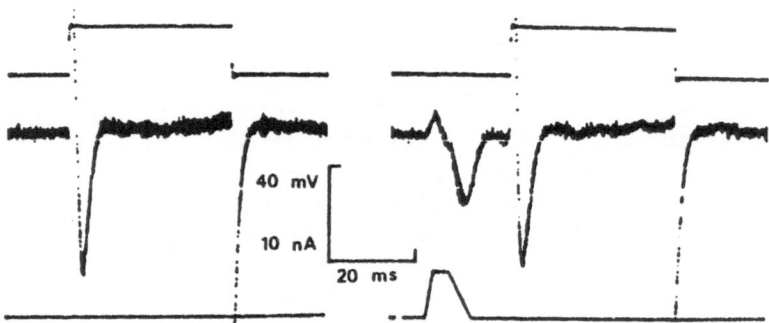

Fig. 21. Absence of interactions between somatic Ca current from
          anterior mechanostimulation (bottom trace) and ciliary Ca
          current from membrane depolarization (top trace). The
          early inward current (middle trace, right) following step
          depolarization under voltage clamp is not affected by the
          preceding receptor current (compare with reference in left
          column). Data from Stylonychia, Deitmer (1983).

     These experiments show that Ca entering across mechanoreceptor
channels cannot directly signal ciliary motor activation, even though
Ca is a prerequisite for DCA. It is the electrotonic spread of the
depolarizing receptor potential which activates voltage-sensitive Ca
channels in the ciliary membrane, thereby supplying Ca to the intraciliary space.

     Effective compartmentalization of ciliary and somatic spaces to
limit Ca fluxes is further underscored by observations showing that
the voltage-sensitive ciliary Ca current is not diminished by a Ca
receptor current immediately preceding that current (Fig. 21). Recent
evidence has suggested that a raised concentration of ionic Ca in the
cilia leads to inactivation of the voltage-sensitive Ca channels
(Brehm and Eckert, 1978b; Deitmer, 1984).

     The receptor potential also communicates to the cilia the signal
following posteriorly applied mechanical stimuli (Fig. 22). No evidence is available, so far, as to how hyperpolarization of the ciliary membrane can induce the cilia to augment their beating frequency
(HCA). Electrophysiological and cinematical analysis suggests, however, that Ca is involved in this regulation. In Paramecium, for
instance, the direction and frequency of ciliary beating are continuous functions of the membrane potential between hyperpolarizations and a depolarization up to +5 mV (Fig. 23). Double-pulse
voltage clamp experiments in Stylonychia showed that a hyperpolarization immediately following a depolarization reduced the amount and
duration of the DCA; at the same time a predepolarization decreased
the HCA (Machemer and de Peyer, 1982).

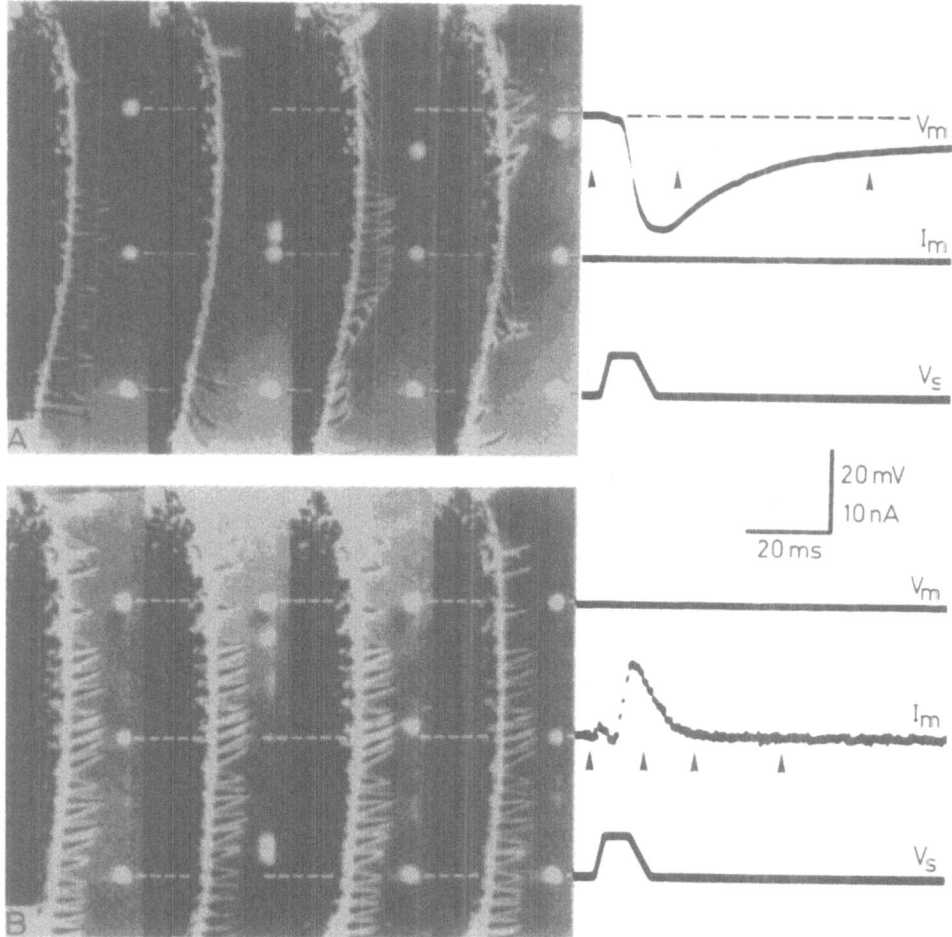

Fig. 22. Sensory-motor coupling in <u>Stylonychia</u> following hyperpolar-
          izing stimulation at the posterior cell end. (A) Beating
          of cirri toward the posterior cell end is activated, while
          the negative receptor potential rises. (B) Under voltage
          clamp the same stimulus generates an outward receptor cur-
          rent without affecting the resting cirri. Consult legend
          of Fig. 20 for further explanations. De Peyer and
          Machemer (1978b).

     Assuming, in accordance with the literature, that depolarization
raises intraciliary Ca, these observations agree with the hypothesis
that negative shifts in membrane potential serve to decrease the
concentration of intracellular ionic Ca (Machemer, 1974) and/or of
Ca bound to axonemal sites.

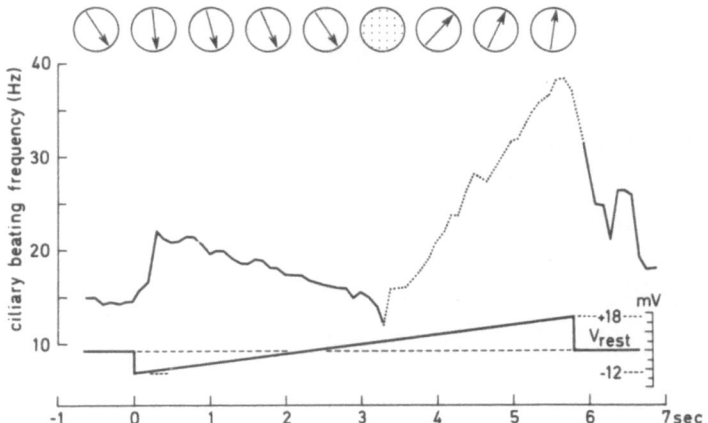

Fig. 23.   Voltage control of ciliary functions. The membraᵤe of
           Paramecium was stimulated under voltage clamp applying a
           slow positive voltage ramp between -12 mV hyperpolarization
           and +18 mV depolarization. It is seen that the frequency
           of ciliary beating rises with large negative or positive
           shifts from the membrane resting level (dashed line). The
           frequency drops with positive-going voltage up to small
           depolarizations. The orientation of the ciliary power
           stroke is turned counterclockwise from "posterior" (during
           hyperpolarization) to "anterior" (during depolarization;
           see arrows in circles). Modified after Machemer (1976).

    Recent biochemical studies in Paramecium have indicated that
cyclic adenosine monophosphate (cAMP) rises in conjunction with the
performance of HCA (Gustin et al., 1983; Schultz et al., 1984). These
data potentially indicate that cAMP is a hyperpolarization-generated
signal for axonemal activation. Injections of cAMP into Paramecium
cells under voltage clamp failed, however, to raise ciliary beating
frequency. Increases in ciliary frequency were seen in unclamped
cells following injection of cAMP only in conjunction with membrane
hyperpolarizations (Hennessey et al., in preparation). Thus, cAMP
is not a second messenger to the ciliary motor of stimulus-induced
negative voltage signals.

CONCLUSIONS

    Mechanoresponses in the ciliate protozoa involve a number of
physiological steps which are now emerging to be comparable to
similar processes in tissue for stimulus reception, excitation and
generation of a motor response. Universality in cellular functions,
being indispensable for the unicellular organism, may have been an
evolutionary basis of metazoan tissue differentiation.

Some of the properties of sensory-motor coupling are unique to the ciliate, such as the topographical difference in cellular mechano-sensitivity and the ability of the cilia to respond to membrane hyper-polarization and depolarization. While ciliates are highly advanced in these respects, there are other properties, such as a long response latency of mechanical stimulation and membrane excitation using voltage-sensitive Ca channels, which are undoubtedly archaic proper-ties. It is the position of protozoans between the worlds of pro-caryotic life and metazoan organization which will continue to define their roles as model systems for stimulus-response coupling in animals.

ACKNOWLEDGEMENTS

This work was supported by the Deutsche Forschungsgemeinschaft, SFB 114, TP A5. I would like to thank Drs. T. Hennessey, J. Deitmer, I. Ivens, S. Machemer-Röhnisch and Mr. K. Sugino for reading the manuscript.

REFERENCES

Afzelius, B., 1959, Electron microscopy of the sperm tail, J. Biophys. Biochem. Cytol., 5:269.

Brehm, P., and Eckert, R., 1978a, An electrophysiological study of the regulation of ciliary beating frequency in Paramecium, J. Physiol., 283:557.

Brehm, P., and Eckert, R., 1978b, Calcium entry leads to inactivation of calcium channel in Paramecium, Science, 202:1203.

Corey, D. P., and Hudspeth, A. J., 1979, Ionic basis of the receptor potential in a vertebrate hair cell, Nature, 281:675.

Deitmer, J. W., 1981, Voltage and time characteristics of the po-tassium mechanoreceptor current in the ciliate Stylonychia, J. Comp. Physiol., 141:173.

Deitmer, J. W., 1982, The effects of tetraethylammonium and other agents on the potassium mechanoreceptor current in the ciliate Stylonychia, J. Exp. Biol., 96:239.

Deitmer, J. W., 1983, Ca-channels in the membrane of the hypotrich ciliate Stylonychia, in: "The Physiology of Excitable Cells," A. Grinnel, and W. J. Moody, eds., A. Riss, New York.

Deitmer, J. W., 1984, Evidence for two voltage-dependent calcium currents in the membrane of the ciliate Stylonychia, J. Physiol., 355: in press.

Deitmer, J. W., Machemer, H., and Martinac, B., 1984, Motor control in three types of ciliary organelles in the ciliate Stylonychia, J. Comp. Physiol. A, 154:113.

Dunlap, K., 1977, Localization of calcium channels in Paramecium caudatum, J. Physiol., 271:119.

Eckert, R., and Brehm, P., 1979, Ionic mechanisms of excitation in
   Paramecium, Annu. Rev. Biophys. Bioeng., 8:353.
Grell, K. G., 1973, "Protozoology," Springer, Berlin.
Gustin, M. C., Bonini, M. N., and Nelson, D. L., 1983, Membrane
   potential regulation of cAMP: Control mechanism for the
   swimming behavior in Paramecium, Society for Neuroscience,
   Abstr. 9:167, 50.11.
Hara, R., Asai, H., and Naitoh, Y., 1980, Dobutsugaku Zasshi
   (Zool. Mag., Tokyo), 89:450.
Hennessey, T., Machemer, H., and Nelson, D. L., 1985, Injected
   cAMP increases ciliary beat frequency in conjunction with
   membrane hyperpolarization (in preparation).
Jennings, H. S., 1899, Studies on reactions in unicellular organisms.
   III. Reactions to localized stimuli in Spirostomum and Stentor,
   Am. Naturalist, 33:373.
Jennings, H. S., 1906, "Behavior of the Lower Organisms," Columbia
   University Press, New York.
Kafka, G., 1914, "Einführung in die Tierpsychologie auf Experimen-
   teller und Ethologischer Grundlage," Vol. I, J. A. Barth,
   Leipzig.
Machemer, H., 1965, Analyse kurzzeitlicher Bewegungserscheinungen
   des Ciliaten Stylonychia mytilus Ehrenberg, Arch. Protistenk.,
   108:153.
Machemer, H., 1972, Temperature influences on ciliary beat and
   metachronal coordination in Paramecium, J. Mechanochem. Cell
   Motil., 1:57.
Machemer, H., 1974a, Ciliary activity and metachronism in Protozoa,
   in: "Cilia and Flagella," M. A. Sleigh, ed., Academic Press,
   New York.
Machemer, H., 1974b, Frequency and directional responses of cilia
   to membrane potential changes in Paramecium, J. Comp. Physiol.,
   92:293.
Machemer, H., 1976, Interactions of membrane potential and cations
   in regulation of ciliary activity in Paramecium, J. Exp. Biol.,
   65:427.
Machemer, H., 1977, Motor activity and bioelectric control of cilia,
   Fortschr. Zool., 24:211.
Machemer, H., and de Peyer, J., 1977, Swimming sensory cells:
   Electric membrane parameters, receptor properties and motor
   control in ciliated Protozoa, Verh. Dtsch. Zool. Ges. Erlangen,
   1977:86.
Machemer, H., and de Peyer, J., 1982, Analysis of ciliary beating
   frequency under voltage clamp control of the membrane,
   Cell Motility Suppl., 1:205.
Machemer, H., and Machemer-Röhnisch, S., 1984, Mechanical and
   electric correlates of mechanoreceptor activation of the ciliated tail in Paramecium, J. Comp. Physiol. A, 154:273.
Machemer, H., and Ogura, A., 1979, Ionic conductances of membranes
   in ciliated and deciliated Paramecium, J. Physiol., 296:49.

Machemer-Röhnisch, S., and Machemer, H., 1984, Receptor current
    following controlled stimulation of immobile tail cilia in
    Paramecium caudatum, J. Comp. Physiol. A, 154:263.
Meech, R. W., 1978, Calcium-dependent potassium activation in
    nervous tissues, Annu. Rev. Biophys. Bioeng., 7:1.
Naitoh, Y., 1983, Mechanosensory transduction in Protozoa, in:
    "Membrane and Sensory Transduction," G. Colombetti, and
    F. Lenci, eds., Plenum Publishing Company, New York, pp. 113-135.
Naitoh, Y., and Eckert, R., 1969, Ionic mechanism controlling be-
    havioral responses in Paramecium to mechanical stimulation,
    Science, 164:963.
Naitoh, Y., and Eckert, R., 1973, Sensory mechanisms in Paramecium.
    II. Ionic basis of the hyperpolarizing mechanoreceptor
    potential, J. Exp. Biol., 59:53.
Naitoh, Y., and Kaneko, H., 1972, Reactivated Triton-extracted
    models of Paramecium: Modification of ciliary movement by
    calcium ions, Science, 176:523.
Ogura, A., and Takahashi, K., 1976, Artificial deciliation causes
    loss of calcium-dependent responses in Paramecium, Nature,
    264:170.
Ogura, A., and Machemer, H., 1980, Distribution of mechanoreceptor
    channels in the Paramecium surface membrane, J. Comp. Physiol.,
    135:233.
Peyer, J. de and Deitmer, J., 1980, Divalent cations as charge
    carriers during two functionally different membrane currents
    in the ciliate Stylonychia, J. Exp. Biol., 88:73.
Peyer, J. de and Machemer, H., 1978a, Hyperpolarizing and depolar-
    izing mechanoreceptor potentials in Stylonychia, J. Comp.
    Physiol., 127:255.
Peyer, J. de and Machemer, H., 1978b, Are receptor-activated ciliary
    motor responses mediated through voltage or current?, Nature,
    276:285.
Sale, W. S., and Satir, P., 1977, Direction of active sliding of
    microtubules in Tetrahymena cilia, Proc. Natl. Acad. Sci. USA,
    74:2045.
Satow, Y., Murphy, A. D., and Kung, C., 1983, The ionic basis of
    the depolarizing mechanoreceptor potential of Paramecium
    tetraurelia, J. Exp. Biol., 103:253.
Schultz, J. E., Grünemund, R., von Hirschhausen, R., and Schönefeld,
    U., 1984, Ionic regulation of cyclic AMP levels in Paramecium
    tetraurelia in vivo, FEBS Lett., 167:113.
Wood, D., 1975, Protozoa as models of stimulus transduction, in:
    "Aneural Organisms in Neurobiology," E. M. Eisenstein, ed.,
    Plenum Press, New York.

SIGNAL PROCESSING IN THE TRANSDUCTION MECHANISMS OF PHOTOTROPISM

W. Shropshire, Jr.

Smithsonian Environmental Research Center
Rockville, MD 20852

INTRODUCTION

Once a stimulus is perceived it produces a signal, chemical or physical in nature, that is processed within an organism and leads to a response. If a light stimulus produces an asymmetric signal with respect to the growth axis of an organism, a response of differential elongation results in a re-orientation of the growth axis. This response is phototropism.

The processes which occur between light impinging on the surface of a phototropically sensitive organism and the observed phototropic response are called transduction. Transduction in an engineering sense is the conversion of energy from one form into another. Transduction may include the change of information in a signal to the information in a response. Thus, incident light quanta contain both energy and information that are transduced into the energy and information that lead to a response.

In general, for sensory systems:

stimulus $\longrightarrow$ black box $\longrightarrow$ response

By measuring relationships between the signal and the response, statements can be made about the transduction processes occurring within the "black box".

For phototropism:

light $\longrightarrow$ transmission $\longrightarrow$ attenuation/concentration $\longrightarrow$ gradient $\longrightarrow$ absorption $\longrightarrow$ photoproduct $\longrightarrow$ amplification $\longrightarrow$ altered metabolism $\longrightarrow$ synthesis $\longrightarrow$ response

211

There are several fundamental differences that must be kept in mind in the light-growth and phototropic responses of fungi and higher plants. In fungi, in general, the light-growth response is positive; i.e., for an increase in light intensity there is an increase in growth rate. For flowering plants the light-growth response is usually negative; i.e., for an increase in light intensity there is an inhibition of growth rate. However, in some systems such as coleoptiles and mesocotyls of grasses there may be a distribution of responses; i.e., red light may stimulate coleoptile growth but inhibit mesocotyl growth. The optical properties of the lower and higher plants are usually quite different. For fungi, where there is no chlorophyll, the responding cells are often nearly transparent and their uniform cylindrical structure allows for dioptrics. In higher plants not only is the tissue very green due to chlorophyll for light-grown plants but even in dark-grown tissue there is the presence of the photomorphogenic pigment phytochrome that complicates the interpretation of light responses in photo-tropism. In addition, the tissues that perceive light are usually multicellular and somewhat irregular in shape, thus making it ex-tremely difficult to determine the light intensity distribution within the tissue with any degree of certainty.

LIGHT STIMULUS

Electromagnetic radiation of the wavelength range 250 nm to 800 nm and collimated as from a distant point source is incident on an ellipsoidal or cylindrical surface. A portion of the incident radiation enters the organism and the remainder is reflected or back scattered, depending upon the angle of incidence and the dif-ference in index of refraction between the exterior and interior media that constitute the surface.

In the case of a cylindrical cell for a collimated beam of blue light incident in a plant normal to the growth axis, it is apparent that many optical processes occur. These include reflection, re-fraction and scattering. The magnitude of the reflection is a function of the angle of incidence and the index of refraction dif-ference at the surface. For an air (external):chitin cell wall (internal) surface in the case of a Phycomyces sporangiophore the indices of refraction at 450 nm are 1.00:1.50. For the cell taken as a whole, without regard to individual differences in the internal components, the index of refraction is 1.38 (Castle, 1933a). For a multicellular coleoptile or hypocotyl, the many interfaces between each cell complicate the internal distribution of light (Parson et al., 1984).

The ray diagram for a homogeneous cylinder in air is relatively complex (Gressel and Horwitz, 1982). For example, consider the intensity distribution on the surface of the distal side. Since

index of refraction is a function of the wavelength, the intensity
distribution at the distal surface will be quite different by refraction for blue light (450 nm) as compared to red light (650 nm).

A second major complication is if the angle of incidence of
the beam of light is not normal to the growth axis. For a cylindrical Phycomyces sporangiophore the focal band moves from external
to the cell, at an angle of incidence of 90°, to internal to the
cell about 25° (Dennison, 1979). If measurements are made of the
bending rate and time of onset for phototropic curvatures of sporangiophores as a function of the angle of incidence it is clear
that the size of the stimulus (difference in effectiveness between
the light on the proximal and distal side of the sporangiophore)
diminishes as the angle of incidence decreases.

Remember that as the beam exits the cylinder there will also
be appreciable internal reflection. Jaffe (1960) has calculated
the magnitude of this effect with reference to plant-polarized beams

Table 1. Phototropism as a Function of Angle of Incidence.

β	Average Bending Rate[*] (°/minute)	Average Time of Onset (minute)
90	2.9 ± 0.4	5.4 ± 0.8
70	3.0 ± 0.3	3.9 ± 0.7
48	3.0 ± 0.6	7.0 ± 1.0
40	2.9 ± 0.8	9.7 ± 1.2
30	2.6 ± 0.5	12.7 ± 4.1
20	1.7 ± 1.4	14.9 ± 6.7
08	-1.2 ± 2.1	8.9 ±10.2
0	(0.4 ± 3.2)[**]	random oscillations

β is the angle of incidence measured from the growth axis
of the sporangiophore (wild type strain). Sporangiophores
adapted to unilateral broad band blue light (0.31 μWcm^{-2})
with rotation of 2 rpm and at angle indicated for one hour.
[*]Longest straight line fit of successive points taken at
one minute intervals for at least thirty minutes after
rotation stopped at time = 0 to induce curvature.
[**]No consistent pattern observed. (Unpublished data of
Shropshire and Gettens, 1963).

of light. The Fresnel differences in reflection of transversely polarized and longitudinally polarized beams of light incident upon the front surface of a cylindrical organism (transversely polarized light enters about 15% more effectively than longitudinally polarized) are balanced by the increased internal reflections within the cylinder at the exiting surface so that both beams would be expected to be equally effective. The residual difference in effectiveness that is observed is due to dichroically oriented photoreceptors (Jesaitis, 1974).

However, organisms are not homogeneous isotropic materials and scattering occurs due to inhomogeneities. In addition, for even a relatively simple system such as the Phycomyces sporangiophore there are multiple layers within the cell having marked differences in index of refraction (Dennison and Foster, 1977); e.g., cell wall (1.50), cytoplasm (1.38) and vacuole (1.36).

TRANSMISSION

The transmission of the light through the cylindrical structure is altered by dioptric effects so that the intensity distribution per unit area of exiting surface is higher than that of the entering surface. That is, there is for an incident collimated beam of light in air a concentration of the number of quanta passing through a region on the distal side as compared to a region on the proximal side. This concentration is called the focussing advantage (Castle, 1933b).

ATTENUATION

There may be attenuation of the light beam by absorption or scattering of the light. Scattering will increase the effective path length of the light as it passes through the organism and will increase the probability for absorption either by the photoreceptors that lead to phototropism or by absorbers that are screens and do not directly lead to phototropism (see Delbrück and Shropshire, 1960, for a detailed quantitative description of such effects).

In the case of only absorption by a phototropic photoreceptor (self-screening) it is clear that for the case of light normally incident to the growth axis the net irradiance entering the distal half of the cell must be less than that entering the proximal half by the amount of absorption. For the distal side to have a greater effectiveness for bending, the product of the number of photoproducts produced times their distance from a midline normal to the incident direction must be greater than for the proximal side. This theory is known as the "moment arm advantage" (Bergman et al., 1969).

CONCENTRATION

The fact that dioptrics is important for the resultant phototropic curvature of Phycomyces sporangiophores has been well
established (Shropshire, 1962). If a cylindrical glass lens is
placed in front of a sporangiophore such that the dioptrics is reversed the phototropic response is reversed. Similarly if cells
are immersed in media which have an index of refraction greater
than 1.29 the cells bend away from an incident beam of blue light.

Balance experiments can be used to determine the effective
index of refraction for the entire cell (1.29) for which the
focussing advantage due to refraction is just balanced by the losses
across the cell due to absorption, scattering and reflection
(Shropshire, 1962). From such measurements the magnitude of the
focussing advantage has been estimated to be about 0.3 for blue
light (450 nm) and the loss due to absorption at 450 nm about 0.1
(Delbrück and Shropshire, 1960). In fluorochemical the focussing
advantage was calculated to be 14% to match the total internal
losses (Shropshire, 1962).

GRADIENTS

In coleoptiles having an ellipsoidal cross-section at the
sensitive tip region (Shropshire, 1975) or for hypocotyls with a
nearly circular cross-section the presence of many cells complicates
the determination of the gradient produced within the tissues
(Parson et al., 1984). Blaauw in 1915 and van Dillewijn in 1927
measured the gradients across hypocotyls and coleoptiles. (See
Parson et al., 1984, for a summary of this work.)

In coleoptiles of Avena data were obtained (Shropshire, 1975)
for unilateral exposures to blue light at fluence levels for first
positive curvature with a cylindrical glass lens inserted in the
light path before the coleoptile and the resultant curvatures were
negative. These data are difficult to interpret for two reasons:
(a) The ray path through the tip could not be symmetrically reversed
as is possible for a cylindrical organism. (b) The concentration
of the light beam on the proximal surface may have increased the
fluence rate at the surface to a value which would be in the negative portion of the fluence response curve (Dennison, 1979). In
any case there is evidence in support of a dioptric effect in
coleoptiles for first positive curvature from the well known fact
(Pohl and Russo, 1984) that coleoptiles as well as sporangiophores
immersed in paraffin oil and irradiated unilaterally with blue
light give negative phototropic curvatures.

However, for exposures well below the tip of coleoptiles
(second positive curvatures) the principal effect producing a

gradient must be attenuation by scattering and absorption with a minimum effect by refraction (Parson et al., 1984).

In sporangiophores there is a large central vacuole that occupies 70% of the cell diameter (Bergman et al., 1969). This vacuole is a repository for gallic acid produced by metabolism. Gallic acid is present in high concentrations (10 mg/ml) and has an absorption spectrum which begins to absorb strongly about 330 nm and shorter wavelengths (Delbrück and Shropshire, 1960). If the wavelength of the incident unilateral light is less than 330 nm negative phototropic curvatures are produced. This internal screen effectively produces an extremely steep gradient across the cell such that absorption occurs mostly on the proximal side of the cell. The distal side of the cell receives very little light. Therefore, the cell bends away from the proximal side with strong negative phototropic curvatures in the ultraviolet.

In multicellular tissues there is a lack of knowledge about the magnitude of the gradient which is required to produce phototropic curvatures (Parson et al., 1984; Poff, 1983). The magnitude of the gradient and the local light intensity distribution is very difficult to calculate theoretically and even more difficult to measure directly (Seyfried and Fukshansky, 1983). Light-piping occurs in dark-grown Avena seedlings (Mandoli and Briggs, 1982) but the importance of this process for phototropic curvatures is not clear (Firn et al., 1983). Small fibre optic probes have been used to measure directly the light gradients within tissues (Vogelmann and Björn, 1984) but disturbance of the tissue properties by the probe and the acceptance angles of the probes limit their accuracy (Seyfried and Fukshansky, 1983).

STIMULUS ABSORPTION

The Grotthus-Draper law of photobiology states that for a quantum of light to have an effect it must be absorbed. Molecules that absorb light are pigments. The pigments that are excited by absorption of light and lead to the photoproducts required to produce a response are called photoreceptors.

Action spectra measurements are used to determine the wavelength sensitivity of the response and from these data the absorption characteristics of the photoreceptor can be derived (Hartmann, 1983). Action spectra for phototropism early led to the consideration of two classes of pigments, carotenoids and flavins (Presti and Delbrück, 1978). It is generally accepted that for phototropism flavins are the photoreceptors.

Mutant strains of Phycomyces, deficient in carotenoids, including double mutants with no detectable β-carotene present remain phototropically sensitive with normal responses (Presti et al., 1977).

Similarly, action spectra for phototropism and the light growth response at red wavelengths (Delbrück et al., 1976) indicate direct optical excitation for flavin triplets at 595 nm.

In corn seedlings carotenoid levels can be reduced to 2% of normal levels by herbicide inhibition (Vierstra and Poff, 1981). Phototropic sensitivity to 380 nm light and the threshold values for 380 nm and 450 nm are unaffected. However, phototropic sensitivity to 450 nm is reduced indicating that carotenoids are involved in phototropism, probably as an internal screen enhancing the light gradient. Data using selective inhibitors of phototropism (Schmidt et al., 1977) and comparing phototropic responses of corn with geotropic curvatures in the dark for similar seedlings treated with these inhibitors also indicate that flavins are the photoreceptors for phototropism.

Genetic analysis of phototropism of the perithecial necks of Neurospora crassa using white collar and albino mutants (Harding and Melles, 1983) shows that when white collar mutants, in which the light induction of enzymes in the carotenoid biosynthetic pathway is blocked, are used as the protoperithecial parent in crosses, the perithecial necks have no phototropic response. However, when either wild type or one of three albino mutants are used as protoperithecial parent, phototropism occurs. It was concluded that the sensory transduction pathways for both photoinduced carotenogenesis and phototropism in Neurospora crassa must have some steps in common.

Although it is generally agreed that flavins are the photoreceptors, Galland (1983), measuring the interaction between geotropism and phototropism (photogeotropic equilibria) near threshold values for phototropism for wild type as well as a number of mutant strains, finds very complex action spectra with evidence for more than one photoreceptor.

Löser and Schäfer (1983) also have evidence for the participation of a photochromic photoreceptor or two independent pigments in phototropism. They observe responses in Phycomyces to interactions between 650 and 450 nm stimuli given simultaneously.

In coleoptiles the fluence-response curves to pulses of blue light give a first positive curvature followed by a first negative curvature as the fluence is increased. For even greater fluences second positive and even third positive curvatures can be produced (Dennison, 1984). It is generally agreed that first positive curvatures are perceived by the extremely sensitive coleoptile tip (less than 350 μm from the tip) and that second positive and third positive curvatures are perceived by tissues below the tip (so called base curvatures). Action spectra for all three regions of the fluence response curves are essentially identical and indicate that the same photoreceptor is responsible for all three fluence regions (Dennison, 1984).

Sensitivity changes to blue light stimuli can be brought about by exposing coleoptiles to red light (Briggs and Chon, 1966). The photoreceptor involved in regulating sensitivity is clearly the red, far red, photochromic pigment, phytochrome. However, the levels of phytochrome which are involved in these sensitivity changes are some four orders of magnitude below the spectrophotometrically observable changes in phytochrome concentration. It is not known how this process functions and one explanation is that there are two distinct populations of phytochrome molecules which are spatially separate. One pool is required for the spectrophotometric obser- vations and the second affects the phototropic sensitivity.

The irradiance range over which phototropism occurs is extremely large (Dennison, 1979). In Phycomyces single pulses for 10 seconds of blue light produce curvatures (Iino and Schäfer, 1984). A maximum curvature occurs for stage I sporangiophores of 30° for a fluence of about 10^{-4} Jm^{-2}, decreases to a minimum of 1 or 2° about 0.1 Jm^{-2}, and then begins to increase in curvature for fluences greater than 1 Jm^{-2}. Similar values have been obtained by Galland for stage IV sporangiophores (personal communication, 1984).

For continuous exposure unilaterally, curvatures of stage IV sporangiophores occur up to a fluence rate of about 10 Wm^{-2}. For greater fluence rates the phototropic curvatures rapidly decrease in size presumably because both sides of the sporangiophore are light saturated and there is no effective response gradient across the cell. An action spectrum of this high irradiance response is essentially the same as that found by a null method for mid-range fluence rates (Bergman et al., 1969).

In flowering plants solar tracking occurs for full sunlight (Ehleringer and Forseth, 1980) at about 500 Wm^{-2} or μmol s^{-1} m^{-2} (400 to 700 nm). This phototropic response is a nastic type movement (Satter, 1979) and not directly connected with growing tissue. The pulvinus is the site of perception of light in Lupinus (Vogelmann and Björn, 1983) and in Lavatera the major leaf veins perceive the light vectorially (Schwartz and Koller, 1978). Only light which strikes the upper leaf surface is effective in activating tracking and the response is blue light mediated. Turgor changes in the pulvinus (Satter, 1979) are brought about by changes in potassium ion balances.

PHOTOPRODUCTS

The initial photoproduct produced after light absorption in phototropism is not known. Changes in the amounts of extractable

ATP and cyclic AMP from Phycomyces sporangiophores have both been
reported following blue stimuli. However, these findings have
been difficult to confirm (Dennison, 1984).

Rapid changes in bioelectric potentials are observed in coleoptiles and hypocotyls after blue light exposures. However, for
coleoptiles the convex side is always found to be electrically
positive regardless of whether the curvature is positive or negative
(Schmidt, 1983). With dwarf bean seedlings a hyperpolarization is
observed with 8.5 Wm^{-2} blue light (437 nm) given for 60 seconds with
a maximum occurring between one and two minutes after the onset of
light (Hartmann and Schmid, 1980). Such hyperpolarizations occur
sufficiently fast prior to the onset of bending that it is plausible
to imagine that they are among the first events occurring after light
absorption. However, there are few data connecting them with phototropic responses.

In coleoptiles and seedlings it has long been argued that the
differential growth required for a phototropic curvature is brought
about by a lateral transport of auxin. However, this idea has been
questioned (Firn and Digby, 1980) for all but first positive curvatures occurring for very low fluences given in brief pulses. In
Phycomyces sporangiophores there is no evidence for auxin involvement in the development of phototropic curvatures. In any case, for
the differences in light distribution in the growing zone of the
sporangiophores there must be some sort of intracellular communication system which allows for every element of the responding region
to be influenced by the rate of growth locally produced. Otherwise
there could not be a smooth curvature developed that integrates the
growth of every part. Nothing is known about the nature of this
process.

ALTERED METABOLISM

The initial photoabsorption event is temperature independent
but subsequent biochemical events would be expected to be temperature
dependent. By the use of low temperature, attempts have been made to
store the initial photoproducts and then to allow responses to occur
with subsequent warming. In Phycomyces unilateral ultraviolet stimuli
can be stored for at least 2 hours at 2° C prior to the response
occurring at room temperature (Petzuch and Delbrück, 1970). In Avena
coleoptiles a 45 minute unilateral blue stimulus persists for 450
minutes at 2° C. However, some transduction steps occur at 2° C
because an exposure given at the low temperature is three fold less
effective than a control exposure given at 25° C (Pickard et al.,
1969). It is concluded that some transduction steps take place
even at the low temperature.

AMPLIFICATION

For threshold stimuli a relatively small number of absorptions must occur. In coleoptiles, where an asymmetric auxin distribution has been shown to occur, a calculation of the number of auxin molecules formed per each quantum absorbed (Galston, 1959) gives a value of 4682. The conclusion is that some sort of amplification process must occur. The mechanism of this process is unknown. However, the sensitivity to red light for phytochrome controlled growth of coleoptile sections cut from dark-grown oat seedlings can be increased 10 000 fold if the sections are incubated in 6 μM indole-3-acetic acid (IAA). It is concluded that the effect of IAA on the sensitivity is on some step of the transduction of the phytochrome signal, rather than on the growth response (Shinkle and Briggs, 1984).

In Phycomyces sporangiophores adaptation (range-adjustment) occurs which regulates the sensitivity of both the light-growth and phototropic responses to stimuli, depending upon the ambient level of light prior to the stimulus. For an increase in light intensity, the light adaptation kinetics depends on the magnitude of the increase as well as the absolute level of the intensity range in which the increase is given (Galland and Russo, 1984a). Dark adaptation kinetics is observed to be bi-phasic with a slow and fast component. The conclusion is that dark and light adaptation are regulated by different mechanisms. A comparison of the dark adaptation kinetics with the time course of the dark-growth response led to the postulate of two adaptation mechanisms. These are: an input adaptation which probably operates at the photoreceptor pigment level and an output adaptation which modulates the growth response directly (Galland and Russo, 1984a).

The threshold for light sensitivity of Phycomyces depends upon the growth conditions of the sporangiophores (Galland and Russo, 1984b). If the wild-type strain is dark-grown its threshold for bending is 10^{-9} Wm^{-2} while under continuous light it is 2 x 10^{-7} Wm^{-2}. Adaptation is also influenced by the growth conditions. The kinetics of dark adaptation is slower in light-grown sporangiophores than in dark-grown ones.

The observed bending rates for maximum response are twice as high in dark-grown sporangiophores as in light-grown ones. In addition the diameter and growth rates for light-grown sporangiophores are greater than for dark-grown sporangiophores (Galland and Russo, 1984b). Thus, Phycomyces regulates its own light sensitivity by its light-sensing apparatus. This light-sensing apparatus has three principal functions (a) absolute sensitivity (threshold), (b) sensory adaptation (range adjustment), and (c) wavelength sensitivity as discussed above under photoreceptors.

RESPONSES

A phototropic curvature develops because of a differential
elongation on the two sides of the organism. However, it is clear
that this differential elongation can develop in several ways (Firn
et al., 1983). There can be a redistribution of growth such that
the overall elongation rate remains constant, or there can be an
inhibition or stimulation of one side with the other remaining
unaffected, or a quantitative difference in the stimulation or
inhibition on each side. In every case, the resultant curvature
could be the same. Therefore, in order to understand the relationship of growth to phototropic curvatures it is not enough to measure
simply the angle of bend after an arbitrary time but rather growth
rates and distribution of growth as a function of time need to be
measured (Badham, 1984; Silk, 1984). Of course, null experiments
have the advantage of eliminating many of these problems (Delbrück
and Shropshire, 1960) if a null response can be devised for the
process being studied.

It is difficult in many cases to correlate the change in
growth rates with the resultant phototropic curvatures. In many
cases it is simply because the kinetic data do not exist. In
Phycomyces sporangiophores the growth distribution in the growing
zone is the same during steady-state adapted growth as during a
light-growth response in which the absolute rate is more than
doubled (Castle, 1959). It is concluded that light affects growth
proportionally in all parts of the light-sensitive growing zone
and that the same pattern of growth is maintained as in the steady
state.

If sporangiophores are allowed to bend phototropically under
conditions where a light-growth response does not occur (Castle,
1961; Shropshire, 1963) it is clear that phototropism is basically
a steady-state process. The growth is redistributed between proximal
and distal sides. For example, helices can be formed in the rigid
wall below the growing zone as a record of the continuous phototropic
response occurring over many hours to a unilateral blue light stimulus when the container in which the sporangiophore is growing is
rotated slowly (one revolution every six hours). Thus the light-
growth response adapts but the phototropic response does not.

One explanation for this observation is that local autonomous
adaptation does not occur and the adaptation level is spread across
the cell. During phototropism the rotation of the cell as it grows
converts the spatial asymmetrical distribution of light into a
temporal one (Dennison and Foster, 1977). Thus, the light sensitive
region is subjected to a continuous temporal stimulus and phototropism never adapts as long as there is a spatial asymmetry in
light intensity. If an external back-rotation (counterclockwise
from above) of 10° per minute is applied to sporangiophores the

phototropic response is reduced by two-thirds. There is no cyto-
logical evidence for any rotating internal structure that might
contain the photoreceptors.

In coleoptiles there is disagreement about whether lateral
redistribution of growth occurs or whether growth of the proximal
side is inhibited while the distal side remains unaffected (Franssen
et al., 1981; Briggs and Iino, 1983; Iino and Briggs, 1984). It
appears clear that first positive phototropic curvature, at least
in maize coleoptiles, is the result of lateral distribution of
growth. This first positive curvature appears to be under control
of the tip of the coleoptile while second positive curvatures and
those produced by larger fluences are local responses and do not
seem to depend upon a simple redistribution of auxin regulated
growth (Firn and Digby, 1980; Franssen et al., 1982).

The growth of stems of dicotyledonous seedlings is inhibited
rapidly by exposure to blue light (Cosgrove, 1983) with a 50% reduction in rate occurring after about three minutes and a lag period
of one minute for a 16 second exposure (2.6 Wm^{-2}). This inhibition
is produced by altering the properties of the cell wall of growing
tissue. Turgor pressure is a key indicator for the mechanism by
which light affects growth (Cosgrove, 1983). Comparable measurements
of hydraulic conductivity and yield threshold have not been measured
for Phycomyces sporangiophores. The wall of Phycomyces is made of
chitin fibrils (Bergman et al., 1969) and a direct effect of light
on chitin synthetase in vitro has been reported (Jan, 1974). It is
believed that the turgor pressure of the central vacuole is the
driving force which increases elongation growth as the wall is
loosened after a light stimulus but there are scant data in support
of this assumption. In coleoptiles and dicotyledonous seedlings
the wall is cellulose and cell expansion occurs through regulation
of the cell wall mechanical properties (Taiz, 1984) principally by
proton (H^{+}) extrusion as the "wall loosening factor" brought about
as an early response to auxin.

For many years it has been a puzzle why unilateral red light
normally does not produce phototropic curvatures since red light
through the phytochrome system clearly affects the elongation
growth rate of both monocotyledonous and dicotyledonous tissues
(Shropshire and Mohr, 1970). The explanation is that the gradient
produced by red light is insufficient to bring about differential
elongation even though there is a sufficient gradient to produce
differential synthesis of anthocyanin through phytochrome control.
However, unilateral red light induces positive curvatures in maize
mesocotyls (Iino et al., 1984). The second positive curvatures
result from a gradient in P_{fr} across the plant axis. First positive
curvatures induced by red light are inhibited by far-red light given
equilaterally from above either before or after phototropic
induction. The fluence response relationship for blue light is

shifted two orders of magnitude to greater fluences than for red
light and the blue light induces classical first positive curvatures
in the coleoptile while red light does not. Because of the inhibition of this red response by far red light it is concluded that
it is a phytochrome mediated response.

DISSECTION OF THE SENSORY TRANSDUCTION PATHWAY

Eight genes are known which affect phototropism in Phycomyces
(Lipson et al., 1980; López-Diaz and Lipson, 1983; Lipson et al.,
1983). These are known as madA to madH genes. By complementation
and recombination analyses they have been shown to be unlinked
genes. Phenotypically they are divided into three groups, "night-
blind" (madA, madB and madC which have reduced photosensitivity),
"stiff" (madD, madE, madF and madG which have reduced tropic responses and are probably affected near the common output of the
sensory transduction pathways), and "hypertropic" (madH which show
enhanced tropisms). In the mycelial responses of light induced
carotenoid synthesis and the formation of sporangiophores, the
"night-blind" madA and madB strains are abnormal while madC strains
are normal. Recently (López-Diaz and Lipson, 1983), the madC gene
has been connected with the photoreceptor at the input to the
sensory pathway. In addition, the madH gene has been shown to be
associated with the growth control output. Double mutants carrying
madC and madH genes are behaviorally intermediate between phenotypes
of the parentals. In heterocaryons carrying madC and madH genes the
expression of the recessive "hypertropic" allele becomes dominant.
It is concluded (López-Diaz and Lipson, 1983) that a physical interaction may occur between the gene products, madH and madC, in a
molecular complex for the photosensory transduction chain.

Little has been done in the way of genetic analysis for phototropism in higher plants. In the rapidly growing seedlings of
Arabidopsis thaliana (Koornneef et al., 1980) the growth of some 41
mutants were compared with wild type. Some mutants were found in
which spectrophotometrically detectable phytochrome in dark-grown
seedlings was greatly reduced or absent. All 41 mutants had
reduced sensitivity to inhibition of hypocotyl elongation by white
light. From recombinant data it was concluded that there is
probably more than one photoreceptor for the high irradiance
response that leads to inhibition of hypocotyl growth. If growth
is directly connected to phototropism then this system may offer a
way to genetically dissect the sensory transduction pathway.

Much is known about the growth of cellulose cell walls in
higher plants (Taiz, 1984) and since the discovery of spiral growth
in Phycomyces (Oort, 1931), much effort has been directed biophysi-
cally to describe the growth patterns of the Phycomyces chitin cell
wall (Gamow, 1980; Gamow and Böttger, 1981). Elongation growth for

each system appears to take place largely by intussusception in which newly synthesized wall polymers of either chitin or cellulose are inserted into the wall to form new wall material (Preston, 1974).

However, it is not certain that there is a direct linkage between light-induced growth and the phototropic response. (See the discussion above on adaptation and growth responses.) Iino and Schäfer (1984) find that stage I sporangiophores responding to 10 second pulses of blue light have an elongation light-growth response which is first stimulated and then inhibited when measured overall for the sporangiophores during the course of phototropic curvatures. They suggest that the primary mechanisms of phototropism and light-growth may be distinct. Galland (personal communication, 1984) has measured both light-growth responses and phototropic responses of stage IV sporangiophores, dark-adapted for three hours, to short unilateral pulses of 30 seconds (450 nm). The threshold for the light-growth response is between 50 and 100 times higher than the phototropic threshold. He concludes that there is no clear correlation between phototropism and the positive and negative peaks of the light-growth response. Redistribution of growth appears to be the mechanism of phototropism.

In the sensory transduction pathway outlined in the first section (Introduction) on the first page of this chapter ("For phototropism.."), it is apparent that there is no inherent reason that the connections indicated should follow a simple, linear pathway. There have been few tests (Delbrück and Reichardt, 1956) of the stimulus-response system of phototropism to determine what the transfer functional is that gives the relationship between the input function (stimulus) and the output function (response). The most extensive tests have been those using an automated tracking system (Foster and Lipson, 1973) which allows for precise measurements of responses to white noise stimuli (Lipson, 1975). The Delbrück-Reichardt model (Delbrück and Reichardt, 1956) is inadequate to accommodate the data obtained and new data continually expand the requirements for a comprehensive model (Galland and Russo, 1984a).

CONCLUSIONS

The transduction processes of phototropism are complex in both fungi and higher plants. However, the recent resurgence of interest in these processes has forced a re-examination of long accepted facts and offers hope of solving many intransigent problems. It is anticipated that the application of new techniques and methodologies will bring about a complete and detailed description of the steps occurring between stimulus and response at the molecular and intracellular levels. It is an exciting time to study phototropism.

No attempt has been made in this chapter to review phototropism completely but rather to treat the process as a whole and stimulate some appreciation for the problems remaining. Numerous exhaustive and critical reviews have been published. Some are: Galston (1959), Briggs (1963), Shropshire (1963), Bergman et al. (1969), Foster (1977), Dennison (1979), Firn and Digby (1980), Russo (1980), Gressel and Horwitz (1982), Dennison (1984) and Pohl and Russo (1984).

REFERENCES

Badham, E. R., 1984, Measuring curvature in cylindrical plant organs, Exper. Mycol., 8:176-178.

Bergman, K., Burke, P. V., Cerdá-Olmedo, E., David, C. N., Delbrück, M., Foster, K. W., Goodell, E. W., Heisenberg, M., Meissner, G., Zalokar, M., Dennison, D. S., and Shropshire, Jr., W., 1969, Phycomyces, Bacteriol. Rev., 33:99-157.

Briggs, W. G., 1963, The phototropic responses of higher plants, Annu. Rev. Plant Physiol., 14:311-352.

Briggs, W. R., and Chon, H. P., 1966, The physiological versus the spectrophotometric status of phytochrome in corn coleoptiles, Plant Physiol., 41:1159-1166.

Briggs, W. R., and Iino, M., 1983, Blue-light-absorbing photoreceptors in plants, Phil. Trans. R. Soc. Lond. B, 303:347-359.

Castle, E. S., 1933a, The refractive indices of whole cells, J. Gen. Physiol., 17:41-47.

Castle, E. S., 1933b, The physical basis of positive phototropism of Phycomyces, J. Gen. Physiol., 17:49-62.

Castle, E. S., 1959, Growth distribution in the light-growth response of Phycomyces, J. Gen. Physiol., 42:697-702.

Castle, E. S., 1961, Phototropism, adaptation, and the light-growth response of Phycomyces, J. Gen. Physiol., 45:39-46.

Cosgrove, D. J., 1983, Photocontrol of extension growth: A biophysical approach, Phil. Trans. R. Soc. Lond. B., 303:453-465.

Delbrück, M., Katzir, A., and Presti, D., 1976, Responses of Phycomyces indicating optical excitation of the lowest triplet state of riboflavin, Proc. Natl. Acad. Sci. USA, 73:1969-1973.

Delbrück, M., and Reichardt, W., 1956, System analysis for the light growth reactions of Phycomyces, in: "Cellular Mechanisms in Differentiation and Growth," D. Rudnick, ed., Princeton University Press, Princeton, pp. 3-44.

Delbrück, M., and Shropshire, Jr., W., 1960, Action and transmission spectra of Phycomyces, Plant Physiol., 35:194-204.

Dennison, D. A., 1979, Phototropism, in: "Physiology of Movements," Encyclopedia of Plant Physiology, New Series, W. Haupt and M. Feinleib, eds., Springer Verlag, Berlin, 7:506-566.

Dennison, D. A., 1984, Phototropism, in: "Advanced Plant Physiology," M.B. Wilkins, ed., Pitman Books, Ltd., London, pp. 144-162.

Dennison, D. S., and Foster, K. W., 1977, Intracellular rotation and the phototropic response of Phycomyces, Biophys. J., 18: 103–123.

Ehleringer, J., and Forseth, I., 1980, Solar tracking by plants, Science, 210:1094–1098.

Firn, R. D., and Digby, J., 1980, The establishment of tropic curvatures in plants, Annu. Rev. Plant Physiol., 31:131–148.

Firn, R., Digby, J., Macleod, K., and Parsons, A., 1983, Phototropism, Patterns of growth and gradients of light, What's New Plant Physiol., 14:29–32.

Foster, K. W., 1977, Phototropism of coprophilous zygomycetes, Annu. Rev. Biophys. Bioeng., 6:419–443.

Foster, K. W., and Lipson, E. D., 1973, The light growth response of Phycomyces, J. Gen. Physiol., 62:590–617.

Franssen, J. M., Cooke, S. A., Digby, J., and Firn, R. D., 1981, Measurements of differential growth causing phototropic curvature of coleoptiles and hypocotyls, Z. Pflanzenphysiol., 103: 207–216.

Franssen, J. M., Firn, R. D., and Digby, J., 1982, The role of the apex in the phototropic curvature of Avena coleoptiles: Positive curvature under conditions of continuous illumination, Planta, 155:281–286.

Galland, P., 1983, Action spectra of photogeotropic equilibrium in Phycomyces wild type and three behavioral mutants, Photochem. Photobiol., 37:221–228.

Galland, P., and Russon, V. E. A., 1984a, Light and dark adaptation in Phycomyces, J. Gen. Physiol., 84:101–118.

Galland, P., and Russon, V. E. A., 1984b, Threshold and adaptation in Phycomyces, their interelation and regulation by light, J. Gen. Physiol., 84:119–132.

Galston, A. W., 1959, Phototropism of stems, roots and coleoptiles, in: "Handbuch der Pflanzenphysiologie," 17: 492–529.

Gamow, R. I., 1980, Phycomyces: Mechanical analysis of the living cell wall, J. Exp. Bot., 31:947–956.

Gamow, R. I., and Böttger, B., 1981, Phycomyces: Irregular growth patterns in stage IVB sporangiophores, J. Gen. Physiol., 77: 65–75.

Gressel, J., and Horwitz, B., 1982, Gravitropism and phototropism, in: "The Molecular Biology of Plant Development," H. Smith and D. Grierson, eds., Blackwell, London, pp. 405–433.

Harding, R. W., and Melles, S., 1983, Genetic analysis of phototropism of Neurospora crassa perithecial beaks using white collar and albino mutants, Plant Physiol., 72:996–1000.

Hartmann, K. M., 1983, Action spectroscopy, in: "Biophysics," W. Hoppe, W. Lohmann, H. Markel and H. Ziegler, eds., Springer Verlag, Berlin, pp. 115–144.

Hartmann, E., and Schmid, K., 1980, Effects of UV and blue light on the bipotential [sic] changes in etiolated hypocotyl hooks of dwarf beans, in: "The Blue Light Syndrome," H. Senger, ed., Springer Verlag, Heidelberg, pp. 221–237.

Iino, M., and Briggs, W. R., 1984, Growth distribution during first positive phototropic curvature of maize coleoptiles, Plant, Cell and Environment, 7:97-104.

Iino, M., Briggs, W. R., and Schäfer, E., 1984, Phytochrome-mediated phototropism in maize seedling shoots, Planta, 160:41-51.

Iino, M., and Schäfer, E., 1984, Phototropic response of the stage I Phycomyces sporangiophore to a pulse of blue light, Proc. Natl. Acad. Sci. USA, in press.

Jaffe, L. F., 1960, The effect of polarized light on the growth of a transparent cell. A theoretical analysis, J. Gen. Physiol., 43:897-911.

Jan, Y. N., 1974, Properties and cellular localization of chitin synthetase in Phycomyces blakesleeanus, J. Biol. Chem., 249: 1973-1979.

Jesaitis, A. J., 1974, Linear dichroism and orientation of the Phycomyces photopigment, J. Gen. Physiol., 63:1-21.

Koornneef, M., Rolff, E., and Spruit, C. J. P., 1980, Genetic control of light-inhibited hypocotyl elongation in Arabidopsis thaliana (L) Heynh, Z. Pflanzenphysiol., 100:147-160.

Lipson, E. D., 1975, White noise analysis of Phycomyces light growth response system. I. Normal intensity range. II. Extended intensity ranges. III. Photomutants, Biophys. J., 15:989-1046.

Lipson, E. D., López-Díaz, I., and Pollock, J. A., 1983, Mutants of Phycomyces with enhanced tropisms, Exp. Mycol., 7:241-252.

Lipson, E. D., Terasaka, D. T., and Silverstein, P. S., 1980, Double mutants of Phycomyces with abnormal phototropism, Mol. Gen. Genet., 179:155-162.

López-Díaz, I., and Lipson, E. D., 1983, Genetic analysis of hyper-tropic mutants of Phycomyces, Mol. Gen. Genet., 190:318-325.

Löser, G., and Schäfer, E., 1983, Photogeotropism of wildtype Phycomyces blakesleeanus sporangiophores, in: "Abstracts of 2nd International Conference on the Effect of Blue Light in Plants and Microorganisms," Marburg, W. Germany, pp. 146-147.

Mandoli, D. F., and Briggs, W. R., 1982, The photoperceptive sites and the function of tissue light-piping in photomorphogenesis of etiolated oat seedlings, Plant, Cell and Environment, 5:137-145.

Oort, A. J. P., 1931, The spiral-growth of Phycomyces, K. Ned. Akad. Wet., 34:564-575.

Parson, A., Macleod, I., Firn, R. D., and Digby, J., 1984, Light gradients in shoots subjected to unilateral illumination-implications for phototropism, Plant, Cell and Environment, 7:325-332.

Petzuch, M., and Delbrück, M., 1970, Effects of cold periods on stimulus response system of Phycomyces, J. Gen. Physiol., 56: 297-308.

Pickard, B. G., Dutson, K., Harrison, V., and Donegan, E., 1969, Second positive phototropic response patterns of the oat coleoptile, Planta, 88:1-33.

Poff, K. L., 1983, Perception of a unilateral light stimulus, Phil. Trans. R. Soc. Lond. B, 303:479-487.

Pohl, U., and Russo, V. E. A., 1984, Phototropism, in: "Membranes and Sensory Transduction," F. Lenci and G. Colombetti, eds., Plenum Press, New York, pp. 231-329.

Presti, D., and Delbrück, M., 1978, Photoreceptors for biosynthesis, energy storage and vision, Plant, Cell and Environment, 1:81-100.

Presti, D., Hsu, W. J., and Delbrück, M., 1977, Phototropism in Phycomyces mutants lacking β-carotene, Photochem. Photobiol., 26:403-405.

Preston, R. D., 1974, "The Physical Biology of Plant Cell Walls," Chapman and Hall, London.

Russo, V. E. A., 1980, Sensory transduction in phototropism: Genetic and physiological analysis in Phycomyces, in: "Photoreception and Sensory Transduction in Aneural Organisms," F. Lenci and G. Colombetti, eds., NATO Advanced Study Institute, Plenum Press, New York, pp. 373-395.

Satter, R. L., 1979, Leaf movements and tendril curling, in: "Physiology of Movements," Encyclopedia of Plant Physiology, New Series, W. Haupt and M. Feinleib, eds., Springer Verlag, Berlin, 7:442-484.

Schmidt, W., 1983, The physiology of blue-light systems, in: "The Biology of Photoreception," D. J. Cosens and D. Vince-Prue, eds., Society for Experimental Biology, London, pp. 305-330.

Schmidt, W., Hart, J., Filner, P., and Poff, K. L., 1977, Specific inhibition of phototropism in corn seedlings, Plant. Physiol., 60:736-738.

Schwartz, A., and Koller, D., 1978, Phototropic response to vectorial light in leaves of Lavatera cretica L., Plant Physiol., 61:924-928.

Seyfried, M., and Fukshansky, L., 1983, Light gradients in plant tissue, Applied Opt., 22:1402-1408.

Shinkle, J. R., and Briggs, W. R., 1984, Indole-3-acetic acid sensitization of phytochrome-controlled growth of coleoptile sections, Proc. Natl. Acad. Sci. USA, 81:3742-3746.

Shropshire, Jr., W., 1962, The lens effect and phototropism of Phycomyces, J. Gen. Physiol., 45:949-958.

Shropshire, Jr., W., 1963, Photoresponses of the fungus, Phycomyces, Physiol. Rev., 43:38-67.

Shropshire, Jr., W., 1975, Phototropism, in: "Progress in Photobiology," Proceedings of the Sixth International Congress on Photobiology, G. O. Schenck, ed., Springer Verlag, Berlin, pp. 1-6.

Shropshire, Jr., W., and Mohr, H., 1970, Gradient formation of anthocyanin in seedlings of Fagopyrum and sinapsis unilaterally exposed to red and far-red light, Photochem. Photobiol., 12: 145-149.

Silk, W. K., 1984, Quantitative descriptions of development, Annu. Rev. Plant Physiol., 35:479-518.

Taiz, L., 1984, Plant cell expansion: Regulation of cell wall me-
    chanical properties, <u>Annu. Rev. Plant Physiol.</u>, 35:585-657.
Vierstra, R. D., and Poff, K. L., 1981, Role of carotenoids in the
    phototropic response of corn seedlings, <u>Plant Physiol.</u>, 68:
    798-801.
Vogelmann, T. C., and Björn, L. O., 1983, Response to directional
    light by leaves of a sun-tracking lupine (<u>Lupinus succulentus</u>),
    <u>Physiol. Plant</u>, 59:533-538.
Vogelmann, T. C., and Björn, L. O., 1984, Measurement of light gra-
    dients and spectral regime in plant tissue with a fiber optic
    probe, <u>Physiol. Plant</u>, 60:361-368.

PHOTOMORPHOGENESIS IN MICROORGANISMS

V.E.A. Russo, J.A.A. Chambers, F. Degli-Innocenti
and Th. Sommer

Max-Planck-Institut
Institut für Molekulare Genetik
Abt. Trautner
IhnestraBe 63/73
D-1000 Berlin 33, W. Germany

INTRODUCTION

Light can stimulate or inhibit many morphogenetic changes in several microorganisms: algae, fungi, slime molds and myxobacteria.

The series of events that begin with the stimulus perception (e.g. light) and ends with the response (e.g. morphogenesis) is called the sensory transduction chain (e.g. of photomorphogenesis).

There are several questions that can be asked about this chain:

(a) What is the photoreceptor?
(b) What are the first biochemical/biophysical changes?
(c) Is there a change in the pattern of protein synthesis and/ or enzyme activity?
(d) If yes, is the regulation at the level of transcription, translation, modification or at other levels?
(e) How many genes in an organism are "specifically" necessary for photomorphogenesis?
(f) In the case of light inducing several responses, is there any part in common between the different sensory trans- duction chains? In particular, is the photoreceptor the same for all?
(g) Do different organisms have the same or similar photore- ceptors?

In order to answer those questions, the problem must be studied
from a physiological, genetical and biochemical point of view, in all
cases where it is possible.

In this lecture, we will critically discuss several organisms
and the ways that were used to get some insight into the problem.
It is impossible to talk about all the organisms studied until now.

The interested reader can consult some of the latest specialized
reviews (Tan, 1978; Russo and Galland, 1980; Kumagai, 1980; Shropshire and Mohr, 1983; Schmidt, 1984; Dring, 1984). In many cases,
cryptochrome is considered the blue light photoreceptor.

"Cryptochrome" is a general term for unknown photoreceptors
that give "similar" action spectra: one or two peaks between 400
and 500 nm, one between 300 and 400 nm. From a look at a series of
cryptochrome spectra (Tan, 1978; Schmidt, 1980; Gressel and Rau,
1983), it is clear that there are no two spectra that are identical.
It is difficult to say if that is an indication of different photoreceptors or of different artifacts in different organisms
(Shropshire, 1972).

PHILOSOPHICAL CONSIDERATIONS ON THE SENSORY TRANSDUCTION CHAIN

The simplest chain is the linear one; an example for photomorphogenesis is given in Fig. 1.

A represents the photoreceptor.

B is an unknown compound, the level of which we assume to be
   influenced by the photoreceptor (either increased or
   decreased).

C is similarly influenced by B and so on.

Our knowledge of photomorphogenesis will be complete when we are able
to give a name to each of these components. As soon as someone has
the courage to propose a function for these components, it will be
possible to perform experiments designed to disprove the model. It
may seem trivial to some, but we want to repeat that with experiments
in vivo, it is much easier to disprove a model than to prove it. A
model is true as long as it is not disproved. In model I of Fig. 1
we say, for example, that the change in the level of D is a conditio
sine qua non for a response to occur; in other words, it is an essential link in the sensory transduction chain. In model II, it is not.

For photoreceptor A, as well as for any other component of the
model, there are simple rules that must be followed by anyone interested in answering in a meaningful way the questions: Is A the

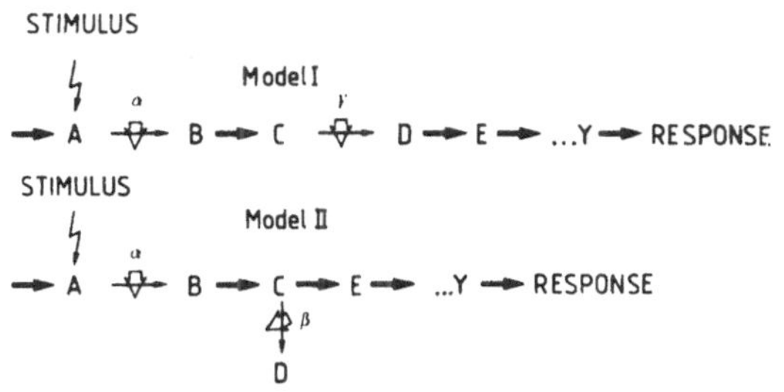

STIMULUS

Model I

→ A ─⊕─ B → C ─⊕─ D → E → ...Y → RESPONSE

STIMULUS

Model II

→ A ─⊕─ B → C → E → ...Y → RESPONSE

β

D

Fig. 1. Hypothetical linear transduction chain for photomorphogenesis. A is per definition the photoreceptor α; β and γ
are inhibitors of the photochange of the level of compound
D. With experiments in vivo, it is possible to distinguish
between the two models only with the inhibitor β. With
experiments in vivo and in vitro, it is possible to distinguish with inhibitor γ also.

photoreceptor? Is a change in D a conditio sine qua non for the
response? Elementary rules for model builders are given elsewhere
(Pohl and Russo, 1984). In Fig. 1, it is shown that an inhibition
of the step from A to B (either by "specific" inhibitors or by mutations) will inhibit both the response as well as the change of the
level of D, and therefore cannot distinguish between the two models.
On the other hand, a block of the step from C to D which does not
inhibit the response is a clear indication that Model I is wrong.

With the genetical and biochemical knowledge that we have today,
there is a theoretical case where we can prove in a quite convincing
way that a change of the level of D is indeed a conditio sine qua non
for the response to occur. However, this requires experiments both
in vivo and in vitro. Listed below is the way to prove it:

    (a) Isolate a mutant which is presumably in the gene coding
        for the enzyme necessary to go from C to D;
    (b) Show that the mutation inhibits "specifically" photo-
        morphogenesis (block γ in Fig. 1);
    (c) Purify the enzyme;
    (d) Make the step from C to D in vitro with the purified enzyme;
    (e) Show that the mutation is indeed in the presumed gene by
        cloning the gene, sequence the gene and the enzyme and
        compare the two sequences.

Until now, no one has done such a complicated series of experiments.

An alternative would be to give the molecule D to the organism
in the dark and see if you obtain a response. This would be possible
in the rare case that molecule D is taken up by the organism.

Here, we want to recall a rule that we will need later in the critical analysis of an experiment regarding the nature of the blue light photoreceptor in <u>Neurospora crassa</u>. What does a fluence-response curve look like in a system where the amount of photoreceptors are artificially lowered? The general form of the mathematical formula of a fluence-response curve can be written

$$R = f(N)$$

where N is the number of absorbed photons which induce a response. For low fluences, it becomes (Shropshire, 1972)

$$R = f(KN_0I)$$

where R is the response, K is a constant which accounts for the absorption cross section, length of the light path in the cell, quantum efficiency, etc., No is the concentration of the photoreceptor and I is the fluence.

If we plot the fluence-response curve on semilog paper, then any time the system has less artificial receptors, the curve will be shifted toward the higher fluence. That is the philosophy that underlies any action spectroscopy (Schäfer et al., 1983).

PHENOMENA

Very often with light, there are other responses besides the photomorphogenic one. As we discussed in the previous section, it is difficult to decide if there is a cause-effect relationship between the two responses. Even in the case where this relationship is not present, there is the possibility that the two sensory transduction chains have the first part in common as in model II of Fig. 1.

For that reason, we list in Table 1 not only the different photomorphogenic effects but also the other photoeffects known in the same organism.

From Table 1, several general statements can be made:

(a) Light can both inhibit or stimulate differentiation;
(b) There is no particular wavelength that is effective. The effective wavelength can be anywhere in the visible spectrum, depending on the organism and the response;
(c) The photon response varies from 16 nmol m^{-2} for the photosporangiogenesis of <u>Phycomyces</u> to 2.25 mol m^{-2} for the induction of 2-D growth in <u>Scytosiphon</u>. There is a difference in sensitivity by a factor of 10^8!
(d) In many organisms, there are several responses;

Table 1.

Genes	Description of Response	Photon[a] Response	Effective Wave Length	Lag Time	Reference
Algae					
Acetabularia mediterranea	Growth	-	blue	-	Clauss, 1963
	Cap formation	-	blue	-	Clauss, 1963
	Hair formation	5 μmol m^{-2}	Cryptochrome	-	Schmid, 1984
	Stimulation of photosynthesis	-	blue	-	Clauss, 1970
	Mobilization of starch reserve	-	blue	-	Clauss, 1970
	Induction of circadian rhythm of photosynthesis	-	blue	-	Clauss, 1979
	Induction of protein synthesis	-	blue	12 h	Schmid, 1984
Dictyota dichotoma	Stimulation of egg release	-	< 500	-	Müller and Clauss, 1976
	Stimulation of hair tufts	-	< 550	-	Müller and Clauss, 1976
	Stimulation of oogonia formation	-	> 600	-	Müller and Clauss, 1976

(continued)

Table 1 (continued)

Genes	Description of Response	Photon[a] Response	Effective Wave Length	Lag Time	Reference
Fucus serratus	Induction of polarity in germinating zygotes	0.17 mmol m^{-2}	Cryptochrome	-	Bentrup, 1963
Laminaria saccharina	Induction of egg formation	1.95 mol m^{-2}	Cryptochrome	2 days	Lüning and Dring, 1975
	Inhibition of egg release	3.8 mmol m^{-2}	Cryptochrome	6 min	Lüning, 1981
Rodochorton purpuream	Night break (tetra-sporangia formation)	0.6 mmol m^{-2}	448,667	-	Dring and West, 1983
Scytosiphon lomentaria	Induction of 2-D growth	2.25 mol m^{-2}	Cryptochrome	2 days	Dring and Lüning, 1975a
	Induction of hair formation	1.95 mol m^{-2}	Cryptochrome	2 days	Dring and Lüning, 1975a
	Night break	20 μmol m^{-2}	Cryptochrome	-	Dring and Lüning, 1975b
Fungi					
Neurospora crassa	Phototropism	-	blue	12 days	Harding and Melles, 1983
	Photocarotenogenesis	32 μmol m^{-2b}	Cryptochrome	30 min	Schrott, 1980

Genes	Description of Response	Photon[a] Response	Effective Wave Length	Lag Time	Reference
Neurospora crassa (continued)	Shift of circadian rhythms	-	Cryptochrome	-	Sargent and Briggs, 1967
	Induction of circadian rhythms	-	blue	-	
	Induction of protoperi-thecia formation	16 μmol m^{-2}[b]	blue	24 h	Innocenti et al., 1983
	Hyperpolarization of membrane potential	-	350 - 500	10 min	Potapova et al., 1984
Phycomyces blakesleeanus	Phototropism	4 pmol m^{-2}[b]	Cryptochrome	5 min	Galland and Russo, 1984
	Carotenogenesis	-	Cryptochrome	20 min	Raugei et al., 1982
	Sporangiophorogenesis	16 nmol m^{-2}[b]	Cryptochrome	24 h	Galland and Russo, 1979b
	Sporangiogenesis	-	Cryptochrome	24 h	Russo and Galland, 1980
Trichoderma harzianum	Conidia induction	20 μmol m^{-2}	Cryptochrome	20 h	Horwitz et al., 1984
	Electric currents around growing hyphae	-	blue	90 min	Horwitz et al., 1984

(continued)

Table 1 (continued)

Genes	Description of Response	Photon[a] Response	Effective Wave Length	Lag Time	Reference
Venturia inequalis	Stimulation ascospore release	-	710 – 770	30 min	Brook, 1969
	Inhibition ascospore release	-	650 – 700	-	Brook, 1969
		-	740 – 800	-	
Slime Molds					
Dictyostelium mucoroides	Inhibition of macrocyst	-	428	-	Chang et al., 1983
Physarum polycephalum	Induction of sporulation	-	350 – 500	-	Daniel and Rusch, 1962
			450, >600		Schreckenbach et al., 1981
	pH changes in the medium	-	Cryptochrome	2 min	Schreckenbach et al., 1981
	Inhibition of glucose consumption	-	Cryptochrome	10 min	Schreckenbach et al., 1981
	Inhibition of sclerotization	-	450	-	Schreckenbach et al., 1981
	Inhibition of growth	-	-	-	Gray, 1938
	Negative phototaxis	-	Cryptochrome	-	Rakoczy, 1980

Genes	Description of Response	Photon[a] Response	Effective Wave Length	Lag Time	Reference
Myxobacteria					
Stigmatella	Sporangiogenesis	–	white, red	–	Qualls et al., 1978

[a]Photon exposures required for a 50% response.
[b]Threshold value.

(e) The lag of the responses varies from 2 min in the pH change in the medium of <u>Physarium</u> to a couple of days in <u>Scytosiphon</u> differentiation.

PHYSIOLOGICAL ANALYSIS

In some cases other factors beside light can influence differentiation. There are examples where light interacts with environmental conditions.

In wild type <u>Phycomyces</u> the lower oxygen threshold for sporangiophorogenesis in the dark is 0.5%, for albino mutants it is up to 15%; in both cases, light can shift the threshold curve toward lower values (Galland and Russo, 1979a). The fact that the albino mutants have a higher oxygen threshold for differentiation in the dark than <u>WT</u> implies that carotenoids have a function in dark reactions. Interestingly, at a higher oxygen concentration (5% for <u>WT</u>, 20% for albino), the dark value (of sporangiophores/tube) is equal to the light value. Another example in <u>Phycomyces</u> is the inhibition of the same differentiation process by CO_2. A pulse of blue light will overcome this inhibition (Russo et al., 1981). Also, here light has little or no effect if there is 0% CO_2 in the atmosphere. In both cases, blue light induces differentiation in "adverse" conditions, too little oxygen or too much carbon dioxide.

In <u>Neurospora crassa</u>, it has been known since Westergaard and Mitchell (1947) that the concentration of nitrogen is important for perithecia formation and Hirsch (1954) found that KNO_3 is important for protoperithecia formation. Recent unpublished results (Sommer et al.) show that in order to have photoinduction of protoperithecia, the nitrogen source concentration (nitrate or ammonium) should be less than 50 mM after illumination. Interestingly enough, too much nitrogen inhibits the photoinduction of protoperithecia but not of carotenoids. Furthermore, it is possible to show that the last step that can be inhibited by nitrogen is at least 5 h after illumination.

In <u>Physarum polycephalum</u> light can induce sporulation and inhibit spherulation only after total starvation (Schreckenbach, 1984). The very interesting results of Wormington and Weaver (1976) indicate that sporulation of <u>Physarum</u> can be induced in the dark by injecting a small molecule (500 Daltons) extracted either from an illuminated plasmodium or from dark grown plasmodium and then illuminated in vitro. The molecules absorb at 380 nm and these results will be discussed further in the section entitled "Photoreceptors."

GENETIC ANALYSIS

Genetic analysis of photomorphogenesis has been conducted essentially in fungi and in particular with Phycomyces, Neurospora and to a lesser extent in Trichoderma.

In Phycomyces, several mutants called mad were isolated which were abnormal for phototropism (Bergman et al., 1973). The mutants in two genes, mad A and mad B are abnormal for all blue light effects known in this organism (Russo et al., 1980; Lipson, 1980) (see Table 1). This is interpreted as evidence that the sensory transduction chains of this blue light effect are the same in the early stages of all responses including the photoreceptor. Presti et al. (1977) used an albino mutant C 173 which has 4×10^{-5} the amount of WT β-carotene and found that it has the same phototropic threshold as WT; they concluded that the bulk of carotenoid cannot be the blue light photoreceptor. Confusion in this matter was created by the results of Galland and Russo (1979b) who have shown that the threshold for photomorphogenesis of C 173 is 2×10^3 times higher than the one of WT. On the basis of these results, they could not decide if β-carotene is the photoreceptor for this type of photomorphogenesis or it is an important molecule in the transduction chain. They argued in favor of the second hypothesis.

We have already mentioned that CO_2 delays (or completely inhibits at high concentration) the production of sporangiophores in Phycomyces. Russo et al. (1981) have isolated two mutants which were more sensitive to CO_2 and two mutants which were more resistant. Using those mutants, Pohl et al. (1983) have shown a correlation between the amount of a b-type cytochrome (not further characterized) and the lag in the differentiation process with 10% CO_2.

In Neurospora crassa, mutants similar to the mad A and mad B were found and called WC-1 and WC-2. They are isolated as mutants which do not have photoinduction of carotenoids and it was also found that:

(a) they are not phototropic (Harding and Melles, 1983);
(b) they do not have photoinduction of protoperithecia
    (Degli Innocenti and Russo, 1984a);
(c) they have less protoperithecia than WT in the dark
    (Degli Innocenti et al., 1984);
(d) the double mutants WC-bd do not make bands in the dark
    and the banding cannot be photoinduced (Russo, unpublished
    results).

Degli Innocenti and Russo (1984b) interpret this pleiotropism in a similar way as in Phycomyces, namely, that the sensory transduction chains of this different effect are the same at the beginning.

Russo (unpublished results) has constructed a triple albino mutant, al- 1, al- 2 and al- 3 which has less than 10^{-5} of the content of carotenoids of WT but the same threshold for the induction of protoperithecia. That seems to exclude a carotenoid as the cryptochrome.

In Neurospora, many mutants are known to be female sterile, i.e., they do not make protoperithecia. From the work of Johnson (1978), it is possible to guess that about 60 - 120 genes are specifically needed for this type of differentiation. Similar values for the conidiation of Trichoderma were given (Gressel and Rau, 1983). This gives an idea of how complex the process of photomorphogenesis is. The genetical analysis will be absolutely necessary if one wants to study the gene regulation in photomorphogenesis at the molecular level.

BIOCHEMICAL ANALYSIS

Very little is known at the biochemical level.

Acetabularia has a property that makes it unique among the organisms considered here; it can be enucleated and still grow and differentiate. Schmid (1984) has shown that the enzyme UDPG pyrophosphorylase is induced almost 4-fold in an enucleate cell in the first 24 h of blue light, while another enzyme, hydroxypyruvate reductase, is not. The overall cytosolic protein synthesis is also enhanced. With density labelling experiments, he came to the conclusion that there is indeed a de novo synthesis, but the difference in photoinduction between the two enzymes is due mainly to the fact that the first one has a half life which is about one-tenth of the one of the second enzyme. As the cell is enucleated, the enhancement of protein synthesis cannot be at the transcriptional level.

In Neurospora crassa, Mitzka-Schnabel et al. (1984) have shown that after 90 min illumination there are (in the membrane fraction) at least four polypeptides which are conspicuously photoinduced. Even more interesting, they have shown that the mRNA isolated after 10 min of illumination produces in an in vitro translation system four more proteins than the corresponding control. If that is reproducible, it is an indication that blue light can regulate either the transcription or the processing of some messengers. Chambers and Russo (unpublished results) have shown from the kinetics of two experiments, labelling between 0 - 30', 30' - 60' or 60' - 90' after beginning the illumination, that there is sequential regulation of about 17 proteins; the synthesis of 12 proteins is stimulated at some time in the first 90', while the synthesis of the other 5 proteins is inhibited (Fig. 2). In the 2- D gels from which those data were obtained, there are about 150 major spots; that indicates a regulation of about 10% of the major proteins.

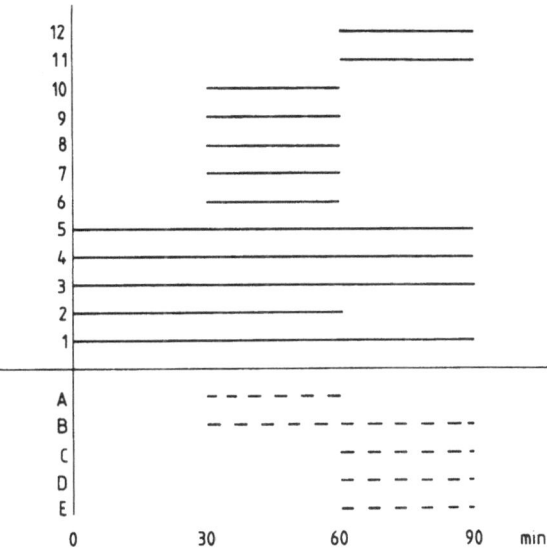

Fig. 2. Summary of changes observed. Proteins that have been
repeatedly shown to be newly synthesized within a given
30 min period are shown as a bar within that period.
Proteins that reproducibly disappear are shown as an open
box.

In Physarum polycephalum, besides the two fast blue light effects
cited in Table 1, there is also a regulation of new translatable mRNA
species 15 h after photoinduction (Schreckenbach, 1984).

HOW TO LOOK FOR THE PHOTORECEPTOR?

(a) One Way to Look

One way to look for, and maybe in principle, the most straightforward, is the one of Otto et al. (1981). They attempted to replace
the putative chromophore flavin with roseoflavin, which has a different absorption spectrum. With some experiments and mathematical
analyses, they were convinced that roseoflavin (which alone cannot
support the growth of a riboflavin auxotroph and it has to be mixed
in a certain ratio with riboflavin) was indeed incorporated into the
photoreceptor of the sensory transduction chain of phototropism. One
criticism is that the data of their Fig. 3 is at the limit of statistical significance.

(b) Another Way

Another way to look for the photoreceptor is the one of
Wormington and Weaver (1976). They extracted a pigment from Physarum

polycephalum that abosrbs around 380 nm. They injected this pigment
into the cell and have shown that there was differentiation if the
pigment was either extracted and partially purified from an illuminated plasmodium or from an unilluminated plasmodium but that was
illuminated in vitro for 6 h with a 40 W fluorescent lamp. The pigment extracted from an unilluminated plasmodium did not induce differentiation if it was not illuminated in vitro.

Unfortunately, there is no action spectrum for photoinduction
of sporulation in Physarum; however, the work of Daniel and Rusch
(1962) and Schreckenbach et al. (1981) indicates that at 450 nm and
even 500 nm, there is sporulation while the isolated pigment does
not absorb at all at 500 nm. This is quite disturbing, although it
is known for other chromophores that binding to a protein can change
the absorption spectrum significantly (Bensasson, 1980).

(c) The Hypothetical Ideal Way

The hypothetical ideal way could be the following. Find out
what the first reaction is in vitro, namely, what is B in Fig. 1.
Find a mutant which does not have a photochange B. Show that the
mutant lacks a pigment which has an absorption spectrum similar to
the action spectrum of photomorphogenesis. Show that it is possible
to have a photochange of B in vitro with the WT but not with the
mutant. Until now, no one has found a system where all this is
possible.

(d) The Wrong Way

The wrong way to look for the photoreceptor is the one of
Paietta and Sargent (1981). They have studied the photosuppression
of circadian conidiation, photoinduction of phase shifting and the
photoinduction of carotenoid synthesis in the Neurospora crassa double
mutants, bd and rib (band mutants which are auxotrophic for riboflavin). They do see a correlation between the light sensitivity
and the amount of flavin in the medium. However, they have no control which indicates that riboflavin is indeed needed for photoreception and not for any dark reaction. This control could have been
done in the case of photoinduction of carotenoids. There is an example where riboflavin is needed for dark reaction of photodifferen-
tiation of a riboflavin autotroph of Trichoderma (Horwitz and Gressel,
1983). In the case of photoinduction of phase shift (Fig. 3 of
Paietta and Sargent, 1981), it is possible to see that at the highest
flavin concentration, the fluence-response curve is already in plateau at 15 s (the shortest time they used). Because there is not a
threshold curve, it is impossible to see if the threshold at low riboflavin concentration is shifted with respect to the high riboflavin
concentration or the fluence-response curve is flatter at low riboflavin concentrations. For the philosophical consideration we presented at the beginning of this chapter, a shift of the threshold

curve could have been an indication that flavin is the chromophore, a flattening of the curve is certainly an indication against it. As it is now, nothing can be learned from this paper.

LAST CONSIDERATION ABOUT THE CRYPTOCHROME

For 35 years, there has been a polarization between carotenoid (α-carotene was suggested by Castle, 1935, as the cryptochrome of Phycomyces, β-carotene by Bünning, 1938a, and b, as the one of Avena) and flavin (suggested by Galston, 1949) as the best candidate for the chromophore. Very little is done to see if other pigments exist in nature which have a similar absorption spectrum than the two previously mentioned compounds. It is, therefore, very interesting that Margraf (1984) in the basidiomycete Pleurotus ostreatus has found a new compound which is neither flavin nor carotenoid but it has a very nice peak at 475 nm.

ACKNOWLEDGMENT

This work was partially supported by the Deutsche Forschungs-gesellschaft.

REFERENCES

Bensasson, R. V., 1980, Molecular aspects of photoreceptor function: Carotenoids and rhodopsins, in: "Photoreception and Sensory Transduction in Aneural Organisms," F. Lenci and G. Colomberti, eds., Plenum Press, New York - London, pp. 211-234.

Bentrum, F.-W., 1963, Vergleichende Untersuchungen zur Polaritäts-induktion durch das Licht an der Equisetumspore und der Fucus-Zygote, Planta, 59:472.

Bergman, K., Eslava, A. P., and Cerda-Olmedo, E., 1973, Mutants of Phycomyces with abnormal phototropism, Mol. Gen. Genet., 123:1.

Brook, P. J., 1969, Stimulation of ascospore release in Venturia inaequalis by far red light, Nature, 222:390.

Bünning, E., 1938a, Phototropismus und Carotenoide. II. Das Carotin der Reizaufnahmezone von Pilobolus, Phycomyces und Avena, Planta, 27:148.

Bünning, E., 1938b, Phototropismus und Carotenoide. III. Weitere Untersuchungen an Pilzen und höheren Pflanzen, Planta, 27:583.

Castle, E. S., 1935, Photic excitation and phototropism in single plant cells, Cold Spring Harbor Symp. Quant. Biol., 3:224.

Chang, M. T., Raper, K. B., and Poff, K. L., 1983, The effect of light on morphogenesis of Dictyostelium mucoroides, Exp. Cell Res., 143:335.

Clauss, H., 1963, Über den Einfluß von Rot- und Blaulicht auf das Wachstum kernhaltiger Teile von Acetabularia mediterranea, Naturwis., 50:719.

Clauss, H., 1970, Effect of red and blue light on morphogenesis and metabolism of Acetabularia mediterranea, in: "Biology of Acetabularia," J. Bracket and S. Bonotto, eds., Academic Press, New York.

Clauss, H., 1979, Auslösung der circadianen Photosynthese-Rhythmik bei Acetabularia durch Blaulicht, Protoplasma, 99:341.

Daniel, J. W., and Rusch, H. P., 1962, Method for inducing sporulation of pure cultures of the myxomycete Physarum polycephalum, J. Bacteriol., 83:234.

Degli Innocenti, F., Chambers, J. A. A., and Russo, V. E. A., 1984, Conidia induce the formation of protoperithecia in Neurospora crassa: Further characterization of white collar mutants, J. Bacteriol., 159: in press.

Degli Innocenti, F., Pohl, U., and Russo, V. E. A., 1983, Photoinduction of protoperithecia in Neurospora crassa by blue light, Photochem. Photobiol., 37:49.

Degli Innocenti, F., and Russo, V. E. A., 1984a, Isolation of white collar mutants of Neurospora crassa and studies on their behavior in the blue light-induced formation of protoperithecia, J. Bacteriol., 159: in press.

Degli Innocenti, F., and Russo, V. E. A., 1984b, Genetic analysis of blue light induced responses in Neurospora crassa, in: "Blue Light Effects in Biological Systems," H. Senger, ed., Springer, Berlin-Heidelberg-New York, pp. 213-219.

Dring, M. J., 1984, Blue light effects in marine macroalgae, in: "Blue Light Effects in Biological Systems," H. Senger, ed., Springer, Berlin-Heidelberg-New York, pp. 509-516.

Dring, M. J., and Lüning, K., 1975a, Induction of two-dimensional growth and hair formation by blue light in the brown alga Scytosiphon lomentaria, Z. Pflanzenphysiol., 75:107.

Dring, M. J., and Lüning, K., 1975b, A photoperiodic response mediated by blue light in the brown alga Scytosiphon lomentaria, Planta, 125:25.

Dring, M. J., and West, J. A., 1983, Photoperiodic control of tetrasporangium formation in the red alga Rhodochorton purpurum, Planta, 159:143.

Galland, P., and Russo, V. E. A., 1979, The role of retinol in the initiation of sporangiophores of Phycomyces blakesleeanus, Planta, 146:257.

Galland, P., and Russo, V. E. A., 1979b, Photoinduction of sporangiophores in Phycomyces mutants deficient in phototropism and in mutants lacking β-carotene, Photochem. Photobiol., 29:1009.

Galland, P., and Russo, V. E. A., 1984, Light and dark adaptation in Phycomyces phototropism, J. Gen. Physiol., in press.

Galston, A. W., 1949, Riboflavin-sensitized photooxidation of indoleacetic acid and related compounds, Proc. Natl. Acad. Sci. USA, 35:10.

Gray, W. D., 1938, Am. J. Bot., 25:511.

Gressel, J., and Rau, W., 1983, Photocontrol of fungal development, in: "Encyclopedia of Plant Physiology," New Series, W. Shropshire and H. Mohr, eds., Springer-Verlag, Berlin-Heidelberg, 16:603.

Harding, R. W., and Melles, S., 1983, Genetic analysis of phototropism of Neurospora crassa perithecial beaks using white collar and albino mutants, Plant Physiol., 72:996.

Hirsch, H. M., 1954, Environmental factors influencing the differentiation of protoperithecia and their relation to tyrosinase and melanin formation in Neurospora crassa, Physiol. Plant., 7:72.

Horwitz, B. A., and Gressel, J., 1983, Elevated riboflavin requirement for postphotoinductive events in sporulation of a Trichoderma auxotroph, Plant Physiol., 71:200.

Horwitz, B. A., Gressel, J., and Malkin, S., 1984, The quest for Trichoderma cryptochrome, in: "Blue Light Effects in Biological Systems," H. Senger, ed., Springer, Berlin, pp.237-249.

Johnson, T. E., 1978, Isolation and characterization of perithecial development mutants in Neurospora, Genetics, 88:27.

Kumagai, T., 1980, Blue and near ultraviolet reversible photoreaction in conidial development of certain fungi, in: "The Blue Light Syndrome," H. Senger, ed., Springer, Berlin, pp. 251-260.

Lipson, E. D., 1980, Sensory transduction in Phycomyces photoresponses, in: "The Blue Light Syndrome," H. Senger, ed., Springer, Berlin-Heidelberg-New York, pp. 110-118.

Lüning, K., 1981, Egg release in gametophytes of Laminaria saccharina: Induction by darkness and inhibition by blue light and UV, Br. Phycol. J., 16:379.

Lüning, K., and Dring, M. J., 1975, Reproduction, growth and photosynthesis of gametophytes of Laminaria saccharina growth in blue and red light, Mar. Biol. (Berl.), 29:195.

Margraf, W., 1984, Orange yellow pigments in the basidiomycete Pleurotus ostreatus (Jacq. Ex. Fr.) Kummer, in: "Blue Light Effects in Biological Systems," H. Senger, ed., Springer, Berlin-Heidelberg-New York, pp. 55-59.

Mitzka-Schnabel, U., Warm, E., and Rau, W., 1984, Light-induced changes in the protein pattern translated in vivo and in vitro accompanying carotenogenesis in Neurospora crassa and Fusarium aquaeductum, in: "Blue Light Effects in Biological Systems," H. Senger, ed., Springer, Berlin-Heidelberg, pp. 264-269.

Müller, S., and Clauss, H., 1976, Aspects of photomorphogenesis in the brown algae Dictyota dichotoma, Z. Pflanzenphysiol., 78:461.

Otto, M. K., Jayaram, M., Hamilton, R. M., and Delbrück, M., 1981, Replacement of riboflavin by an analogue in the blue light photoreceptor of Phycomyces, Proc. Natl. Acad. Sci. USA, 78:266.

Paietta, J., and Sargent, M. L., 1981, Photoreception in Neurospora crassa: Correlation of reduced light sensitivity with flavin deficiency, Proc. Natl. Acad. Sci. USA, 78:5573.

Pohl, U., Degli Innocenti, F., and Russo, V. E. A., 1983, Effect of carbon dioxide on differentiation and on the level of a soluble b-type cytochrome in Phycomyces blakesleeanus, Planta, 158:51.

Pohl, U., and Russo, V. E. A., 1984, Phototropism, in: "Membranes
    and Sensory Transduction," G. Colombetti and F. Lenci, eds.,
    Plenum, New York, pp. 231-329.

Potapova, T. V., Levina, N. N., Belozerskaya, T. A., Kritsky, M. S.,
    and Chailakhian, L. M., 1984, Investigation of electrophysio-
    logical responses on Neurospora crassa to blue light, Arch.
    Microbiol., 137:262.

Presti, D., Hsu, W. J., and Delbrück, M., 1977, Phototropism in
    Phycomyces mutants lacking β-carotene, Photochem. Photobiol.,
    26:403.

Qualls, G. T., Stephens, K., and White, D., 1978, Light-stimulated
    morphogenesis in the fruiting myxobacterium Stigmatella
    aurantiaca, Science, 201:444.

Rakoczy, L., 1980, Effect of blue light on metabolic processes,
    development and movement in true slime molds, in: "The Blue
    Light Syndrome," H. Senger, ed., Springer, Berlin-Heidelberg-
    New York, pp. 570-583.

Raugei, G., Dohrmann, U., Pohl, U., and Russo, V. E. A., 1982,
    Kinetics of photoaccumulation of β-carotene in Phycomyces
    blakesleeanus, Planta, 155:296.

Russo, V. E. A., and Galland, P., 1980, Sensory physiology of
    Phycomyces blakesleeanus, in: "Structure and Bonding," Vol. 41,
    P. Hemmerich, ed., Springer, Berlin-Heidelberg.

Russo, V. E. A., Galland, P., Toselli, M., and Volpi, L., 1980,
    Blue light induced differentiation in Phycomyces blakesleeanus,
    in: "The Blue Light Syndrome," H. Senger, ed., Springer,
    Berlin-Heidelberg-New York, pp. 563-569.

Russo, V. E. A., Pohl, U., and Volpi, L., 1981, Carbon dioxide
    inhibits phorogenesis in Phycomyces and blue light overcomes
    this inhibition, Photochem. Photobiol., 34:233.

Sargent, M. L., and Briggs, W. R., 1967, The effects of light on a
    circadian rhythm of conidiation in Neurospora, Plant. Physiol.,
    42:1504.

Schäfer, E., Fukshansky, L., and Shropshire, Jr., W., 1983, Action
    spectroscopy of photoreversible pigment systems, in: "Photo-
    morphogenesis," W. Shropshire, Jr., and H. Mohr, eds., Springer,
    Berlin-Heidelberg-New York-Tokyo, pp. 39-68.

Schmid, R., 1984, Blue light effects on morphogenesis and metabolism
    in Acetabularia, in: "Blue Light Effects in Biological Systems,"
    H. Senger, ed., Springer, Berlin-Heidelberg, pp. 419-432.

Schmidt, W., 1980, Physiological blue light reception, in: "Structure
    and Bonding," Vol. 41, P. Hemmerich, ed., Springer, Berlin-
    Heidelberg-New York.

Schreckenbach, T., 1984, Phototaxis and Photomorphogenesis in Physarum
    polycephalum plasmodia, in: "Blue Light Effects in Biological
    Systems," H. Senger, ed., Springer, Berlin, pp. 463-475.

Schreckenbach, T., Walckhoff, B., and Verfuerth, C., 1981, Blue light
    receptor in a white mutant of Physarum polycephalum mediates
    inhibition of spherulation and regulation of glucose metabolism,
    Proc. Natl. Acad. Sci. USA, 78:1009.

Schrott, E. L., 1980, Dose response and related aspects of caroteno-
    genesis in Neurospora crassa, in: "The Blue Light Syndrome,"
    H. Senger, ed., Springer, Berlin-Heidelberg, pp. 309-318.

Shropshire, Jr., W., 1972, Action spectroscopy, in: "Phytochrome,"
    K. Mitrakos, and W. Shropshire, Jr., eds., Academic Press,
    London-New York.

Shropshire, Jr., W., and Mohr, H., 1983, "Photomorphogenesis,"
    Vol. 16B, Springer, Berlin-Heidelberg-New York-Tokyo.

Tan, K. K., 1978, Light-induced fungal development, in: "The
    Filamentous Fungi," J.E. Smith, and D. R. Berry, eds.,
    Edward Arnold Ltd., London, 3:334.

Westergaard, M., and Mitchell, H. K., 1947, A synthetic medium
    favoring sexual reproduction, Am. J. Bot., 34:573.

Wormington, W. M., and Weaver, R. F., 1976, Photoreceptor pigment
    that induces differentiation in the slime mold Physarum poly-
    cephalum, Proc. Natl. Acad. Sci. USA, 73:3896.

MECHANISMS FOR THE MEASUREMENT OF LIGHT DIRECTION

Kenneth L. Poff

Michigan State University
United States Department of Energy
Plant Research Laboratory
Michigan State University
East Lansing, MI 48824

INTRODUCTION

To measure light direction, an organism must experience a gradient in light intensity which can be translated into some difference in light absorption. If this gradient is to be established inside of the organism as is most frequently the case, then an optical dissymmetry must also exist within that organism. The difference in light absorption may be measured at two points in time--a temporal measurement, or at two points in space--a spatial measurement. However, for either the temporal or the spatial measurement, it is the difference in light absorption which is measured and translated into light direction. Thus, for either measurement, some optical dissymmetry is required.

This discussion will describe three mechanisms used to establish light gradients: screening, refraction, and interference. Dichroism may be a factor involved in the measurement of light direction. However, by itself, dichroism can be used only to determine the plane of propagation of light, and must be combined with an additional mechanism to establish a light gradient in order for the direction of non-polarized light to be measured. (Dichroism can be used to measure the light direction of plane polarized light, but is not sufficient by itself to measure the direction of non-polarized light.)

Any investigation directed toward an understanding of a photoresponse includes, at some point, the measurement of fluence-response relationships and action spectra. Each is designed to yield information concerning the nature of the photoreceptor pigment, and each

251

is based on the assumption that the response is dependent solely upon the limiting primary photochemistry. Unfortunately, in those instances where light direction is measured, the response is a complex function of both the photoreceptor pigment and of the mechanism used to establish the light gradient. The purpose of this discussion will be to describe the three mechanisms used to establish a light gradient and to describe the way in which the mechanism might interfere with a measurement of fluence response relationships or action spectra. Finally, a few selected biological examples will be discussed to illustrate these mechanisms. The intention is not to present an exhaustive analysis of the measurement of light direction, but rather to give an overview of our current knowledge in hopes of stimulating further work in this area.

SCREENING

Screening decreases the intensity of light beyond the screen relative to the intensity of light before the screen. Screening, which may occur as a result of absorption or scatter, must be assumed to occur in all organisms. However, the extent of screening may vary from one organism to another and may, in fact, be manipulated to some extent experimentally. By definition, any photoresponsive system must contain a photoreceptor pigment, and, although the absorption of that pigment may be small, it must be finite. Similarly, all organisms are particulate or optically non-homogenous. All organisms must therefore scatter light.

Absorptive-type screening may be from self absorption in which the photoreceptor pigment itself also serves as the screen, or from foreign absorption in which some pigment other than the photoreceptor pigment serves as the screen. It is evident that two conditions must be satisfied for light direction to be measured: photoreception or the measurement of light, and a light gradient from which direction is perceived. It follows from this that light direction can only be measured at wavelengths where the photoreceptor pigment absorbs. Thus, there could not exist a case in which the absorptive-type screening resulted solely from foreign absorption.

In a typical fluence-response relationship, the magnitude of the response increases log-linearly with increasing fluence above some threshold, and saturates at some still higher fluence (Fig. 1). Such a relationship illustrates a basic assumption for photobiology-- that the response depends upon the number of absorbed quanta. Let us assume for a hypothetical organism that the detection of light direction is accomplished by calculating the difference in the response induced on either side of the organism. If the difference in light intensity across the organism were infinitely small, the same response would be induced on both sides and light direction could not be measured. Given screening across the organism, the

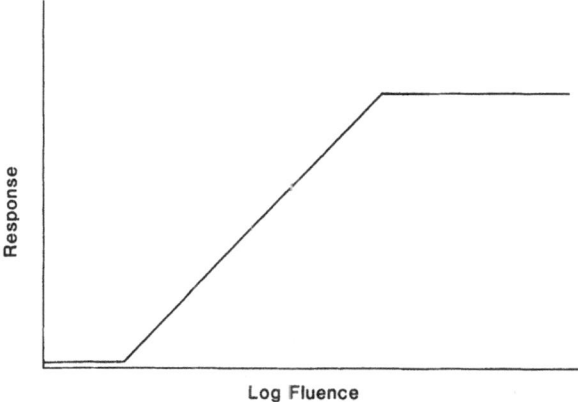

Fig. 1. Hypothetical fluence response curve with response as a function of log fluence.

light intensity will be highest on the proximal side, and the hypothetical fluence response relationship for the distal side will be displaced along the fluence axis toward higher fluence, where the extent of displacement is equal to the attenuation caused by the screening (Fig. 2). Since the response on both sides of the organism depends on the fluence, and since the fluence on the distal side of

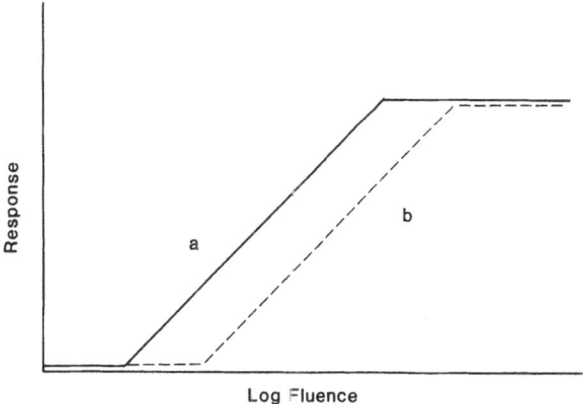

Fig. 2. Hypothetical fluence response curve for an organism using screening to establish an internal light gradient. Curve "a" represents the response on the proximal or lighted side of the organism, while "b" represents the response on the distal side.

the organism depends on the extent of screening, it follows that
the difference in response, proximal to distal, depends both on
fluence and on the extent of screening.

Other methods could be used to detect light direction other
than calculating the difference in the response on either side of
the organism. For example, one can imagine a system in which the
response is related to the difference in the amount of photoproduct
(P^*), on the proximal and distal sides. These other methods for
detecting light direction might possibly result in different mathematical relationships between fluence and the ultimate response
(detection of light direction), but this response will still depend
on both the fluence and on the extent of screening. (Note, however,
that the response would be independent of the screening mechanism
at the "threshold" fluence at which no response would be induced on
the distal side of the organism.)

Since the number of absorbed quanta depends upon both the
absorption coefficient of the photoreceptor pigment and the fluence,
a change in absorption coefficient will result in a shift of the
fluence-response relationship along the fluence axis. A series of
fluence-response relationships for different wavelengths of mono-
chromatic light can thus be used to calculate an action spectrum
which should approximate the absorption spectrum of the photoreceptor
pigment. However, as mentioned above, given a measurement of light
direction based on screening, the ultimate response will also depend
on the extent of screening. Therefore, the action spectrum will
represent not just the absorption spectrum of the photoreceptor
pigment but some complex product of the photoreceptor pigment and
the screen. Since we do not know the precise nature of the dependenc
of the response on fluence, we cannot predict the nature of this
"complex product", but can conclude that the action spectrum will
involve both the wavelength dependence of the photoreceptor pigment
absorption and the wavelength dependence of the screen.

The influence on the action spectrum of screening by scatter
is the least troublesome of the mechanisms for measuring light direction. The extent of scatter is inversely proportional to some
power of the wavelength. Scatter would, therefore, increase the
screen (and the response) at low wavelengths more than at high wavelengths. Peaks at lower wavelengths in the action spectrum would
thus be higher in amplitude relative to peaks at higher wavelengths.
Over the relatively narrow wavelength range of the usual action
spectrum, the distortion due to scatter-type screening would be
minimal. "Back scatter" would increase the light intensity just
inside of the proximal outer surface magnifying the light gradient
expected from scattering (Seyfried and Fukshansky, 1983).

A more difficult problem is presented by the case of screening
via absorbance. If the primary screening pigment be the photorecepto

pigment itself, the action spectrum should represent not only the photoreceptor pigment, but the photoreceptor pigment multiplied by itself in some manner. Since the precise nature of the multiplication is unknown, one cannot predict the exact effect on the action spec-trum, nor can one correct for that effect. One can assume that, in general, this would considerably sharpen the peaks. For example, assuming that the response is proportional to the simple product of the photoreceptor pigment and the screen, and since the screen is the photoreceptor pigment itself, the response would be proportional to the photoreceptor pigment squared. The action spectrum peak sould have the same wavelength maximum as the photoreceptor pigment but would be considerably sharper (Fig. 3). If the ultimate response of the organism (measurement of light direction) is the difference in the response on the proximal and distal sides, then the response on the distal side will become more significant at fluences above the saturation of the response on the proximal side. Thus, with increasing fluence, the response of the organism will decrease until the responses on both sides are saturated.

REFRACTION

Light can be focused within an organism by refraction at a curved air-organism interface (Fig. 4). It is evident from a

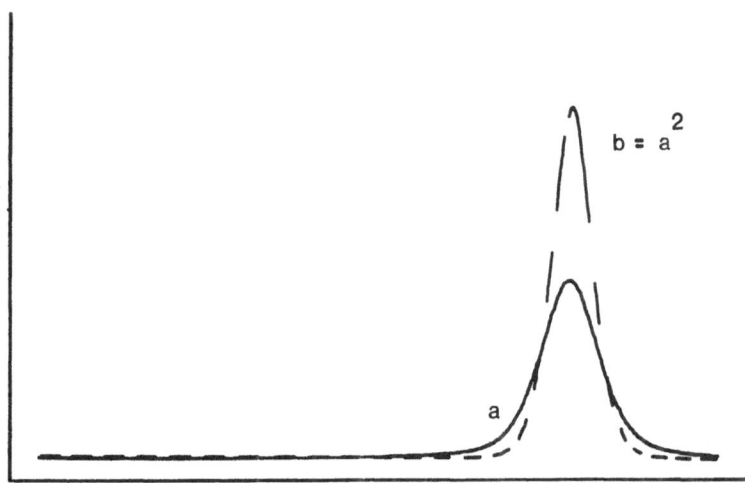

Fig. 3. Computer-generated curves demonstrating the sharpening which occurs when one curve is multiplied by itself. Curve "a" represents the original curve and "b" the product of "a" times "a".

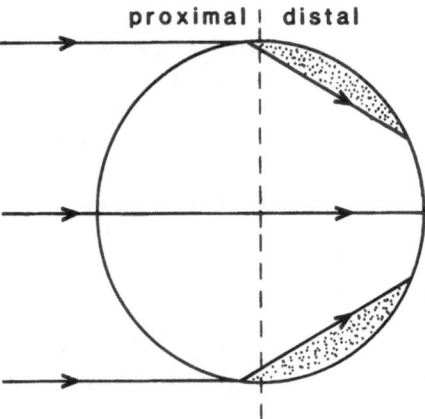

Fig. 4.   Schematic drawing of the cross section of an organism ex-
          posed to unilateral light. The lines represent light rays
          from a light source to the left of the organism. Light is
          focused onto the distal side of the organism by refraction
          at the air-organism interface.

diagram of an organism with a circular cross section that a larger
volume of the proximal side is illuminated than of the distal side,
and that the average pathlength on the distal side is greater than
that on the proximal side. Moreover, given relatively low losses
of light due to screening, the light intensity is greater on the
distal side than on the proximal side. Making assumptions similar
to those above for a hypothetical organism, fluence response relationships can be constructed for the proximal and distal sides
(Fig. 5). Given focussing and minimal absorption, the light intensity
will be higher on the distal side than on the proximal side so the
fluence response relationship for the proximal side will be displaced
along the fluence axis toward higher fluence.

     Since a greater volume is illuminated on the proximal side than
on the distal side, if the photoreceptor pigment is evenly distributed
and the primary photoproducts not readily diffusible, the fluence
response relationship for the proximal side will saturate at a higher
response than will that for the distal side. Based on these factors,
a refraction mechanism used to measure light direction might be expected to produce a response, the sign of which would be fluence
dependent (e.g., movement at low fluences toward the light when the
response is greatest on the distal side, and away from the light at
higher fluences where the response on the distal side is saturated
and less than that on the proximal side with its greater volume).

     Screening would decrease the light intensity as it passed through
the organism, and would decrease the focussing advantage. A

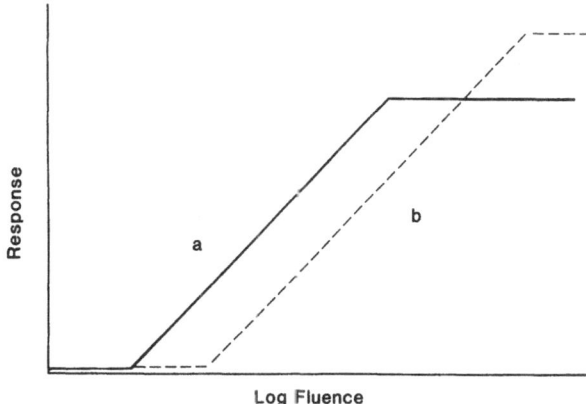

Fig. 5. Hypothetical fluence-response curve for an organism using
         refraction to establish an internal light gradient.
         Curve "a" represents the response on the distal side of the
         organism, and curve "b" the response on the proximal side.

refraction mechanism can only work therefore with a relatively transparent organism. This provides the potential for distinguishing between the screening and refraction mechanisms -- decreasing the
absorptive-type screening should decrease the final response if the
screening mechanism is operative and increase the final response if
the refraction mechanism is operative. Scatter-type screening would
also act to decrease the dark areas on the distal side of the organism
and would decrease the fluence-dependent sign reversal referred to
above.

INTERFERENCE

    Recently, a third mechanism used to establish light gradients
based on interference has been proposed (Foster and Smyth, 1980).
If one alternates layers of high and low refractive indices, of
which the optical thickness (thickness times refractive index) is
approximately one-fourth of the wavelength of the incident light
(a quarter-wave stack), reflection will occur at each surface such
that the reflected waves will be in phase (Fig. 6). Constructive
interference occurs, and the reflected waves become additive. The
light intensity in this reflected wave is then the sum of the incident light intensity and the reflected light intensity; the transmitted light is reduced by the amount of reflected light. Thus,
an optical dissymmetry is established producing a difference in
light intensity on the two sides of the quarter-wave stack. Since
the light intensity on the proximal side of the organism is increased
over the incident intensity by an amount due to reflection and the

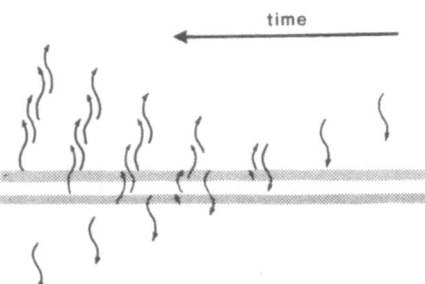

Fig. 6. Diagramatic representation of an interference reflector.
The lines with arrows represent segments of a light wave
at successive intervals of time during which the wave
advances one-half wavelength. The shaded bars represent
layers of high refractive index. (after Foster and Smyth,
1980).

intensity on the distal decreased by the same amount, the fluence
response relationship for the proximal side will be shifted along
the fluence axis toward lower fluence and that for the distal side
will be shifted toward higher fluence.

The efficiency of the quarter-wave stack can be increased by
adding absorption in the layers, increasing the number of layers,
modifying the shape of the stack, etc. However, these are not
strictly relevant to the basic mechanism - the establishment of an
optical dissymmetry using a quarter-wave interference reflector.

It directly follows from this discussion that the quarter-wave
stack is optimum for a particular wavelength or range of wavelengths.
The half-bandwidth for the Chlamydomonas quarter-wave stack, for
example, has been calculated to be 113 nm (Foster and Smyth, 1980).
This means that one particular wavelength is reflected most effectively and that wavelength plus or minus 56.5 nm are reflected at
50% of that effectiveness. The absorption spectrum would, in effect,
be multiplied by the effectiveness of the quarter-wave stack at each
wavelength to give the action spectrum. Unless the absorption
spectrum of the photoreceptor pigment were extremely narrow, a
half-bandwidth of 113 nm for the optical dissymmetry mechanism would
easily modify the measured action spectrum such that it would not
accurately represent the absorption spectrum of the pigment. The
effect on the action spectrum would become even more complex given
absorption.

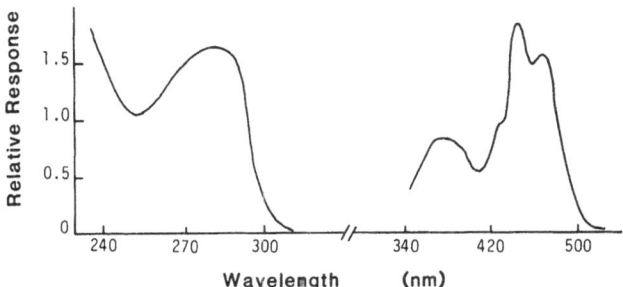

Fig. 7.   Action spectrum for phototropism in Phycomyces sporangio-
          phores. The curve on the left is for negative phototropism;
          the curve on the right is for positive phototropism.
          (after Curry and Gruen, 1959).

BIOLOGICAL EXAMPLES

     No attempt will be made in this section to include all of the
known biological examples for each of the mechanisms for measuring
light direction. Rather, a few select examples will be discussed
to illustrate particular problems likely arising because of the
mechanism used for establishing the optical dissymetry across the
organism.

Phototropism in Phycomyces sporangiophores

     Sporangiophores of Phycomyces show positive phototropism toward
visible light but a strong negative phototropism to ultraviolet
light (Fig. 7). A considerable amount of data supports the conclusion
that refraction is used to establish an optical dissymmetry across
the sporangiophore, but that this is possible only for visible light
to which the sporangiophore is relatively transparent (Dennison,
1979). The sporangiophore is virtually opaque to ultraviolet light
and this screening establishes the optical dissymmetry at these
wavelengths. Since the two mechanisms result in a higher light intensity on opposite sides of the organism, changing from one mechanism
to the other changes the sign of the response.

Phototaxis by Amoebae of Dictyostelium discoideum

     Amoebae of the cellular slime mold, Dictyostelium discoideum,
move toward or away from the direction of the unilateral light
dependent upon the intensity of that light (Häder and Poff, 1979 a, b;
Hong et al., 1981). The action spectra for these phototactic responses show peaks at 405 nm, 440-520 nm, 580 nm, and 640 nm (Fig. 8).
This last peak is of interest for this discussion since none of the
mechanisms discussed is obviously sufficient to establish the optical
dissymmetry required for phototaxis.

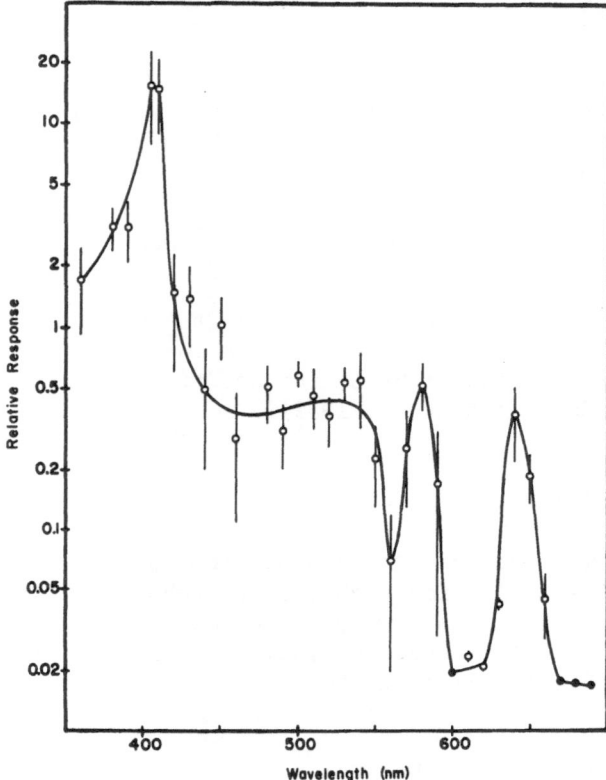

Fig. 8. Action spectrum for phototaxis by <u>Dictyostelium</u> amoebae.
Response was based on accumulation of amoebae in "light
traps." (after Hader and Poff, 1979).

1. At 640 nm, the ability of the 1 μ pseudopodium or even the
entire 10 μ amoeba to act as a lens is doubtful. The ability of a
lens to focus light decreases as the diameter of the lens approaches
the wavelength of the light used. At 640 nm, refraction should be a
relatively ineffective mechanism for establishing an optical dissymmetry.

2. At 640 nm, scattering, which is inversely proportional to
some power of the wavelength, should be relatively ineffective.

3. Given the very low absorbance of cells at 640 nm, and the
short 10 μ pathlength, absorbance should be relatively ineffective
in establishing a significant gradient in light intensity.

4. Given the broad wavelength range over which the amoebae
are capable of phototaxis (< 405 nm to > 640 nm), it seems unlikely
that the relatively wavelength specific interference mechanism
would be used.

The extremely sharp peak in the action spectrum at 640 nm is
compatible with screening as the mechanism for establishing optical
dissymmetry in these amoebae. If the absorption band of the photoreceptor pigment served both as the receptor and also to establish
the light intensity gradient, the peak in the action spectrum would
be expected to be sharp. Thus, it is suggested that the absorptive-
type screening probably assisted by scatter-type screening is used
for optical dissymmetry. Any lens effect would tend to diminish
the gradient established by screening, but any reflective type
interference would tend to increase that gradient.

Phototropism by Etiolated Plant Seedlings

The fluence-response relationship for plant phototropism is
extremely complex showing at least three different components which
have been called first positive phototropism, first negative phototropism, and second positive phototropism. Recently, Iino et al.
(1984) have demonstrated that the so-called negative response is a
geotropic compensation for the phytochrome-mediated positive phototropic curvature of the mesocotyl and not a true photoresponse.
This still leaves a complex fluence-response relationship with first
and second positive phototropism separated by a fluence range over
which the response decreases with increasing fluence.

Either refraction or screening as an optical dissymmetry mechanism might play a role in the decreased phototropism with increasing fluence between the maximum for first positive phototropism
and second positive phototropism. In either case, one can predict a
decreased response with increasing fluence over some fluence range.
In the case of refraction, this would occur between saturation on
the distal side and saturation on the proximal side of the organism.
In the case of screening, this would occur between saturation on
the proximal side and saturation on the distal side of the organism.

Vierstra and Poff (1981) have shown using carotenoid inhibitors
that screening is the mechanism used to establish the light gradient
in second positive phototropism. More recently, Piening (1984) has
confirmed these conclusions and extended them to first positive
phototropism using both inhibitors and carotenoidless mutants.
These results leave one with the possibility that the region of decreasing response with increasing fluence results from a saturated
response on the proximal side of the shoot at the same incident

fluence where the response is increasing on the shaded distal side.*

Phototaxis in Chlamydomonas

The role of interference in establishing an optical dissymmetry has been most carefully studied by Foster and Smyth and Morel-Laurens and Feinleib. The results of these two groups were discussed by Feinleib at the beginning of this conference.

SUMMARY

This discussion has presented the three mechanisms whereby an optical dissymmetry can be introduced within an organism enabling the measurement of light direction. The response of the organism is a function both of photoreception and of the mechanism used for measuring light direction. Thus, each of these mechanisms, screening, refraction, and interference, can cause perturbations in an action spectrum for the response, such that the action spectrum may not accurately represent the absorption spectrum for the photo-receptor pigment.

REFERENCES

Curry, G. M., and Gruen, H. E., 1959, Action spectra for the positive and negative phototropism of Phycomyces sporangiophores, Proc. Natl. Acad. Sci. USA, 45:797-804.

Dennison, D. S., 1979, Phototropism, in: "Encyclopedia of Plant Physiology, New Series," W. Haupt and M. E. Feinleib, eds., Vol. 7, Springer-Verlag, Berlin, Heidelberg, pp. 506-566.

Foster, K. W., and Smyth, R. D., 1980, Light antennas in phototactic algae, Microbiol. Rev., 44:572-630.

Häder, D.-P., and Poff, K. L., 1979a, Light-induced accumulations of Dictyostelium discoideum amoebae, Photochem. Photobiol., 29: 1157-1162.

Häder, D.-P., and Poff, K. L., 1979b, Photodispersal from light traps by amoebae of Dictyostelium discoideum, Exp. Mycol., 3:121-131.

---

*The student is encouraged to follow the work from the groups of Shäfer and of Fukshansky at Freiburg, W. Germany. These groups are evaluating the gradients used in plant phototropism using both theoretical and empirical approaches. Recent data from Shäfer's group (personal communication) support the conclusion that a gradient is used as the mechanism whereby the plant measures light direction and underscore the contribution of that mechanism to the measured response of the organism.

Hong, C. B., Häder, M., Häder, D.-P., and Poff, K. L., 1981,
    Phototaxis in Dictyostelium discoideum amoebae, Photochem.
    Photobiol., 33:373-377.
Iino, M., Briggs, W. R., and Schäfer, E., 1984, Phytochrome-mediated
    phototropism in maize seedling shoots, Planta, 160:41-51.
Piening, C. J., 1984, Mechanisms of establishing a light gradient
    for first positive phototropism in Zea mays (L.), M. S. Thesis,
    Michigan State University, E. Lansing, MI.
Seyfried, M., and Fukshansky, L., 1983, Light gradients in plant
    tissue, Appl. Opt., 22:1402-1408.
Vierstra, R. D., and Poff, K. L., 1981, Role of carotenoids in the
    phototropic response of corn seedlings, Plant Physiol., 68:
    798-801.

PHYTOCHROME REGULATION OF PLANT DEVELOPMENT AT THE WHOLE PLANT,

PHYSIOLOGICAL, AND MOLECULAR LEVELS

Winslow R. Briggs, Dina F. Mandoli, James R. Shinkle,
Lon S. Kaufman, John C. Watson and William F. Thompson

Department of Plant Biology
Carnegie Institution of Washington
290 Panama Street
Stanford, CA 94305

INTRODUCTION

Of the many environmental factors which play a role in the
survival, growth, and reproduction of green plants, light is among
the most crucial. Not only does it drive the process of photosynthesis, through which its energy is transduced through the photosynthetic pigments into usable chemical form, but it also provides
vital environmental information which can affect seed germination,
leaf growth, stem growth, flowering, and a host of other processes.
Such effects of light, in which it provides an environmental cue to
trigger a given response, rather than providing a direct energy
source for the response itself, collectively define the field of
photomorphogenesis. An excellent series of review articles on
photomorphogenesis was recently edited by Shropshire and Mohr (1983).

Light can cue plant responses through several different photoreceptors. At the blue end of the spectrum, there is almost certainly more than one photoreceptor pigment, regulating processes
as diverse as chloroplast movement, leaflet movement, phototropic
curvature, and stem elongation (Briggs and Iino, 1983). In the red
and near infrared regions, the principal photoreceptor is phytochrome.
The enormous range and diversity of processes regulated by phytochrome, and the complexity of this regulation, are reviewed in detail
by Satter and Galston (1976) and by Smith and Morgan (1983). The
intricacies encountered in higher plant photomorphogenesis are no-
where better illustrated than in chloroplast development. Both
phytochrome (Mohr, 1977) and protochlorophyllide (Boardman et al.,
1978) serve as photoreceptors for this process, and it is probable
that a blue light photoreceptor is involved as well (Senger, 1982).

Of the several photoreceptors involved in photomorphogenesis, phytochrome has attracted by far the most effort. Since the first spectrophotometric detection and preliminary purification of phytochrome (Butler et al., 1959), much work has been devoted to its purification and characterization (Pratt, 1982; Smith, 1983). Phytochrome is a chromoprotein with a bilitriene chromophore (see Rüdiger and Scheer, 1983), and can exist in either of two spectrally distinct and photointerconvertible forms: P_r, with an absorption maximum near 667 nm, and P_{fr}, with an absorption maximum near 730 nm. It is generally accepted that P_{fr} is the biologically active form of the molecule in most systems. Red light, which drives P_r to P_{fr}, normally potentiates a response; far red light, which drives P_{fr} back to P_r, will usually cancel the effect of red, provided it is given sufficiently soon after the red (prior to any irreversible transduction steps). In most cases, the relative effectiveness of far red light in cancelling the consequences of red diminishes with increasing time between red and far red treatments. It is this escape from photoreversibility after red, but not after far red light, that led to the notion that P_{fr} was the active form. These are classical properties of the phytochrome system; indeed red, far red reversibility is the standard criterion by which phytochrome involvement in a process is assessed.

As will become clear, below, however, there are some responses for which photoreversibility (and hence escape therefrom) cannot be demonstrated, but for which phytochrome involvement is nonetheless relatively certain (see below). Included in this category are certain extremely sensitive responses (Blaauw et al., 1968; Mandoli and Briggs, 1981) and certain responses for which light-driven cycling of the pigment may be required for action (Hartmann, 1966). A more detailed consideration of the various types of phytochrome responses will be presented below.

In unravelling any sensory transduction chain, one can start with the receptor molecule itself, examining the immediate biochemical or biophysical consequences of its excitation; one can study events as early as possible elsewhere following receptor excitation; one can probe events deep in the transduction chain; or one can study the final display itself, working back from there into the transduction chain. All of these approaches have been applied in the case of phytochrome, though only rarely have more than a couple been applied for any given system.

Elucidation of the physicochemical changes occurring upon phytochrome phototransformation has proved difficult and frustrating. In the 25 years since the initial physical detection of this photochromic molecule, great effort has been devoted to purifying and characterizing the molecule itself, and in particular to comparing the physical and chemical properties of P_r with those of P_{fr}, in hopes of finding some clues as to its primary action. These efforts

have been severely hampered by the unfortunate coincidence that
aqueous plant extracts are usually rich in protease activity, and
phytochrome is highly susceptible to proteolytic cleavage (Gardner
et al., 1971). Quail et al. (1983) document the history of efforts
to combat this limitation.

Only very recently (Vierstra and Quail, 1982) have phytochrome
preparations been obtained which were unambiguously undegraded.
Although these preparations are only slightly larger than those
obtained by earlier workers (124 kDa, as compared to 114 or 118
kDa), the polypeptide fragment(s) missing from the earlier preparations is obviously of considerable importance to the chromophore
environment, affecting the absorption spectrum, the quantum efficiency for photoconversion of P_r to P_{fr}, and a number of other
physical and chemical properties (see Quail et al., 1983). The importance of this fragment is underlined by the observation by Hahn
et al. (1984) that degradation of the 124 kDa molecule to a mixture
of the 114 and 118 kDa forms leads to an eight-fold increase in the
relative susceptibility of the chromophore to oxidation by tetranitromethane when the pigment is in the P_{fr} form. Thus, for the first
time, one may be able to compare P_r and P_{fr} with the assurance that
any differences observed might be of biological consequence. The
recent cloning of an mRNA for phytochrome (Hershey et al., 1984)
should also lead to significant advances in our knowledge of the
phytochrome protein moiety. However, there is as yet insufficient
information from the undegraded molecule to move confidently to
the immediate next step in the signal transduction chain, and it is
necessary to turn to other approaches.

Failing demonstration of direct interaction of undegraded P_{fr}
with another cellular component, one can nonetheless study events
occurring as rapidly as measurable following the initial photoreceptor photoexcitation. Quail (1983) summarizes the many studies
of rapid responses to phytochrome photoconversion. Many of these
events involve membranes, although in no case has a direct interaction between phytochrome and a biological membrane been definitively shown. Quail (1983), Hendricks and VanDerWoude (1983), and
Raven (1983) deal quite specifically with the evidence for early
involvement of membrane phenomena in phytochrome response.

In some rapid responses such as rapid leaf closure, membrane
events may be very close to the end of the transduction chain.
However, there may be a great deal more involved in longer term
responses. Hock and Mohr (1964) proposed two decades ago that P_{fr}
might act through a signal chain to activate certain genes; the
products of these activated genes would then work through a further
reaction chain to invoke the response (see also Mohr, 1966). Regulation at the gene level was visualized as being either positive
or negative. Within a relatively few years, evidence began to
accumulate for the light regulation of the levels of specific

enzymes, and there are now dozens of such photoregulated enzymes known (Mohr, 1974; Lamb and Lawton, 1983). More recently a number of studies have demonstrated phytochrome control of the level of specific mRNA species (see Jenkins et al., 1983; Thompson et al., 1983; Kaufman et al., 1984, 1985). In three cases (Gallagher and Ellis, 1982; Silverthorne and Tobin, 1984; Mösinger and Schäfer, 1984), regulation at the level of transcription has been demonstrated (although regulation of other steps, such as transcript processing or degradation, cannot be ruled out).

It has been argued in the past that many rapid responses to phytochrome phototransformation could not be mediated through changes in gene activity simply because too much time would be required from initial photoexcitation, through the signal chain to the gene, and then out again through the reaction chain, with the signal ultimately affecting the level of some protein. Indeed, this argument is undoubtedly valid for many rapid responses. However, it is beginning to lose force for others, since phytochrome-mediated changes in the level of specific mRNAs can occur within as little as 30 minutes of photoexcitation (Colbert et al., 1983; Stiekema et al., 1983; see also below).

The above-mentioned studies have concentrated on events occurring relatively soon following phytochrome photoexcitation - and indeed are beginning to provide some hints as to the nature of some of the primary steps in the transduction chain. However, as mentioned above, another valid approach is to begin with the final display and then work back through the physiology to earlier events.

In a photosensory system, the first steps are normally to characterize the photobiology of the response. What wavelengths of light are required to potentiate it? If the action spectrum suggests phytochrome involvement, how much red light is required? Is the effect far red reversible? If so, what are the kinetics of escape from far red reversibility? Is the reciprocity law (which states that a given fluence is equally effective given at a high fluence rate for a short time or at a low fluence rate for a long time) valid? If it is valid, what are the limits? In the studies to be described below, this approach has been used to probe back from whole plant responses through physiological changes to molecular events which might possibly underlie the responses.

WHOLE PLANT STUDIES

As mentioned above, three kinds of phytochrome responses have been suggested (see Fig. 1 for a schematic illustration). The first of these is classically photoreversible, shows a fluence-response relationship not dissimilar to the fluence-response relationship for phytochrome phototransformation itself, shows far-red

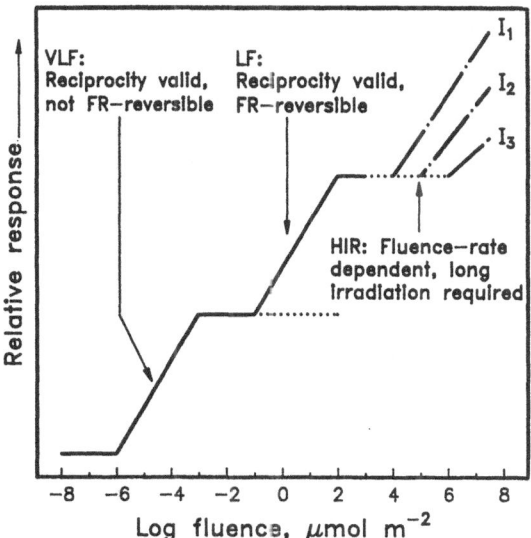

Fig. 1. Summary of the three kinds of phytochrome responses discussed in the text. The relative magnitudes of repre-
sentative responses are plotted against increasing red
light fluence. For the VLF and LF responses, short pulse
irradiations (e.g. 1 s) will potentiate the response.
For the HIR, three fluence rates given over long periods
are shown: $I_1 < I_2 < I_3$.

reversibility, and obeys the reciprocity law. Saturation of such
responses requires that a major fraction of the phytochrome present
be phototransformed into the P_{fr} form. This type of response has
recently been designated the low fluence (LF) response (Mandoli
and Briggs, 1982). The second kind of response is orders of magnitude more sensitive to red light than the LF response and has been
designated the very low fluence (VLF) response (Mandoli and Briggs,
1982). As a VLF response may require less than 1% of the total
phytochrome in the P_{fr} form for saturation, and far red light alone
produces a photostationary equilibrium value near 3% (Pratt and
Briggs, 1966), it is clear a) that VLF responses should not be far-
red reversible, and b) that far red light itself should potentiate
them. Since photoreversibility cannot normally be demonstrated
(but see Briggs and Chon, 1966), the principle evidence that a VLF
response is mediated by phytochrome is the similarity of its action
spectrum to that of a LF response and correspondingly to the absorption spectrum of P_r (Blaauw et al., 1968; Mandoli and Briggs,
1981). Where tested, VLF responses appear to obey the reciprocity
law (e.g. Blaauw-Jansen, 1974; Mandoli and Briggs, 1981) at least
over a limited range. Blaauw-Jansen (1983) has recently reviewed
VLF and LF responses.

The third type of response, called the high irradiance response (HIR), requires irradiations on the order of hours (or frequent brief irradiations over an extended period of time, see Mancinelli and Rabino, 1978), shows fluence-rate dependence, and does not obey the reciprocity law. Though there is good evidence that some HIR responses are mediated by phytochrome (e.g. Hartmann, 1966), there may well be HIR responses mediated by other pigments as well. The reader is referred to reviews by Mancinelli and Rabino (1978), Fukshansky and Schäfer (1983) and Hendricks and VanDerWoude (1983) for details.

It is not uncommon for a particular system to show more than one response type. Thus, suppression of mesocotyl growth in etiolated oat (Goodwin, 1941; Blaauw et al., 1968; Mandoli and Briggs, 1981) or maize (Vanderhoef et al., 1979; Iino, 1982a, b) seedlings may show VLF, LF, and HIR responses. Seed germination in lettuce (Blaauw-Jansen and Blaauw, 1975; Small et al., 1979; VanDerWoude, 1983), Kalenchoe blossfeldiana (Fredricq et al., 1983), and Rumex obtusi-folius (Hand et al., 1982) all can show both VLF and LF responses. Anthocyanin synthesis in Sinapsis alba also shows both a VLF and a LF response (Blaauw-Jansen, 1974) as do both chlorophyll and fresh weight increases in etiolated peas (Kaufman et al., 1983).

Dose-response curves showing more than one response type are by no means an invariant property of a given system. Seed germination provides a number of examples. High or low temperature treatments (Blaauw-Jansen and Blaauw, 1975; Small et al., 1979; Hand et al., 1982; VanDerWoude, 1983) can increase seed sensitivity to light by several orders of magnitude. Seeds showing only a LF response before treatment may show both LF and VLF responses subsequent to treatment. Osmotic stress (0.25 M mannitol) can also bring about the appearance of a VLF where none was present before treatment (Blaauw-Jansen, 1983), as can ethanol (VanDerWoude, 1983), or gibberellin (Fredricq et al., 1983).

Characteristically, in systems showing both VLF and LF responses, the thresholds are four or more orders of magnitude apart, and there often is an intermediate range of doses covering two or more orders of magnitude in which the VLF response is saturated, but LF response is far below threshold. Over this range, the response is essentially indifferent to dose. Since VLF responses require such minute amounts of light, even very brief exposures to dim green safelight can elicit VLF responses (e.g. Blaauw-Janse, 1974; Mandoli and Briggs, 1981). (Light of 530 nm is about one-tenth as effective as light of 667 nm in phototransformation of P_r to P_{fr}, and yields virtually the same photoequilibrium mixture of P_r and P_{fr}, see Pratt and Briggs, 1966.) Hence, to work with VLF responses, one must perform all manipulations under complete darkness (Mandoli and Briggs, 1981) or make use of some kind of infrared image converter (Iino and Carr, 1981). Theo-retically, one should be able to probe LF effects in the presence of

dim green safelight. However, where there is not a distinct break
between VLF and LF responses (Iino, 1982a, b), the use of safelights must be avoided even if just LF phenomena are being studied.
In any case in which safelights are used, their potential effect
must be carefully evaluated.

PHYSIOLOGICAL AND CELLULAR STUDIES

     Given that a system shows more than one response type, it is
appropriate to ask what the physiological and cellular bases of the
responses might be. Two good responses to which such questions can
be addressed are light stimulation of coleoptile growth and inhibition
of mesocotyl growth in grass seedlings. In oats, both responses have
VLF and LF components (Blaauw et al., 1968; Mandoli and Briggs, 1981),
and in fact are virtually mirror images of each other (see Fig. 3 in
Mandoli and Briggs, 1981) showing identical fluence-response relationships. In examining these systems more closely, it is tempting
to begin with several assumptions. The first is that one would expect the greatest light sensitivity where there is the most phytochrome. For example, the coleoptile tip and the coleoptilar node
contain far more phytochrome than either the elongating regions of
the coleoptile or the mesocotyl, and the mesocotyl itself contains
far less phytochrome than the coleoptile (Briggs and Siegelman,
1965). Thus, one might expect irradiation of the coleoptile tip or
node to be more effective in coleoptile growth promotion or mesocotyl
growth inhibition, respectively, than irradiation in other regions.
Mandoli and Briggs (1982) have tested this assumption. Maximum
sensitivity both for LF and VLF responses was found several millimeters _below_ the coleoptilar node in both cases, with a second region
of sensitivity for the coleoptilar response several millimeters _above_
the node. Thus, the phytochrome-rich tips and nodes are relatively
insensitive, and the first assumption is false.

     With the failure of the first assumption, a second (testable)
assumption is that it is the responding tissues themselves which
show maximal photosensitivity. Such is evidently the case in the
rapid suppression of hypocotyl growth in Cucumis sativa seedlings
in response to blue light (Cosgrove, 1981). In the present case,
however, the correlation again does not survive close examination.
For example, in the mesocotyl, the VLF is by far the most dramatic
immediately below the node, while the LF extends throughout the
entire mesocotyl (Schaer et al., 1983). Similarly, the percent increase in length is relatively constant throughout the growing region
of the coleoptile (Schaer et al., 1983) rather than being localized
to the putative sites of photosensitivity (Mandoli and Briggs, 1982).
Hence the second assumption is false.

     A third assumption (also testable) is that the VLF and LF re-
sponses might occur through different mechanisms - for example,

one might be the consequence of a change in cell division frequency,
while the other might be an effect on cellular elongation alone.
Indeed, the mesocotyl VLF does involve primarily suppression of cell
division, while the LF involves primarily inhibition of cell elongation (Schaer et al., 1983). However, the coleoptile presents a different story: both VLF and LF responses involve a promotion of cell
elongation despite a slight suppression of cell division in the
coleoptile base (Schaer et al., 1983). Thus, the third assumption
is false.

Iino (1982a, b) has carefully examined both the auxin relationships and the growth responses of corn mesocotyls to red light.
He concludes (1982b) that a portion of the growth suppression obtained must be through suppression of auxin production by the coleoptile for subsequent delivery to the mesocotyl. However, this mechanism can only be a part of the story, since red light clearly suppresses the elongation of isolated mesocotyl segments in the absence
of any direct influence of the coleoptile. Similarly, Shinkle and
Briggs (1984) have demonstrated red light promotion of the elongation
both of oat and maize coleoptile sections in the absence either of
the auxin-supplying coleoptile tip or the putative photosensitive
sites below. In the coleoptile, a correlative mechanism involving
the effect of red light on auxin delivery is clearly impossible in
any case, since the effect of red light is to reduce the diffusible
auxin supply (Briggs, 1963; Iino, 1982a, b), while the tissue responds with an increase in the elongation rate.

Even in isolated sections where the only growth process affected
by light is probably cell elongation, the situation is complex.
Shinkle and Briggs (1984) have examined the fluence-response relationships for the growth of isolated sections cut from oat coleoptiles grown and harvested in complete darkness. If the sections are
incubated in buffered sucrose, they show only a LF response for
growth stimulation by red light. However, with 6 μM indole-3-acetic
acid in the growth medium, they show only a VLF response. Lower
auxin concentrations cause sections to show responses very similar
to those seen in intact plants: both VLF and LF responses occur,
separated by a dose-indifferent range of about two orders in magnitude. As expected, the VLF component, when present, was not far-
red reversible, while the LF component was fully reversible. Thus,
hormonal status can determine relative light sensitivity, just as it
did in the case of lettuce seeds and gibberellic acid (Fredricq et
al., 1983).

In the case of coleoptile sections, the auxin need only be present shortly before red light treatment to bring about the dramatic
increase in red light sensitivity, and can be removed two hours after
irradiation without loss of the effect (Shinkle, unpublished observations). Under these conditions, there is no significant effect
of auxin on overall growth, despite its dramatic effect on light

sensitivity. Hence, the effects of auxin on light sensitivity and on growth per se must be quite separate. This conclusion is underscored by the observation that auxin sensitizes corn mesocotyl sections to red light <u>inhibition</u> of growth, while simultaneously <u>promoting</u> section growth (Shinkle, unpublished observations).

A well-known effect of IAA is the stimulation of proton pumping across the plasma membrane (see Cleland, 1982). Shinkle (unpublished observations) has shown that a significant conversion of a LF response in oat coleoptile sections to a VLF response can, in fact, be accomplished simply by incubating the sections in a low pH (pH 4) buffer before irradiation. Hence, the auxin effect on red light sensitivity is almost certainly related to its effect on proton pumping. These results also have a more general significance. They demonstrate that a simple change in cell physiology, in this case a decrease in the extracellular pH, can have a profound effect on the transduction of the phytochrome signal.

It should be clear from the above discussion that even with an apparently simple response such as promotion or inhibition of elongation growth, a single transduction chain from photoexcitation of phytochrome to final display is highly unlikely. It seems more likely, rather, that phytochrome triggers a complex program of events, some of which may play a role in certain responses only, with others playing roles in other responses. We shall explore the complexity of such a program at the molecular level in the following section.

MOLECULAR STUDIES

As was mentioned in the Introduction, a number of recent papers have addressed the regulation of specific mRNA levels by phytochrome. However, none of these studies has explored the photobiological properties of the respective system in any great detail. Light fluences sufficient to saturate the LF response have usually been used and the involvement of phytochrome deduced from far-red reversibility studies.

Recently Kaufman et al. (1984, 1985) have explored the photobiological properties of phytochrome regulation of specific mRNA levels in etiolated pea buds. Building on the observations of Thompson et al. (1983) that a large number of nuclear transcripts are under phytochrome control, they obtained fluence-response curves for 13 different mRNAs, both for red light-induced changes in steady state mRNA level and for far-red reversibility of the red effect. The protocol involved growing pea seedlings for six days from planting in total darkness, giving them a single red light treatment, and then measuring the levels of particular mRNAs after 24 hours of darkness or after 24 hours of darkness followed by 24 hours of white light.

Both bud fresh weight and chlorophyll levels, as measured at the
end of the 24 hour light period, were significantly increased by red
light treatment (Kaufman et al., 1983) with both increases displaying
both LF and VLF responses.

When transcript abundance was assayed immediately after the 24
hours of darkness, four distinct fluence-response relationships
emerged. One transcript showed a VLF response only, one showed both
VLF and LF responses, nine showed LF responses only, and two showed
no response at all. When the assay was performed after the additional 24 hours of white light, there were several distinct differences. Three of the transcripts showing a LF response prior to
white light no longer showed any response afterwards. Both the
transcript showing a VLF response only and that showing both responses showed only LF responses after white light. By contrast,
one transcript which displayed only an LF response before white
light, gained a VLF component after. Not unexpectedly, all LF responses showed full far-red reversibility. Far red did not reverse
the VLF responses, and in fact was fully effective in inducing them
(Kaufman et al., 1984, 1985).

Another aspect of the photoregulation of the level of these
gene products emerges from comparison of their relative amount at
the end of the dark period (without red pretreatment) and their
levels after the 24-hour white light treatment. Three of the transcripts showed a distinct decrease following white light treatment,
while the remainder all showed increases. These increases ranged
from about two-fold to almost ten-fold.

Tow of the thirteen mRNA species can be associated with specific
protein products. These are the small subunit of ribulose-1,5-bis-
phosphate carboxylase and the principal chlorophyll $\underline{a/b}$ binding protein from thylakoids. From a developmental point of view, it is
interesting that it is the pigment-binding protein which shows
greater light sensitivity (both VLF and LF) than the small subunit
(LF only) (Kaufman et al., 1984). Although much more needs to be
learned before the biological significance of this difference in
sensitivity will be clarified, we may speculate that as a germinating
plant moves through the soil into an increasingly lighter environment, it could be advantageous to start building the light-harvesting
machinery before building the enzymatic capacity to carry out dark
CO_2 fixation.

Kaufman (unpublished) has also carried out detailed time-course
studies for mRNA accumulation during the 24 hours of darkness following a saturating red light pulse and obtained information on reciprocity relationships for a single subsaturating (LF) light dose
as measured at the end of the 24-hour period. The kinetic studies
show four distinct patterns for increase in mRNA level for the
transcripts showing a light-induced increase. Seven of them began

to increase immediately and climbed almost linearly over the next
16-24 hours. Two of the transcripts began climbing dramatically
right away, reaching a plateau level within two hours. (One of
these almost reached the plateau within the first hour.) Two other
transcripts showed no significant change for a full sixteen hours,
beginning to climb steeply only after that lag. Finally, one transcript which climbed steadily in the dark, simply climbed more
rapidly after red light treatment, levelling out after about 12
hours while the dark control continued to climb to meet it (at 24
hours). For reference, a significant increase in bud fresh weight
could only be seen eight hours after the red light treatment. The
fresh weight increase, and all of the transcripts save for the one
which increased in the dark, obeyed the reciprocity law from exposure times of 50 through 4000 seconds' duration.

It is not possible to derive any simple relationship between
fluence-response relationships and time courses. For example, both
the transcript for the chlorophyll a/b binding protein (VLF and LF)
and for the small subunit of carboxylase (LF only) showed similar
time courses - a gradual climb over the first 20 hours. On the
other hand, the transcript showing the highest sensitivity (VLF
only) didn't even begin to increase until 16 hours after red light
treatment.

Genes encoding the chlorophyll a/b binding protein and the
small subunit of carboxylase are well known to consist of small
multigene families, with on the the order of 5-10 copies per haploid
nuclear genome, both in pea (Coruzzi et al., 1983; J. C. Watson,
unpublished observations) and in petunia (Dunsmuir et al., 1983a,
b). In these cases, it is possible that different genes are responding differently. For example, some chlorophyll a/b binding
protein genes might be active already in the dark while others
respond only to light in the VLF range and still others only to
light in the LF range. In marked contrast, several of the other
genes under study appear, on the basis of Southern blotting experiments, to be present in only a single copy per haploid genome
(J. C. Watson, unpublished data). These genes are noteworthy since
most higher plant gene systems studied so far have turned out to be
multigene families. Clearly, differential expression of different
gene family members cannot be postulated in the case of single copy
genes.

Among the transcripts which represent single copy genes we find
some belonging to each of the major time course classes and at least
one for which both VLF and LF responses can be demonstrated. Thus,
the presence of more than one gene is not required in order to
explain complex fluence-response curves.

The complexity of these molecular responses is clearly comparable
to that encountered at higher levels of organization, and indeed may

underlie the complexities previously observed in whole plant and cellular studies.

CONCLUDING REMARKS

A wide range of processes in higher plants are known to be regulated by phytochrome. From the above discussion, a few tentative conclusions emerge. First, there are at least three different kinds of phytochrome responses, VLF, LF, and HIR (Fig. 1). They are clearly distinguishable on the basis of their photobiological properties, and cover an enormous dynamic range - over ten orders of magnitude. Second, both hormonal and environmental status can affect the relative contributions of the various systems to a given biological display. Third, the VLF and LF responses can be distinguished at the mRNA level, with different transcripts showing quite distinct fluence-response relationships and time courses.

We are left with a number of open questions. First, is there an interaction between hormones and light that is detectable at the molecular level? Second, is there an interaction between photoexcitation of phytochrome and any other photoreceptor detectable at the molecular level? Third, do hormones or other photoreceptors bring about mRNA changes independent of phytochrome status? Fourth, what is the nature of the regulation by phytochrome? Fifth, is it different for transcripts showing different fluence-response relationships and/or time courses? The biological function of these (and other) photoregulated genes will also be of interest, although it is worth noting that understanding the mechanisms by which gene activity is regulated does not require a knowledge of the final gene products.

The picture which is emerging is complicated, but all of these questions are approachable. Addressing them should greatly help in unravelling the various interactions between light, other environmental factors, and hormones on plant growth and development.

ACKNOWLEDGEMENTS

Part of the work described above was supported by USDA grant 82-CRCR-1-1081 to W. F. T. and W. R. B. The authors are grateful for this aid. This is Carnegie Institution of Washington Department of Plant Biology Publication No. 867.

REFERENCES

Blaauw, O. H., Blaauw-Jansen, G., and van Leeuwen, W. J., 1968, An irreversible red-light-induced growth response in Avena, Planta, 82:87-104.

Blaauw-Jansen, G., 1974, Dose-response curves for phytochrome-mediated anthocyanin synthesis in the mustard seedling, Acta Botanica Neerlandica, 23:513-519.

Blaauw-Jansen, G., 1983, Thoughts on the possible role of phytochrome destruction in phytochrome-controlled responses, Plant Cell Environ., 6:173-179.

Blaauw-Jansen, G., and Blaauw, O. H., 1975, A shift in the response threshold to red irradiation in dormant lettuce seeds, Acta Botanica Neerlandia, 24:199-202.

Boardman, N. K., Anderson, J. M., and Goodchild, D. J., 1978, Chlorophyll-protein complexes and structures of mature and developing chloroplasts, Curr. Top. Bioenerget. B 8:35-109.

Briggs, W. R., 1963, Red light, auxin relationships, and the phototropic responses of corn and oat coleoptiles, Am. J. Botan., 50:196-207.

Briggs, W. R., and Chon, H. P., 1966, The physiological versus the spectrophotometric status of phytochrome in corn coleoptiles, Plant Physiol., 41:1159-1166.

Briggs, W. R., and Iino, M., 1983, Blue-light-absorbing photoreceptors in plants, Phil. Trans. Roy. Soc. Lond. B., 303:347-359.

Briggs, W. R., and Siegelman, H. W., 1965, Distribution of phytochrome in etiolated seedlings, Plant Physiol., 40:934-941.

Butler, W. L., Norris, K. H., Siegelman, H. W., and Hendricks, S. B., 1959, Detection, assay, and preliminary purification of the pigment controlling photoresponsive development of plants, Proc. Natl. Acad. Sci. USA, 45:1703-1708.

Cleland, R. E., 1982, The mechanism of auxin-induced proton efflux, in: "Plant Growth Substances 1982," Proceedings of the 11th International Conference on Plant Growth Substances, P. F. Wareing, ed., Academic Press, London, pp. 23-31.

Colbert, J. T., Hershey, H. P., and Quail, P. H., 1983, Autoregulatory control of translatable phytochrome mRNA levels, Proc. Natl. Acad. Sci. USA, 80:2248-2252.

Coruzzi, G., Broglie, R., Cashmore, A., and Chua, N.-H., 1983, Nucleotide sequences for two pea cDNA clones encoding for the small subunit of ribulose-1,5-bisphosphate carboxylase and the major chlorophyll a/b binding thylakoid polypeptide, J. Biol. Chem., 258:1399-1402.

Cosgrove, D. J., 1981, Rapid suppression of growth by blue light. Occurrence, time course, and general characteristics, Plant Physiol., 67:584-590.

Dunsmuir, P., Smith, S. M., and Bedbrook, J., 1983, The major chlorophyll a/b binding protein of petunia is composed of several polypeptides encoded by a number of distinct nuclear genes, J. Mol. Appl. Gen., 2:285-300.

Dunsmuir, P., Smith, S. M., and Bedbrook, J., 1983, A number of different nuclear genes for the small subunit of RuBCase are transcribed in Petunia, Nucl. Acid Res., 11:4177-4183.

Fredricq, H., Rethy, R., van Onckelen, H., and de Greef, J. A., 1983, Synergism between gibberellic acid and low P_{fr} levels inducing germination of Kalenchöe seeds, Physiol. Plantarum, 57:402-406.

Fukshansky, L., and Schäfer, E., 1983, Models in photomorphogenesis, in: "Encyclopedia of Plant Physiology, New Series," W. Shropshire, Jr., and H. Mohr, eds., Springer-Verlag, Berlin, pp. 69-95.

Gallagher, T. F., and Ellis, J. R., 1982, Light-stimulated transcription of genes for two chloroplast polypeptides in isolated pea leaf nuclei, EMBO J., 1:1493-1498.

Gardner, G., Pike, C. S., Rice, H. V., and Briggs, W. R., 1971, 'Disaggregation' of phytochrome in vitro - a consequence of proteolysis, Plant Physiol., 48:686-693.

Goodwin, R. H., 1941, On the inhibition of the first internode of Avena by light, Am. J. Botan., 28:325-332.

Hahn, T.-R., Song, P.-S., Quail, P. H., and Vierstra, R. D., 1984, Tetranitromethane oxidation of phytochrome chromophore as a function of spectral form and molecular weight, Plant. Physiol., 74:755-758.

Hand, D. J., Craig, G., Takaki, M., and Kendrick, R. E., 1982, Interaction of light and temperature on seed germination of Rumex obtusifolius L., Planta, 156:457-460.

Hartmann, K. M., 1966, A general hypothesis to interpret high energy phenomena of photomorphogenesis on the basis of phytochrome, Photochem. Photobiol., 5:349-366.

Hendricks, S. B., and VanDerWoude, W. J., 1983, How phytochrome acts - perspectives on the continuing quest, in: "Encyclopedia of Plant Physiology, New Series," 16A, W. Shropshire, Jr., and H. Mohr, eds., Springer-Verlag, Berlin, pp. 3-23.

Hershey, H. P., Colbert, J. T., Lissemore, J. L., Barker, R. F., and Quail, P. H., 1984, Molecular cloning of cDNA for Avena phytochrome, Proc. Natl. Acad. Sci. USA, 81:2332-2336.

Hock, B., and Mohr, H., 1964, Die Regulation der O_2 Aufnahme von Senfkeimlingen (Sinapis alba L.) durch Licht, Planta, 61:209-228.

Iino, M., and Carr, D. J., 1981, Safelight for photomorphogenic studies: Infrared radiation and infrared-scope, Plant Sci. Lett., 23:263-268.

Iino, M., 1982a, Action of red light on the indole-3-acetic acid status and growth in coleoptiles of etiolated maize seedlings, Planta, 156:21-32.

Iino, M., 1982b, Inhibitory action of red light on the growth of maize mesocotyl: Evaluation of the auxin hypothesis, Planta, 156:338-395.

Jenkins, G. I., Hartley, M. R., and Bennett, J., 1983, Photoregulation of chloroplast development: Transcriptional, translational and post-translational controls? Phil. Trans. Roy. Soc. Lond. B, 303:419-431.

Kaufman, L. S., Briggs, W. R., and Thompson, W. F., 1985, Phytochrome control of specific mRNA levels in developing pea buds: The presence of both very low fluence and low fluence responses. Plant Physiol., submitted.

Kaufman, L. S., Thompson, W. F., and Briggs, W. R., 1983, Phytochrome control of specific mRNA levels in developing pea buds: The presence of low and very low fluence responses, Carnegie Institution of Washington Year Book, 82:15-17.

Kaufman, L. S., Thompson, W. F., and Briggs, W. R., 1984, Phytochrome induced accumulation of RNA encoding the small subunit of RuBPcase requires ten thousand fold more red light than does the RNA for the chlorophyll a/b binding protein, Science, in press.

Lamb, C. J., and Lawton, M. A., 1983, Photocontrol of Gene Expression, in: "Encyclopedia of Plant Physiology, New Series," 16A, W. Shropshire, Jr., and H. Mohr, eds., Springer-Verlag, Berlin, pp. 213-287.

Mancinelli, A. L., and Rabino, E., 1978, The "high irradiance response" of plant photomorphogenesis, Bot. Rev., 44:129-180.

Mandoli, D. F., and Briggs, W. R., 1981, Phytochrome control of two low-irradiance responses in etiolated oat seedlings, Plant Physiol., 67:733-739.

Mandoli, D. F., and Briggs, W. R., 1982, The photoperceptive sites and the function of tissue light piping in photomorphogenesis of etiolated oat seedlings, Plant Cell Environ., 5:137-145.

Mohr, H., 1966, Differential gene activation as a mode of action of phytochrome 730, Photochem. Photobiol., 5:469-483.

Mohr, H., 1974, The role of phytochrome in controlling enzyme levels in plants, in: "Biochemistry of Cell Differentiation," J. Paul, ed., Butterworth, London, pp. 37-81.

Mohr, H., 1977, Phytochrome and chloroplast development, Endeavor (n.s.), 1:107-114.

Mösinger, E., and Schäfer, E., 1984, In-vivo phytochrome control of in vitro transcription rates in isolated nuclei from oat seedlings, Planta, 161:444-450.

Pratt, L. H., 1982, Phytochrome: The protein moiety, Annu. Rev. Plant Physiol., 33:557-582.

Pratt, L. H., and Briggs, W. R., 1966, Photochemical and nonphotochemical reactions of phytochrome in vivo, Plant Physiol., 41:467-474.

Quail, P. H., 1983, Rapid action of phytochrome in photomorphogenesis, in: "Encyclopedia of Plant Physiology, New Series," 16A, W. Shropshire, Jr., and H. Mohr, eds., Springer-Verlag, Berlin, pp. 178-212.

Quail, P. H., Colbert, J. T., Hershey, H. P., and Vierstra, R. D., 1983, Phytochrome: Molecular properties and biogenesis, Phil. Trans. Roy. Soc. Lond. B., 303:387-402.

Raven, J. A., 1983, Do plant photoreceptors act at the membrane level, Phil. Trans. Roy. Soc. Lond. B., 303:403-417.

Rüdiger, W., and Scheer, H., 1983, Chromophores in photomorphogenesis, in: "Encyclopedia of Plant Physiology, New Series," 16A, W. Shropshire, Jr., and H. Mohr, eds., Springer-Verlag, Berlin, pp. 119-151.

Satter, R. L., and Galston, A. W., 1976, The physiological functions of phytochrome, in: "Chemistry and Biochemistry of Plant Pigments," 2nd edit., T. W. Goodwin, ed., Academic Press, London, pp. 681-735.

Schaer, J. A., Mandoli, D. F., and Briggs, W. R., 1983, Phytochrome-mediated cellular photomorphogenesis, Plant Physiol., 72:706-712.

Senger, H., 1982, The effect of blue light on plants and microorganisms, Photochem. Photobiol., 35:911-920.

Shinkle, J. R., and Briggs, W. R., 1984, IAA sensitization of phytochrome-controlled growth of coleoptile sections, Proc. Natl. Acad. Sci. USA, 81:3742-3746.

Shropshire, Jr., W., and Mohr, H., ed., 1983, "Encyclopedia of Plant Physiology, New Series," 16A, 16B, Springer-Verlag, Berlin, p. 832.

Silverthorne, J., and Tobin, E., 1984, Demonstration of transcriptional regulation of specific genes by phytochrome action, Proc. Natl. Acad. Sci. USA, 81:1112-1116.

Small, J. G. C., Spruit, C. J. P., Blaauw-Jansen, G., and Blaauw, O. H., 1979, Action spectra for light-induced germination in dormant lettuce seeds, Planta, 144:125-131.

Smith, H., and Morgan, D. C., 1983, The function of phytochrome in nature, in: "Encyclopedia of Plant Physiology, New Series," 16A, W. Shropshire, Jr., and H. Mohr, eds., Springer-Verlag, Berlin, pp. 491-517.

Smith, W. O., 1983, Phytochrome as a molecule, in: "Encyclopedia of Plant Physiology, New Series," 15A, W. Shropshire, Jr., and H. Mohr, eds., Springer-Verlag, Berlin, pp. 96-118.

Stiekema, W. J., Wimpee, C. F., Silverthorne, J., and Tobin, E. M., Phytochrome control of two nuclear genes encoding chloroplast proteins in Lemna gibba L. G3, Plant Physiol., 72:717-724.

Thompson, W. F., Everett, M., Polans, N. O., Jorgensen, R. A., and Palmer, J. D., 1983, Phytochrome control of RNA levels in developing pea and mung bean leaves, Planta, 158:487-500.

Vanderhoef, L. N., Quail, P. H., and Briggs, W. R., 1979, Red light-inhibited mesocotyl elongation in maize seedlings, Plant Physiol., 63:1062-1067.

VanDerWoude, W. J., 1983, Mechanisms of photothermal interactions of phytochrome control of seed germination, in: "Strategies of Plant Reproduction," Beltsville Symposium on Agricultural Research, Vol. 6, W. J. Meudt, ed., Allanneld Osmun, Totawa, NJ, USA, pp. 234-244.

Vierstra, R. D., and Quail, P. H., 1982, Native phytochrome: Inhibition of proteolysis yields a homogeneous monomer of 124 kilodaltons from Avena, Proc. Natl. Acad. Sci. USA, 79:5272-5276.

MOLECULAR MECHANISMS OF PHOTOINDUCED CHLOROPLAST MOVEMENTS

Gottfried Wagner and Franz Grolig

Botanisches Institut I
der Justus-Liebig-Universität
Senckenbergstrasse 17-21
D-6300 Giessen
W. Germany

Cell metabolism requires that substrates, intermediates, co-factors, messengers, and enzymes must be able to move from one part of the cytoplasm to another. In small cells, such as bacteria or even most animal cells, diffusion is sufficient for small solutes to move over distances comparable to the size of the cell in fractions of a second. However, plant cells, because of their cell walls, vacuoles, and turgor, are able to grow very large. They are commonly more than 100 μm long, while some are few millimeters or even centimeters long. Diffusion is relatively ineffective over such distances, as the time needed for a molecule to reach its destination by diffusion alone depends on the square of the distance involved. It is not surprising, therefore, that large plant cells display an extensive cytoplasmic streaming that stirs their cytoplasm and moves material including chloroplasts around.

Examination of living plant cells reveals that the larger the cell, the more extensive are the movements of its cytoplasm. Small cells exhibit agitated movements of their organelles known as saltations by which cells and particles stop and go, being suddenly propelled at a high speed in a particular direction. In larger plant cells, there is more directionality to the cytoplasmic movements; and in cells with a thin rim of cytoplasm surrounding a huge central vacuole, it is common for the cytoplasm to rotate almost continuously, at a rate of several micrometers per second. Such cytoplasmic movements promote not only intracellular traffic but also intercellular transport by delivering solutes to the openings of plasmodesmata connecting adjacent cells (Hope and Walker, 1975; Alberts et al., 1983).

Many plant cells can respond to changes in irradiance and direction of light by altering the position of their chloroplasts. At low irradiances, i.e. light at low intensities, the chloroplasts tend to become aligned in a monolayer perpendicular to the incident light, thereby maximizing their exposure to the light. High irradiances induce a protective response in which the chloroplasts become aligned against the cell walls that are parallel to the incident light, thereby minimizing their exposure (Haupt, 1982, 1983).

Cytoplasmic streaming and transport of cell materials including chloroplasts is based on a complex array of structural proteins, notably the microfilaments and microtubules, collectively referred to as the plant cytoskeleton. A third class of structural proteins known from animal cells, i.e. intermediate filaments of thicker diameter than that of actin but thinner than that of myosin, has not yet been reported in plants.

TUBULIN AND DYNEIN ATPase

Microtubules are the main structures underlying movement of chromosomes, e.g. during M-phase in the life cycle of a cell; movement of cytoplasmic vesicles during cellulose cell-wall formation also is a good example. For chloroplast reorientation, however, their function seems to emerge only recently. (Microfilaments of actin and myosin, on the other hand, appear to be the common structural basis for cytoplasmic streaming and chloroplast movement in most cases investigated so far; see below).

Microtubules comprise a class of structural proteins of strikingly similar morphology in the kingdom of eukaryotes, including cilia and flagella. A microtubule is built from 13 individual protofilaments in parallel rows to form a hollow cylinder of about 25 nm in diameter. Each protofilament consists of globular dimers of tubulin (α and β, 55 000 daltons each) connected end on.

Microtubules in the cytoplasm prove to be far less stable than those in the orderly array of parallel microtubules, e.g. in cilia. Cytoplasmic microtubules disintegrate upon cleavage of the two anhydride bonds in GTP; disintegration may be triggered in vivo by external factors, such as exposure to low temperatures, high pressure, or antimitotic drugs. The drug colchicine binds rather specifically to unpolymerized microtubule protein and finally leads to microtubule disintegration. Hence, blockage or alteration of a movement by colchicine is taken as strong evidence that the movement is dependent on microtubules (Filner and Yadav, 1979).

Rhythmic chloroplast movement in Ulva is sensitive to colchicine (Britz, 1979). While submillimolar concentrations of colchicine shift Ulva chloroplasts more into profile position without inhibiting the extent of movement, millimolar concentrations of the drug actually stop profile-to-face movement. This inhibition is not reversed by washing, but (as expected for an effect on microtubules) the inhibition is reversed by irradiation with UV light, which forms the non-tubulin-binding isomer of colchicine, lumicolchicine.

In 1967, Sabnis and Jacobs showed that Caulerpa prolifera, a coenocytic marine green alga, comprises large bundles of microtubules in the cytoplasm. Chloroplasts in the vicinity of microtubule bundles lie in parallel to them. No structures corresponding to the actin microfilaments were observed. In the endoplasmatic layer, which is adjacent to the stationary ectoplasmatic layer, numerous two-way streams are oriented. Chloroplasts move bidirectionally along multistriated tracks, much the same as in Acetabularia; the rate of streaming is about 3 µm/s. Kuroda and Manabe (1983) report the streaming in Caulerpa being insensitive to cytochalasin B in concentrations of 1 mmol/l present for 80 min. These results suggest that the streaming in Caulerpa might be caused by a microtubule-associated system. This conclusion is supported by the inhibitory effect of erythro-9-[3-(2-hydroxynonyl)]adenine which has been reported as a specific inhibitor of dynein ATPase activity (Penningroth et al., 1982).

With regard to the circadian chloroplast movement in Ulva, the multistriate cytoplasmic streaming in Caulerpa and possibly light-induced chloroplast accumulation in the marine green alga Bryopsis, involvement of microtubuli in chloroplast movement seems to be recognized only recently and deserves further studies.

ACTIN AND MYOSIN ATPase

Long before protoplasm existed as a concept, Corti (1774, 1776) observed and reported cytoplasmatic streaming in charophyte cells. This turned out to be the first report of microfilaments active in cytoplasmic movement.

The cylindrical cells of the green algae Chara and Nitella are tremendously large, being 2-5 cm in length. These giant, multinucleated cells provide the most dramatic examples of cytoplasmic streaming. A continuous ribbon of cytoplasm streams along a gentle helical path down one side of each cell and back across the other side in an endless belt, sweeping internal membranes, mitochondria, nuclei and cytosol around and around the cell at speeds of up to 75 µm per s. Not all of the cytoplasm, however, flows in these giant cells. The cell cortex is static; it consists of the cytoplasm just beneath the plasma membrane, which contains a monolayer of chloroplasts that are aligned in rows parallel to the direction of streaming. In the light microscope, thin fibrils 0.2 µm in diameter can be seen just under the rows of chloroplasts. The electron microscope reveals that each of these fibrils consists of a bundle of actin filaments in such that the movement of myosin filaments along them could produce the observed cytoplasmic streaming. It seems possible, therefore, that organelles in the moving cytoplasm are indirectly attached to the actin filaments by myosin molecules, which use the energy of ATP hydrolysis to slide along the actin filaments, pulling the organelles with them (Williamson, 1975; Nagai and Hayama, 1979).

It has been proposed that microfilaments found in animal and plant cells are the less highly evolved ancestors of skeletal actin because of their fibrous appearance, their size (4-10 nm in diameter), and their presence in regions of cellular movement (Goldman, 1975). Further substantiation of this hypothesis was derived from effects on movement by certain drugs like cytochalasin B (Carter, 1967), N-ethylmaleimide and p-chloromercuribenzoate (Dreizen and Gershman, 1970) or phalloidin (Wieland, 1977) which are known to interact with actomyosin-like microfilaments.

The most convincing and direct evidence for these microfilaments being actin, however, comes from the use of the heavy meromyosin (HMM) binding technique originally described by Ishikawa et al. (1969). This technique is based on H. E. Huxley's pioneering studies (1963) on the binding of rabbit skeletal muscle HMM to isolated rabbit F-actin to form "arrowhead" configurations or "decorated" actin strands. Arrowheads along individual strands point in one direction due to a polarity in F-actin (Wagner, 1979). This distinctive pattern is most convincingly visualized in thin sections or after negative staining. Due to the amazingly accurate conservation of the actin molecule throughout evolution, actin decoration by HMM works using HMM even from a distantly divergent source, no matter whether animal or plant.

There are several clear cases in which actin is rapidly assembled into filaments and broken down to globular actin in animal and plant cells (Isenberg and Wohlfarth-Bottermann, 1976; Korn, 1982). The globular actin monomers (G-actin) do not actually require energy for polymerization; assembly occurs even in the presence of nonhydrolyzable nucleotide triphosphate analogues. But nucleotide triphosphate hydrolysis during polymerization allows the polymers to assemble in filamentous polar structures (F-actin) that grow at different rates at their two ends. Two of these actin strands are twisted around each other in helical double stranded arrangement. It is probable that a dynamic equilibrium exists between actin molecules and actin filaments in all cells and that this plays a part in various cell movements such as chloroplast reorientation.

Drugs that affect the state of actin polymerization can be shown to disrupt many of these cell movements. For example, the cytochalasins, a family of metabolites excreted by various species of molds, paralyze many cytoplasmic and organellar movements in a

wide range of lower and higher plants (Haupt and Wagner, 1984).
They do not, however, inhibit mitosis, which principally involves
microtubules, nor do they affect muscle contraction, which does not
involve the assembly and disassembly of actin filaments. The cytochalasins act by binding specifically to one end of actin filaments
(corresponding to the "barbed" end of an actin filament decorated
with heavy meromyosin or its subfragment S1), thereby preventing
the addition of actin molecules to that end.

Phalloidin is a highly poisonous alkaloid produced by the
toadstool Amanita phalloides. By contrast with the cytochalasins,
it stabilizes actin filaments and inhibits their depolymerization.
Phalloidin is found to block the migration of tissue culture cells,
amoebae and Physarum (Stockem et al., 1978; Wehland et al., 1980)
suggesting that the dynamic assembly and disassembly of actin filaments is crucial for these movements. Because phalloidin stabilizes
actin filaments by binding to them in a highly specific fashion along
their lengths, fluorescent derivatives of the drug have been useful
for staining actin filaments inside cells (Haupt and Wagner, 1984).

The second major component of the microfilamenteous network,
i.e. the myosin molecule consists of two very large polypeptide
chains of molecular weight of about 200 000 daltons (the "heavy
chains") and two pairs of smaller polypeptide chains having molecular
weights in the 20 000-dalton range (the "light chains"). Two of
these light chains appear to be required for myosin ATPase activity,
while the second pair has been shown in many cases to play a regulatory role in the actin myosin interaction (the "regulatory myosin
light-chains"). Along part of their length, the two heavy chains of
myosin are coiled around each other to form a double-stranded structure about 140 nm in length and 2 nm in diameter. The rest of each
heavy chain is folded up to form a globular "head," together with
some or all of the light chains. By proteolytic attack, the large
polypeptide chains of myosin can be split into well-characterized sub-
fragments, the size of which is dependent on the splitting enzyme
used; e.g. papain cleavage tears off the two globular myosin heads,
each of this is commonly defined as subfragment S1 (molecular weight
120 000); the remainder is the myosin rod (molecular weight 260 000).
Trypsin digestion, by contrast, yields heavy meromyosin, which consists of the two S1 "heads" (S1 units) and part of the myosin rod,
the S2 unit, which clamps the two S1 heads together (HMM; molecular
weight 350 000). The remaining part of the myosin rod is called
light meromyosin (LMM; molecular weight 140 000).

ATPase activity, and the ability to interact with actin, are
present only in those fragments that contain globular heads of
myosin, i.e. HMM or S1 itself. Stimulation or inhibition of myosin
ATPase activity depends on the presence of actin and regulatory
factors the best known of which is the myosin light chain kinase
(MLCK).

Nonmuscle cells contain much less myosin, relative to actin, than muscle cells, and their myosin filaments are very much smaller and more labile than the thick filaments of skeletal muscle cells - not surprisingly, since the forces they must generate are much weaker.

The activation of nonmuscle myosin, as with smooth muscle myosin, depends on the phosphorylation of the regulatory myosin light-chains; phosphorylation induces the assembly of nonmuscle myosin molecules into small bipolar aggregates in which 10 to 20 myosin molecules (compared with 500 or so myosin molecules in a skeletal muscle thick filament) are held together by the aggregation of their tail regions. Phosphorylation is catalyzed by a myosin kinase that is stimulated by Ca^{2+}, so that both the state of aggregation of nonmuscle myosin and its ability to interact with actin filaments are altered by small changes in the concentration of free Ca^{2+} in the cytosol. Such changes in intracellular Ca^{2+} commonly occur in response to extracellular signals.

In order to exert a mechanical force, a contractile assembly must be anchored to other cellular components. Thus, many, if not most, actin filaments are anchored at one end to cell membranes.

Since the free Ca^{2+}-concentration in the cytosol is usually less than 10^{-7} molar and usually does not rise much above 10^{-5} molar even when the cell is activated by an influx of Ca^{2+}, any structure in the cell that is to serve as a direct target for Ca^{2+}-dependent regulation must have a binding affinity for Ca^{2+} of around 10^{-6} M. Moreover, since the concentration of free Mg^{2+} in the cytosol is relatively constant at 10^{-3} M, the Ca^{2+}-binding sites must have a selectivity for Ca^{2+} over Mg^{2+} of at least 1000-fold. Several specific Ca^{2+}-binding proteins fulfill these criteria.

The first such protein to be discovered, and the best characterized, is troponin C in skeletal muscle cells. A related Ca^{2+}-binding protein, known as calmodulin, has been identified in all animal and plant cells that have been examined (Cheung, 1980; Cormier et al., 1981). It has been highly conserved during evolution and appears to be an ubiquitous intracellular Ca^{2+} receptor, playing a part in the majority of the Ca^{2+}-regulated processes that have been studied in eucaryotic cells. Calmodulin is a single polypeptide chain of 148 amino acid residues (MW 16 700), whose sequence is related to that of troponin C, suggesting that the latter is a specialized form of calmodulin. Like troponin C, calmodulin has four high-affinity Ca^{2+}-binding sites and undergoes a large conformational change when it binds Ca^{2+}.

Among the increasing numbers of cellular proteins known to be regulated by calmodulin in a Ca^{2+}-dependent manner are the myosin light-chain kinase of both muscle and nonmuscle cells and plant NAD kinase (Marmé and Dieter, 1983).

Since calmodulin can exist in several different conformations, depending on the number of calcium ions bound per molecule, it is possible that the different conformations interact with different target proteins in a cell. In this way, calmodulin could translate quantitative differences in the concentration of intracellular free Ca^{2+} into different cellular responses.

All calmodulins, no matter whether from animals or plants, stimulate Ca^{2+}-dependent phosphodiesterase (Wagner et al., 1984), and the stimulation is blocked by trifluoperazine, a phenothiazine derivative often used as an antipsychotic agent. Calmodulin binds two moles of trifluoperazine with a dissociation constant of 10^{-6} M; upon binding of trifluoperazine, calmodulin becomes biologically inactive. Myosin light-chain kinase appears peculiar among calmodulin-regulated proteins in that its activity is absolutely dependent on calmodulin and catalyzes the phosphorylation of the regulatory myosin light-chains. Phosphorylation of the light chains activates the actomyosin ATPase, leading to the hydrolysis of ATP and shearing of the myosin along filamentous actin in smooth muscle or nonmuscle cells.

PHOTOINDUCED CHLOROPLAST MOVEMENTS AND ACTOMYOSIN

Chloroplasts can stay motionless in the cytoplasmic rim or can be swept away by the cytoplasmic stream or, in highly elaborate cases, can perform autonomous movements within the cell induced by light (Seitz, 1979; 1982). In the characean algae rows of motionless chloroplasts are found, while the cytoplasm streams. Accordingly, light shows a minor effect on this type of cyclosis which, however, is extremely sensitive to membrane action potentials. Upon firing, intracellular Ca^{2+}-concentration was calculated from aequorin measurements to increase 30-fold to 6.7 μM (the mean of six cells) during a Chara action potential and 42-fold for Nitella to 43 μM (the mean of four cells) (Williamson and Ashley, 1982).

In the siphonaceous alga Vaucheria, the blue-light-induced chloroplast aggregation is not entirely hooked onto the bulk stream of cytoplasm; in darkness or when the alga is fully exposed to light, the chloroplasts swim with the stream of cytoplasm. Upon local irradiation of the alga, however, cortical fibers can be seen to reticulate, and concomitantly chloroplasts aggregate in this region. Local light, therefore, acts as a trap to the passing chloroplasts, while part of the cytoplasm continues to move.

Several lines of evidence indicate that the cortical fibers enclose actin filaments and filament bundles of actin (Blatt et al., 1980). Cytochalasin B (CB) was found both to inhibit streaming and to alter the organization of the cortical fibers, while treatment of the cells with phalloidin, which stabilizes the F-actin, protects the organellar movement from the action of CB. When decorated by a preparation of purified subfragment 1 (S1) from rabbit muscle myosin, the filament bundles appear fuzzy; in single filaments, the distinct arrowhead pattern characteristic of F-actin bound S1 is clearly visible.

The organelles may move along the cortical actin fibers in much the same way as proposed for characean algae (Williamson, 1975), but light apparently disrupts the cable assembly. Thus, the organelles are trapped in the irradiated region, where actin-myosin interaction is interrupted (Blatt et al., 1980).

With a new method, the vibrating electrode system, it has been found that locally illuminated areas of Vaucheria reveal an electrical outward current, which results from an efflux of protons (Blatt et al., 1981). As a result, a membrane hyperpolarization is generated. This proton efflux shows the same fluence-response dependency and the same spectral sensitivity as the chloroplast aggregation, and it slightly preceded the latter. Moreover, when the light is switched off, the former resting potential is re-established.

It seems reasonable to attribute the function of a transduction step to the light-induced proton efflux. Nevertheless, several problems remain unanswered. First, we have no knowledge of how light absorption in the blue-light photoreceptor affects proton transport through the cell membrane, although we may be aware that such effects are well known for other light-mediated responses (e.g. vectorial electron/proton transport in photosynthesis according to the Mitchell theory, or vectorial proton transport in light-harvesting Halobacterium cells). Second, the question of how actin is inactivated as a consequence of proton efflux has not yet been answered. It is suggested that this proton efflux begins a redistribution of other ions; in particular, a change in local calcium concentration has been discussed as the missing link (Blatt et al., 1981). Since calcium is known to control actin-myosin interaction, it seems reasonable to assume that this ion is an integral part of the transduction mechanism.

Finally, the most advanced mechanism of chloroplast reorientation, permanently independent of cytoplasmatic streaming, is realized in Mougeotia. The cell of this organism has been shown to orient its chloroplast in a tetrapolar gradient of phytochrome. This seems true both for the low and the high irradiance response (Haupt and Schönbohm, 1970). After phytochrome phototransformation in the cell has reached a certain threshold value, e.g., in response to a light pulse even as short as microseconds (Haupt, 1982), light is no longer needed, and the chloroplast reorients in darkness in a low irradiance response. Thus, the gradient of the metastable active form of phytochrome (P_{fr}) seems to enable the cell to memorize the direction of light. Indeed, persistence of this memory may be judged from inhibitor experiments with CB. When movement is blocked by CB over

a certain period of time, the information proper of chloroplast reorientation is lost very slowly by the cell. The time constant of
this first-order reaction of $t/2 \approx 90$ min is comparable to the half-
life of dark reversion or of dark destruction of P_{fr} in other organisms (Wagner and Klein, 1981).

These experiments with CB (Haupt and Wagner, 1984) are also
consistent with the inhibitory action of p-chloromercuribenzoate
(Schönbohm, 1980) and of N-ethylmaleimide (Haupt and Wagner, 1984).
Altogether they indicate the involvement of actomyosin in chloroplast movement in Mougeotia. Furthermore, filamentous structures
of 5 - 10 nm in diameter, running between the chloroplast edge and
the plasmalemma, have been shown by transmission electron microscopy
(Wagner and Klein, 1978). Additionally actin was identified by
HMM decoration in a cell homogenate (Marchant, 1976) and in spread
protoplasts of Mougeotia (Klein et al., 1980). Finally, fluorescent
phallotoxin heavily stains the chloroplast edge merging in the
cortical cytoplasm (Haupt and Wagner, 1984).

There apparently is little doubt as to the role of actin and
the recently isolated myosin (Doris Altmüller, unpublished results)
in the movement of Mougeotia chloroplasts. The most exciting question, however, remains of the molecular mechanism of actin-myosin
regulation by phytochrome and interaction with membranes.

Chloroplast movement in Mougeotia decreases much in parallel to
the concentration of an H_2O-insoluble fraction of intracellular
calcium, which can be manipulated by growing the cells in calcium-
free or in calcium-containing medium (Wagner and Klein, 1978).

The electron microprobe and vital fluorescence staining have
demonstrated that significant amounts of calcium are restricted to
particular membrane vesicles (Wagner and Rossbacher, 1980). These
organelles are most abundant in the cytoplasm close to the chloroplast edge, where the plastid merges into the cortical cytoplasm.
In addition, this is precisely the region in which actin filaments
have been demonstrated to be present.

Thus, a hypothetical photosensory reaction network has been
proposed, consisting of phytochrome-regulated release of calcium
from vesicular calcium stores, calcium-activation of actin microfilaments and transformation of the tetrapolar phytochrome gradient
into a gradient of membrane anchorage sites for actin (Wagner and
Klein, 1981). In this network, three steps have to be considered
in more detail: phytochromic control of calcium transport through
membranes, the effect of calcium on cytoplasmic actin microfilaments
and the phytochrome control of membrane anchorage sites for actin.

Phytochromic control of calcium transport has been tested in

Mougeotia by Dreyer and Weisenseel (1979), who measured the uptake of ^{45}Ca from the medium after different irradiations. Red-light pretreated cells accumulate much more radioactivity than do unir-radiated cells, but this increase is not observed when red light is followed by far-red light. On the other hand, Roux and coworkers (1981) report on a red/far-red reversible calcium efflux from Mougeotia under comparable conditions. Assuming a Ca^{2+}-regulated Ca^{2+} export pump, Roux (1984) suggested a model which could reconcile the data from both laboratories as follows: Red light promoted Ca^{2+}-uptake into Mougeotia may rapidly lead to the activation of the discussed Ca^{2+} efflux pump. The active efflux of Ca^{2+} should occur at a rate faster than the rate of passive uptake stimulated prior to red light; thus, the net flux of Ca^{2+} would be efflux. The con-clusion of this hypothetical scenario would be that the autoradio-graphic method of Dreyer and Weisenseel (1979) would show that red light promotes Ca^{2+} uptake, whereas the murexide method of Hale and Roux (1980) would show that the red light response was net efflux of Ca^{2+}.

Even accepting these arguments in light of new data from the same laboratory (Serlin and Roux, 1984), the general questions re-mains whether or not red/far red reversibility of a photoenergetic process such as calcium export (Wagner and Bellini, 1976) is suf-ficient criterion to prove involvement of phytochrome.

Montavon and coworkers (1983) have only recently started a promising action spectroscopic study to look for phytochrome-specific effects on membrane potential and ion movements clearly different from chlorophyll effects; no final answer, however, can be given as yet. Thus, primary mechanism of calcium redistribution in Mougeotia upon induction of chloroplast reorientational response remains un-known so far.

Membrane-coated calcium vesicles are abundant close to the chloroplast edge where actin filaments have been demonstrated to be present. As revealed by fluorescence microscopy in the presence of the in vivo calcium stain chlorotetracycline, the count of calcium vesicles relative to cell length sharply decreases upon cell division possibly for wall synthesis, with reappearance only during the late G-1 phase. This is the period of time when chloroplast movement is diminished. A similar correlation is found in extensively growing cells, reaching cell lengths up to 300 μm, when vesicle restoration is not enabled to keep pace with the extension growth rate. Evidence in favor of the functional relevance of these vesicles in chloroplast reorientational movement seems to come from the vesicular aggregation observed during the lag phase preceding the light-induced chloroplast reorientation response (Fig. 1). These unpublished data (Rogler, 1980) gained from light microscopic observations in our laboratory are in support of data of Klein (1981) after time-lapse electron microscopy (see also Wagner and Klein, 1981).

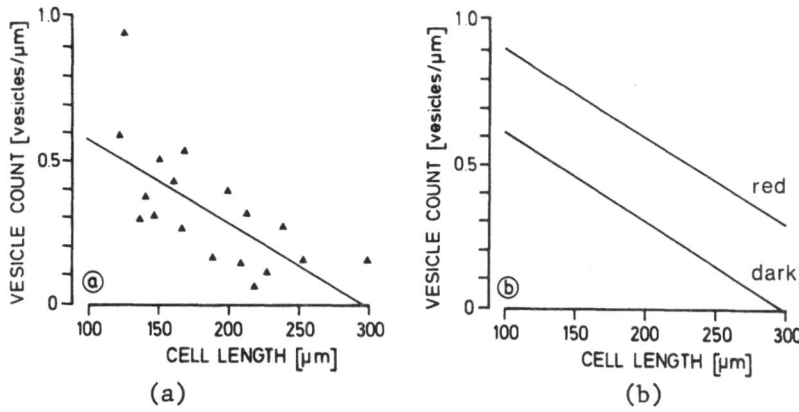

Fig. 1.   Normalized number of calcium-vesicles in Mougeotia (vesicle
          count) as a function of cell length. Mougeotia cells were
          preincubated with 400-500 µmol l^{-1} chlorotetracycline in
          MXS medium for 5-10 min, and fluorescent micrographs were
          taken as given by Wagner and Rossbacher (1980).
          (a) For each determination, the number of fluorescent ves-
          icles along the chloroplast in edge on position was counted,
          the cell length measured and the normalized vesicle count
          (vesicles/µm) indicated with the graph as a function of
          given cell length. Cell length probably is a measure of
          cell age. The data, therefore, indicate a decrease in the
          measured calcium-vesicular density during cell growth.
          (b) Calcium vesicular density along the Mougeotia chloroplast
          was increased after induction of chloroplast reorientational
          response. The number of fluorescent vesicles along the
          chloroplast in edge on position was determined 1-5 min after
          induction of chloroplast reorientational response (red) and
          in darkness as in Fig. 1a (dark); see also legend of Fig. 2.
          The regression lines run in parallel indicating an appreci-
          able increase in chloroplast calcium vesicular density soon
          after the red light pulse (unpublished results).

     Chloroplast reorientational response allowed to proceed to its
maximal value after a light pulse in darkness (Fig. 2) is only
slightly diminished in the presence of LaCL$_3$ (Fig. 3) in concentrations up to disintegration of the Ca^{2+}-pectinate middle lamella
interconnecting the Mougeotia cells to filaments. Under these conditions, La^{3+} certainly has penetrated unstirred layers of the cell
wall (Doughty and Diehn, 1982; Saunders and Hepler, 1983), to reach
the plasmalemma and possibly block transmembrane calcium fluxes
(Wagner and Bellini, 1976).

     This indirect evidence, taken together with earlier findings of
a decline of the reorientational response much in parallel to an

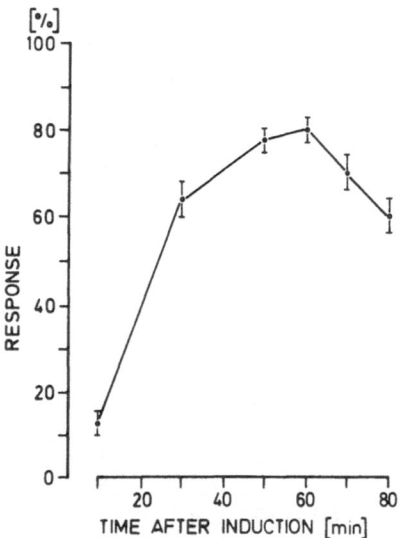

Fig. 2. Reorientational response of the Mougeotia chloroplast as a function of time after red light induction. Preoriented chloroplasts were induced to perform low irradiance response by a pulse of red light equivalent to a quantum flux of 8.5 mmol m^{-2} x s^{-1} gained from a double interference line pass filter with λ_{max} = 683 nm (Haupt, 1970). Percentage of chloroplast response at room temperature in darkness was evaluated as given by Wagner et al. (1984) (unpublished results).

intracellular fraction of bound calcium over a period of up to 7 days in daily exchanged calcium-deficient medium (Wagner and Klein, 1978) renders the Mougeotia intracellular calcium pools as a probable calcium source to activate the Mougeotia actomyosin. Such an acto-myosin activation could lead to a fastening of the chloroplast with-standing centrifugal forces as observed by Schönbohm (1973) in red-irradiated Mougeotia cells. Accordingly, loosening of the chloro-plast after far-red irradiation would be explained by a depletion of cytoplasmic calcium (Wagner and Bellini, 1976). However, for the chloroplast reorientational movement to be precise within the cylin-drical cells, plasmalemma anchorage sites for actin reflecting the tetrapolar gradient of phytochrome (Fig. 4) have been proposed to exist (Wagner and Klein, 1981).

Phytochrome upon phototransformation into its physiologically active form of Pfr may release a proton (Rüdiger, 1983; Song, 1983). This acidification close to the cell membrane (and possibly other intracellular membranes) could either open a calcium channel in the Mougeotia plasmalemma as concluded from the findings in Weisenseel's laboratory. Alternatively, P_{fr} may cause a shift in the binding

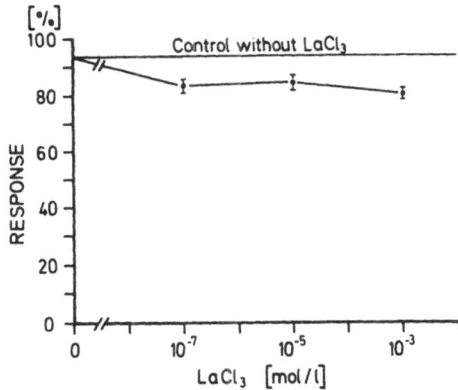

Fig. 3.   Reorientational response of the <u>Mougeotia</u> chloroplast in
the presence of LaCl₃. <u>Mougeotia</u> cells were grown for two
days in Na-MXS medium (Wagner and Bellini, 1976) and kept
in darkness 12 h before the start of the experiment. Cells
were then preincubated 60 min before red light induction
with given LaCl₃ concentrations in modified Na-MXS medium,
deficient of Ca^{2+} (Wagner and Rossbacher, 1980) and the
anions being exchanged for chloride or nitrate. The percentage of the chloroplast reorientational response in the
absence (control) and in the presence of La^{3+} was evaluated
60 min after red light induction as given in the legend of
Fig. 2 (unpublished results).

equilibrium of receptor-bound Ca^{2+} ions in internal sequestering
vesicles as supposed by Rossbacher (1980). The favorite candidates
certainly would be membrane-coated calcium vesicles in the <u>Mougeotia</u>
cell detected as in <u>Mesotaenium</u> (unpublished results) by X-ray microanalysis and by the <u>in vivo</u> calcium stain chlorotetracycline
(Rossbacher and Wagner, 1980; Rossbacher et al., 1984).

   Thus, the sequence of the photosensory reorientational chloroplast response in <u>Mougeotia</u> established so far includes phytochrome,
calcium, actin and myosin. These data found its significant extension
by the recent identification of calmodulin (Wagner et al., 1984) and
a calcium-regulated myosin light-chain kinase in <u>Mougeotia</u> (Doris
Altmüller, unpublished results; see also Altmüller et al., 1984).

   Consistently, the calmodulin-inhibitors trifluoperazine and W-7
inhibit the reorientational response of the <u>Mougeotia</u> chloroplast
(Wagner et al., 1984); the data was recently confirmed by Serlin and
Roux (1984). Thus, the hypothetical network of events (Fig. 4) could
lead from light-absorption via phytochrome to calcium as the signal
and then to calmodulin which upon calcium activation regulates myosin
light-chain kinase and interaction of myosin with anchored actin in
this model organism of chloroplast photo-orientational response.

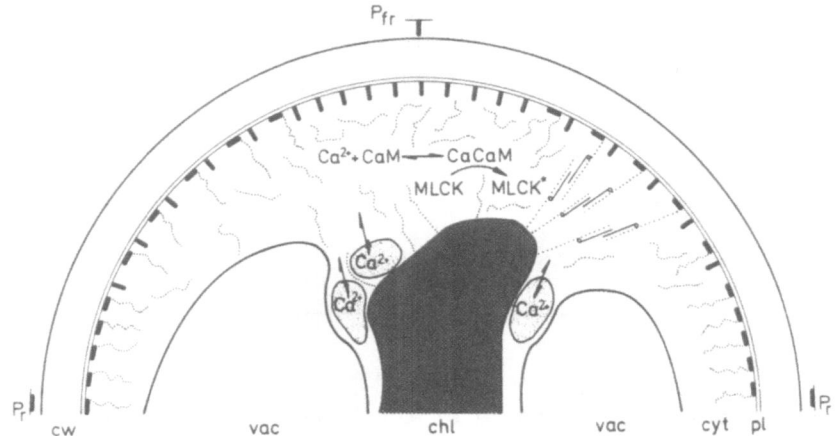

Fig. 4.   Scheme of the upper half of a Mougeotia cell in cross
section. Phytochrome in its state as Pfr ($\mathord{\uparrow}\mathord{\uparrow}$) and Pr
($\mid$, $\mid$), respectively, is proposed to control plasmalemma
(pl) anchorage sites for actin and thus reorientational
movement of the chloroplast after actomyosin activation
via Ca^{2+}, Ca^{2+}-calmodulin (CaCaM) and myosin light-chain
kinase (MLCK). Headed bars symbolize myosin molecules,
the dotted lines actin filaments. The chloroplast (chl)
is shown as turning away from high concentrations of Pfr
to seek high concentrations of Pfr. Further details are:
cw = cell wall; vac = vacuole; cyt = cytoplasm. (After
Wagner and Klein, 1981.)

ACKNOWLEDGEMENT

The work in the authors' laboratory has been supported finan-
cially by the Deutsche Forschungsgemeinschaft.

REFERENCES

Alberts, B., Bray, D., Lewis, J., Raff, M., Roberts, K., and Watson,
       J. D., 1983, "Molecular Biology of the Cell," Garland, New York.
Altmüller, D., Grolig, F., and Wagner, G., 1984, Calcium sequestra-
       tion and protein phosphorylation in the green alga Mougeotia
       sp., Eur. J. Cell Biol., Suppl., 5:3.
Blatt, M. R., Weisenseel, M. H., and Haupt, W., 1981, A light-de-
       pendent current associated with chloroplast aggregation in the
       alga Vaucheria sessilis, Planta, 152:513.
Blatt, M. R., Wessells, N. K., and Briggs, W. R., 1980, Actin and
       cortical fiber reticulation in the siphonaceous alga Vaucheria
       sessilis, Planta, 147:363.

Britz, S. J., 1979, Chloroplast and nuclear migration, in: "Encyclo-
    pedia of Plant Physiology," New Ser., Vol. 7, W. Haupt and
    M. E. Feinleib, eds., Springer, Berlin, pp. 170-205.
Carter, S. B., 1967, Effects of cytochalasins on mammalian cells,
    Nature, 213:261.
Cheung, W. Y., 1980, Calmodulin plays a pivotal role in cellular
    regulation, Science, 207:19.
Cormier, M. J., Charbonneau, H., and Jarrett, H. W., 1981, Plant
    and fungal calmodulin: Ca^{2+}-dependent regulation of plant
    NAD kinase, Cell Calcium, 2:313.
Corti, B., 1774, Osservationi microscopiche sulla tremelle e sulla
    circulazione del fluido in una pianata aquajuola, Lucca.
Corti, B., 1776, Sur la circulation d'un fluide, de'couverti en
    diverses plantes, Rosier obs. sur la Physique, sur l'Histoire
    Nat., 8:232.
Doughty, M. J., and Diehn, B., 1982, Photosensory transduction in
    the flagellated alga, Euglena gracilis, Biochim. Biophys. Acta,
    682:32.
Dreizen, P., and Gershman, L. C., 1970, Molecular basis of muscular
    contraction: Myosin, Transact. NY Acad. Sci., Ser. 11, Vol.
    32:170.
Dreyer, E. M., and Weisenseel, M. H., 1979, Phytochrome-mediated
    uptake of calcium in Mougeotia cells, Planta, 146:31.
Filner, Ph., and Yadav, N. S., 1979, Role of microtubules in intra-
    cellular movements, in: "Encyclopedia of Plant Physiology,"
    New Ser., Vol. 7, W. Haupt and M. E. Feinleib, eds., Springer,
    Berlin, pp. 95-113.
Goldman, R. D., 1975, The use of heavy meromyosin binding as an
    ultrastructural cytochemical method for localizing and determin-
    ing the possible functions of actin-like microfilaments in
    nonmuscle cells, J. Histochem. Cytochem., 23:529.
Hale, II, C. C., and Roux, S. J., 1980, Photoreversible calcium
    fluxes induced by phytochrome in oat coleoptile cells, Plant
    Physiol., 65:658.
Haupt, W., 1970, Hellrot- und Dunkelrot-Wechselwirkungen bei der
    Chloroplastendrehung von Mougeotia, Wiss. Zeitschrift Ernst-
    Moritz-Arndt-Universität Greifswald, 19:47.
Haupt, W., 1982, Light-mediated movement of chloroplasts, Annu. Rev.
    Plant Physiol., 33:205.
Haupt, W., 1983, Movement of chloroplasts under the control of light,
    Prog. Phycol. Res., 2:228.
Haupt, W., and Schönbohm, E., 1970, Light-oriented chloroplast move-
    ments, in: "Photobiology of Microorganisms," P. Halldal, ed.,
    Wiley-Interscience, London, pp. 283-307.
Haupt, W., and Wagner, G., 1984, Chloroplast movement, in: "Membranes
    and Sensory Transduction," G. Colombetti and F. Lenci, eds.,
    Plenum, New York, pp. 331-375.
Hope, A. B., and Walker, N. A., 1975, "The Physiology of Giant Algal
    Cells," University Press, Cambridge.

Huxley, H. E., 1963, Electron microscope studies on the structure of natural and synthetic protein filaments from striated muscle, J. Mol. Biol., 7:281.

Isenberg, G., and Wohlfarth-Bottermann, K. E., 1976, Transformation of cytoplasmic actin. Importance for the organization of the contractile gel reticulum and the contraction-relaxation cycle of cytoplasmic actomyosin, Cell. Tiss. Res., 173:495.

Ishikawa, H., Bishoff, R., and Holtzer, H., 1969, Formation of arrowhead complexes with heavy meromyosin in a variety of cell types, J. Cell Biol., 43:312.

Klein, K., 1981, Feinstrukturelle Untersuchungen zur Bewegung des Mougeotia-Chloroplasten, Ph.D. Thesis, Univ. Erlangen-Nürnberg.

Klein, K., Wagner, G., and Blatt, M. R., 1980, Heavy-meromyosin-decoration of microfilaments from Mougeotia protoplasts, Planta, 150:354.

Korn, E. D., 1982, Actin polymerization and its regulation by proteins from nonmuscle cells, Physiol. Rev., 62:672.

Kuroda, K., and Manabe, E., 1983, Microtubule-associated cytoplasmatic streaming in Caulerpa, Proc. Jpn. Acad., 59:131.

Marchant, H. J., 1976, Actin in the green algae Coleochaete and Mougeotia, Planta, 131:119.

Marmé, D., and Dieter, P., 1983, Role of Ca^{2+} and calmodulin in plants, in: "Calcium and Cell Function," Vol. 4, W. Y. Cheung, ed., Academic Press, New York.

Montavon, M., Horwitz, B. A., and Greppin, H., 1983, Far-red light-induced changes in intracellular potentials of spinach mesophyll cells, Plant Physiol., 73:671.

Nagai, R., and Hayama, T., 1979, Ultrastructure of the endoplasmic factor responsible for cytoplasmic streaming in Chara internodal cells, J. Cell Sci., 36:121.

Penningroth, S. M., Cheung, A., Bonchard, Ph., Gagnon, C., and Bardin, C. W., 1982, Dynein ATPase is inhibited selectively in vitro by erythro-9-[3-2(hydroxynonyl)] adenine, Biochem. Biophys. Res. Commun., 104:234.

Rogler, S., 1980, Fluoreszenzmikroskopische Untersuchungen über die Calcium-Verteilung in Mougeotia, Diploma, Univ. Erlanger-Nürnberg.

Rossbacher, R., 1980, Röntgenmikroanalytische Untersuchungen zur Kompartimentierung von Calcium und anderer Ionen bei der Grünalge Mougeotia spec., Diploma, Univ. Erlangen-Nürnberg.

Rossbacher, R., Wagner, G., and Pallaghy, Ch. K., 1984, X-ray microanalysis of calcium in fixed and in shock-frozen hydrated green algal cells: Mougeotia, Spirogyra and Zygnema, Nucl. Instr. Meth. Phys. Res., B3:664.

Roux, S. J., 1984, Ca^{2+} and phytochrome action in plants, BioScience, 34:25.

Roux, S. J., McEntire, K., Slocum, R. D., Cedel, T. E., and Hale, C. C., 1981, Phytochrome induces photoreversible calcium fluxes in a purified mitochondrial fraction from oats, Proc. Natl. Acad. Sci. USA, 78:283.

Rüdiger, W., 1983, Chemistry of the phytochrome photoconversion, Phil. Trans. R. Soc. London, B303:377.

Sabnis, D. D., and Jacobs, W. P., 1967, Microtubules in the Coenocytic marine alga, Caulerpa prolifera, J. Cell Sci., 2:465.

Saunders, M. J., and Hepler, P. K., 1983, Calcium antagonists and calmodulin inhibitors block cytokinin-induced bud formation in Funaria, Developmental Biol., 99:41.

Schönbohm, E., 1973, Die lichtinduzierte Verankerung der Plastiden in cytoplasmatischem Wandbelag: Eine phytochromgesteuerte Kurzzeitreaktion, Ber. Deutsch. Bot. Ges., 83:423.

Schönbohm, E., 1980, Phytochrome and non-phytochrome dependent blue light effects in intracellular movements in fresh water algae, in: "The Blue Light Syndrome," H. Senger, ed., Springer, Berlin.

Seitz, K., 1979, Cytoplasmic streaming and cyclosis of chloroplasts, in: "Encyclopedia of Plant Physiology," New Ser., Vol. 7, W. Haupt and M. E. Feinleib, eds., Springer, Berlin, pp. 150-169.

Seitz, K., 1982, Chloroplast motion in response to light in aquatic vascular plants, in: "Studies on Aquatic Plants," J. J. Symoens, S. S. Hooper, and P. Compère, eds., Roy. Soc. Belg., Brüssel.

Serlin, B. S., and Roux, S. J., 1984, Modulation of chloroplast movement in the green alga Mougeotia by the Ca^{2+}-ionophore, A 23187, and by calmodulin antagonists, Proc. Natl. Acad. Sci. USA, in press.

Song, P.-S., 1983, Protozoan and related photoreceptors: Molecular aspects, Annu. Rev. Biophys. Bioeng., 12:35.

Stockem, W., Weber, K., and Wehland, J., 1978, The influence of micro-injected phalloidin on locomotion, protoplasmic streaming and cytoplasmic organization in Amoeba proteus and Physarum polycephalum, Cytobiologie, 18:114.

Wagner, G., 1979, Actomyosin as a basic mechanism of movement in animals and plants, in: "Encyclopedia of Plant Physiology," New Ser., Vol. 7, W. Haupt and M.E. Feinleib, eds., Springer, Berlin, pp. 114-126.

Wagner, G., and Bellini, E., 1976, Light-dependent fluxes and compartmentation of calcium in the green alga Mougeotia, Z. Pflanzenphysiol., 79:283.

Wagner, G., and Klein, K., 1978, Differential effect of calcium on chloroplast movement in Mougeotia, Photochem. Photobiol., 27:137.

Wagner, G., and Klein, K., 1981, Mechanism of chloroplast movement in Mougeotia, Protoplasma, 190:169.

Wagner, G., and Rossbacher, R., 1980, X-ray microanalysis and chlorotetracycline staining of calcium vesicles in the green alga Mougeotia, Planta, 149:298.

Wagner, G., Valentin, P., Dieter, P., and Marmé, D., 1984, Identification of calmodulin in the green alga Mougeotia and its possible function in chloroplast reorientational movement, Planta, 162:62.

Wehland, J., Osborn, M., and Weber, K., 1980, Phalloidin associates with microfilaments after microinjection into tissue culture cells, Eur. J. Cell. Biol., 21:188.

Wieland, Th., 1977, Modifications of actins by phallotoxins, Naturwissenschaften, 64:303.

Williamson, R., 1975, Cytoplasmic streaming in Chara: A model activated by ATP and inhibited by cytochalasin B, J. Cell Sci., 17:655.

Williamson, R. E., and Ashley, C. C., 1982, Free Ca^{2+} and cytoplasmic streaming in the alga Chara, Nature, 296:1.

TEMPERATURE SENSING IN MICROORGANISMS

Kenneth L. Poff

Michigan State University
United States Department of Energy
Plant Research Laboratory
Michigan State University
East Lansing, MI 48824

INTRODUCTION

Perhaps the most common aspect of an organism's environment is heat or temperature. Certainly every living organism responds in some way to temperature if only through thermal effects on enzymatically regulated reactions. In addition, many organisms are known which sense their thermal environment (gain some particular piece of information concerning their thermal environment). A few of the organisms in the latter category have received attention recently. Special problems are encountered in a study of thermosensing, and these may account for the relatively slow progress in this area. This discussion will emphasize those problems, and then briefly cover examples of thermosensing in several specific organisms. The emphasis will be on the evidence for specificity - that a specific thermoreceptor molecule is perturbed by heat energy and interacts with a specific transduction pathway, resulting in a specific behavioral response.

Some of the difficulties encountered in a study of thermosensing can best be described by comparing a hypothetical thermosensory transduction sequence with the transduction sequence for another stimulus. Photosensing will be used for comparison, although any one of several different sensory transduction sequences could have been used.

In photosensing, light energy is absorbed by a specific pigment and transduced into chemical energy. The activated pigment initiates or participates in a reaction sequence culminating in some observable, typically behavioral response. A particular pigment which is

one out of thousands of compounds in the cell or organism absorbs
the light. Thus, specificity is achieved at least in part by the
absorption step. One can assume negligible activation of the
photoreceptor pigment in darkness which is relatively easy to impose on the organism. One also can assume an increased amount of
pigment activation (and therefore, an increased biological response)
with an increasing number of absorbed quanta. Thus, a fluence
response relationship can be measured and this has a solid basis
for interpretation. One can calculate an action spectrum from
stimulus response curves measured at different wavelengths, and,
given a series of assumptions, know that the action spectrum is
similar to the absorption spectrum for the photoreceptor pigment.
Finally, the interaction of a photoreceptor pigment with light can
be directly examined by means of absorption spectroscopy and fluorescence spectroscopy. Frequently, such an examination can even
be accomplished in vivo if the absorption is at a wavelength where
few other compounds absorb.

In thermosensing, heat energy "affects" every molecule in the
organism. Any specificity must be provided by the fact that only
one "sensor" is connected to the transduction sequence. In the
absence of specificity of the initial interaction with the stimulus,
it is not possible to identify the sensor based on its alteration
by the stimulus; any or all molecules might be altered. If the
temperature (heat energy) itself is being measured, one cannot
impose the condition of zero stimulus which would be the absence
of heat energy or a temperature of absolute zero under any meaningful physiological conditions. However, one can impose the absence
of a temperature gradient which is fortunate since a temperature
gradient is the usual parameter measured by a thermosensing micro-
organism. For a photosensing system, only the number of quanta
absorbed is significant, irrespective of the time involved. No
similar concept has yet been developed for a thermosensing system.
If a similar relationship were to apply in a thermosensing system,
a large stimulus (e.g., a steep temperature gradient) applied for
a short time would be equivalent to a small stimulus (shallow
gradient) over a long time. That is, a constant dT/dX times time
would induce a constant response.

If an organism detects a temperature gradient, stimulus response
curves may be measured at different absolute temperatures. Given a
series of assumptions similar to those for action spectroscopy (the
absence of other limiting steps), the physiological capability to
sense the gradient is a representation of the ability of the thermotransducer to change its state over that temperature range. Thus,
a temperature response curve should be similar to the temperature
response curve for the transducer itself.

Differential scanning calorimetry will directly measure the
interaction of a molecule with heat energy as a function of

temperature. However, since all compounds in the organism inherently interact with the applied heat energy, this technique as a probe for a thermosensor would be of marginal value.

As a partial result of these difficulties, research has centered on a description of the phenomena. The behavioral response is typically one of altered movement, and is assumed to involve the same motor response network as movement responses to other stimuli. Because of this, the goal with each biological system seems to have been the identification of the thermoreceptor or "biothermometer", and integration of thermoreception and other sensory responses for that organism.

BIOLOGICAL EXAMPLES

Bacteria

The bacterium, _Escherichia coli_, if grown at one particular temperature, when exposed to a temperature gradient, will accumulate at that growth temperature (Maeda et al., 1976). This accumulation involves negative thermotaxis at higher temperatures and positive thermotaxis at lower temperatures, resulting in an "optimum-seeking" system. Tumbling frequency increases for a short period of time if the cells are exposed to a decreasing temperature below the growth temperature, but decreases for a comparable period if the cells are exposed to an increasing temperature. These responses would result in smooth swimming toward the growth temperature but tumbling if in the opposite direction. Although comparable experiments have not been performed at temperatures above the growth temperature because of other thermal effects on swimming behavior, it seems clear that thermotaxis in this organism uses the same general mechanism as other sensory responses (Maeda et al., 1976; Miller and Koshland, 1977).

Thermoreception in _E. coli_ is closely tied to chemoreception. L-Serine is an inhibitor of thermotaxis (Maeda and Imae, 1979). Mutants blocked in the serine transduction pathway (Tsr) show little response to temperature while mutants blocked in the parallel L-aspartate pathway (Tar) exhibit normal thermotaxis. Based on these results and those with other mutants, Maeda and Imae (1979) have suggested that the high-affinity serine receptor is the thermosensor. Whether or not a temperature change affects the protein directly or _via_ a membrane lipid has not been determined. However, it is clear that the interaction between the serine receptor and a thermal stimulus must occur at or very close to the receptor site.

Ciliates

The protozoan, _Paramecium_, similar to _E. coli_, will accumulate at or near some optimum temperature in a temperature gradient, and this temperature can be shifted by shifting the growth temperature (Tawada and Oosawa, 1972). In addition, the cells exhibit an avoidance response at higher temperatures, swimming away from the potentially harmful temperatures (Hennessey and Nelson, 1979). Ciliary reversal, and therefore direction changes, increases as the cells swim away from this optimum temperature or given an avoidance response (Nakaoka and Oosawa, 1977).

Ciliary reversal, induced by membrane depolarization via an increase in the internal calcium concentration is defective in a class of mutants known as Pawns which are also defective in their avoidance response (Hennessey and Nelson, 1979). Other data are consistent with a model that lipids of the ciliary membrane serve as the thermotransducer for the avoidance response with a phase transition of the lipids surrounding the calcium channels causing a change in the calcium conductance and thereby a change in the internal calcium concentration. This causes membrane depolarization and ciliary reversal.

Cellular Slime Molds

Pseudoplasmodia of the cellular slime mold, Dictyostelium discoideum exhibit thermotaxis on a temperature gradient. At higher temperatures, the organisms move toward the warmer side of the gradient (positive thermotaxis), and at lower temperatures toward the cooler side of the gradient. Higher and lower are roughly defined as temperatures above and below the growth temperature of the amoebae from which the pseudoplasmodia were formed (Fig. 1). Thus, in contrast with most other microorganisms, this is an "optimum leaving" system.

Two obvious possibilities exist to explain the positive and negative thermotaxis: there may be one sensor measuring temperature and a switching mechanism changing the sign of the response at the transition temperature between the positive and negative responses; or there could be two sensors, one regulating positive thermotaxis and one regulating negative thermotaxis, with the response of the organism the sum of the two individual responses. The dependence of the transition temperature on the steepness of the gradient suggests that the second model is more accurate (Whitaker and Poff, 1980). In addition, at temperatures several degrees below the growth temperature, the sign of the thermotactic response depends on the strength of the temperature gradient (Fontana, 1982). At least near the transition temperature, the response of the organism appears to be a combination of the positive response and competing negative response. It would appear then that at least three thermosensors are present in this organism: one regulating positive thermotaxis, one regulating negative thermotaxis, and a third sensor detecting the growth temperature and regulating thermal adaptation.

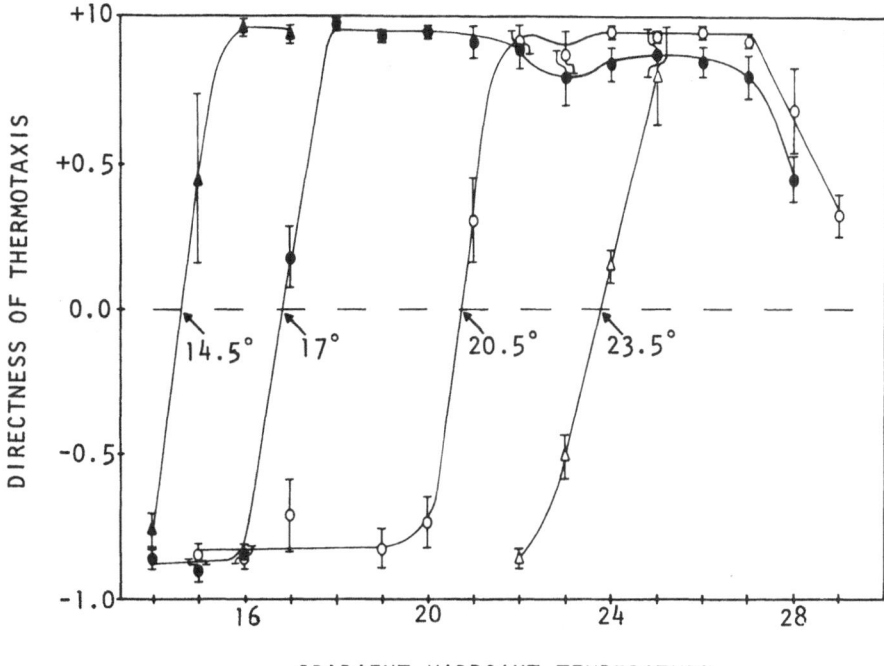

Fig. 1. Temperature dependence of thermotaxis on 0.22 C/cm
gradients for ▲, 16 C, ●, 18 C, O, 23.5 C, and △, 27.5 C
pseudoplasmodia. The vertical bars represent +/- 1 S. D.
(after Whitaker and Poff, 1980).

Whitaker and Poff (1980) have suggested that this third sensor
regulating thermal adaptation might be a temperature controlled
enzyme regulating fatty acid desaturation, and that a change in the
fluidity of the membrane lipids might be the mechanism for adaptation. However, Fontana et al. (1983) have found the fluidity to be
independent of the temperature at which the amoebae grew and developed. Clearly then, adaptation does not involve changes in bulk
lipid fluidity. However, this does not rule out the possibility
that adaptation results from a change in the fluidity of a specific,
minor membrane lipid, as for example, one surrounding the hypothetical positive and negative sensors.

Mutant strains of Dictyostelium have been isolated with altered
thermotactic responses (Schneider et al., 1982). One class of phenotypes shows a decreased response in both positive and negative
thermotaxis. A second class shows normal negative thermotaxis and
decreased positive thermotaxis. A third class shows only positive
thermotaxis over the entire temperature range for migration with

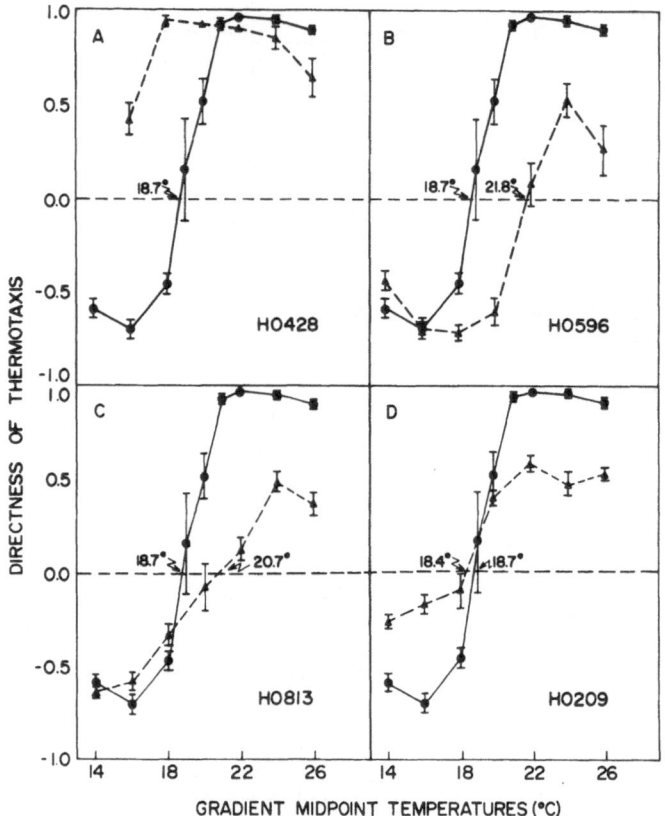

Fig. 2.   Temperature dependence of HL50 (wild type) and mutants of
thermotaxis on a gradient of 1 C across 9 cm. ●-●, HL50;
▲-▲, mutants as indicated on the figure. (after Schneider
et al., 1982).

no negative thermotaxis at any temperature (Fig. 2). These strains
support the idea that positive and negative thermotaxis have separable
pathways, but do not necessarily demonstrate that there are different
sensors.

Amoebae of Dictyostelium also exhibit thermotaxis on a temperature gradient. In fact, this amoebal thermotaxis appears to be the
basis for thermotaxis by the multicellular pseudoplasmodia based on:
the same temperature range, sensitivity, and adaptation for negative
and positive thermotaxis for both amoebae and for pseudoplasmodia.
In addition, mutant strains show the same alterations for both
amoebal and pseudoplasmodial thermotaxis (Hong et al., 1983). The
organizational complexities of the pseudoplasmodium can therefore
be ignored in trying to understand thermotaxis in Dictyostelium.

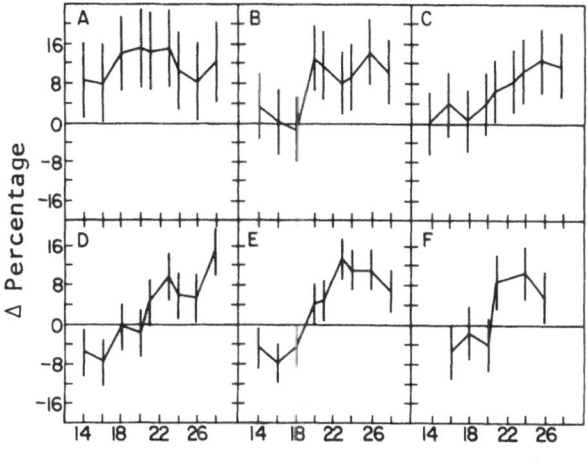

Fig. 3.   Temperature dependence for amoebae exposed for different
          times on the gradient. The exposure was A, 3 h; B, 6 h;
          C, 9 h; D, 12 h; E, 16 h; F, 20 h. (after Hong et al.,
          1983).

     Positive and negative thermotaxis have a different temporal
dependence in amoebae (Fig. 3), with negative thermotaxis dependent
on development of the amoebae (Hong et al., 1983). This is additional evidence for separable pathways for positive and negative
thermotaxis.

OVERVIEW

     Although it is extraordinarily difficult to establish speci-
ficity for a system such as thermotaxis, the term thermosensing
should be reserved for instances where such specificity exists --
a specific thermoreceptor molecule is perturbed by heat energy and
interacts with a specific transduction pathway resulting in a
specific behavioral response. Until we have isolated and identified
the biothermometer molecule itself, we will not be certain about the
specificity of the thermoreceptor. However, with that exception,
the overwhelming evidence for the biological systems discussed above
supports the conclusion that thermotaxis is indeed a specific sensory
response and not just a manifestation of very general thermal effects
on the organism.

SUMMARY

     Several of the problems in thermosensory physiology have been
presented, and three biological examples discussed. The evidence

supports a model for bacterial thermotaxis in which thermoreception occurs at or very close to the high-affinity serine receptor, and that subsequent transduction steps use the classical motor apparatus. In Paramecium, an increase in temperature is theorized to cause a change in phase of one or more lipids surrounding a calcium channel, causing an influx in calcium which would induce membrane depolarization and cause ciliary reversal. This reversal would result in a "tumble" and inhibited swimming in that direction. Thermotaxis by the amoebae of Dictyostelium is the basis for thermotaxis by the multicellular pseudoplasmodia. Several lines of evidence support the suggestion that the behavioral response is a combination of positive thermotaxis and the competing negative thermotaxis. In all three systems, at least one step at or close to the sensor has been shown to be mutable.

REFERENCES

Fontana, D. R., 1982, Thermotactic and phototactic responses of Dictyostelium discoideum pseudoplasmodia. Ph.D. thesis, Michigan State University, East Lansing, MI.

Fontana, D. R., and Poff, K. L., 1983, Effect of stimulus strength and adaptation on the thermotactic response of Dictyostelium discoideum pseudoplasmodia, Exp. Cell Res., 150:250-257.

Fontana, D. R., Poff, K. L., and Haug, A., 1983, Role of bulk lipid fluidity in the thermal adaptation of Dictyostelium discoideum thermotaxis, Exp. Mycol., 7:278-282.

Hennessey, T., and Nelson, D. L., 1979, Thermosensory behavior in Paramecium tetraurelia: A quantitative assay and some factors that influence thermal avoidance, J. Gen. Microbiol., 112:337-347.

Hong, C. B., Fontana, D. R., and Poff, K. L., 1983, Thermotaxis of Dictyostelium discoideum amoebae and its possible role in pseudoplasmodial thermotaxis, Proc. Natl. Acad. Sci. USA, 80:5646-5649.

Maeda, K., Imae, Y., and Oosawa, F., 1976, Effect of temperature on motility and chemotaxis of Escherichia coli, J. Bacteriol., 127:1039-1046.

Maeda, K., and Imae, Y., 1979, Thermosensory transduction in Escherichia coli: Inhibition of the thermoresponse by L-serine, Proc. Natl. Acad. Sci. USA, 76:91-95.

Miller, J. B., and Koshland, Jr., D. E., 1977, Membrane fluidity and chemotaxis: Effects of temperature and membrane lipid composition on the swimming behavior of Salmonella typhimurium and Escherichia coli, J. Mol. Biol., 111:183-201.

Nakaoka, Y., and Oosawa, F., 1977, Temperature sensitive behavior of Paramecium caudatum., J. Protozool., 24:575-580.

Schneider, M. J., Fontana, D. R., and Poff, K. L., 1982, Mutants of thermotaxis in Dictyostelium discoideum, Exp. Cell Res., 140:411-416.

Tawada, K., and Oosawa, F., 1972, Responses of _Paramecium_ to
    temperature change, _J. Protozool._, 19:53-57.
Whitaker, B. D., and Poff, K. L., 1980, Thermal adaptation of
    thermosensing and negative thermotaxis in _Dictyostelium_,
    Exp. Cell Res., 128:87-93.

ENZYMATIC AMPLIFICATION MECHANISM OF VISUAL TRANSDUCTION SIGNAL IN RETINAL RODS

Marc Chabre

Laboratoire de Biologie Moléculaire et Cellulaire
ER CNRS 199
D.R.F. - C.E.N.G., 85 X
38041, Grenoble Cedex
France

I. INTRODUCTION: PHOTOSENSORY PIGMENTS VERSUS PHOTOENERGETIC PIGMENTS

The name "rhodopsin" is currently used for two classes of retinal proteins: the visual pigments of higher organisms, whose archetype is the rhodopsin of vertebrate retinal rod outer segments, and the retinal pigments of halophilic bacteria, whose archetype is the bacteriorhodopsin of <u>Halobacterium</u> <u>rubrum</u>. Much is known on the structure and function of these two proteins, and apparent similarities of their three dimensional structures have often led to speculation of a possible common origin and eventual functional analogies. Both pigment molecules are intrinsic membrane proteins whose hydrophobic core is constituted by a 7 α helices transmembrane barrel. There are, however, no significant analogies between their primary sequences, and their functions are totally different. Bacteriorhodopsin is a photoactivable ion pump, and rhodopsin a receptor molecule, which, in its photoexcited state, catalyses the activation of a biochemical cascade. Indeed, the two proteins may have evolved independently and converged towards a similar structure defined mainly by their intrinsic membrane protein character and their identical chromophore. The α helix is the predominant structure for the intrinsic region of membrane proteins, and helices usually cross membranes perpendicularly to their surface plane. For light catching proteins, there is an advantage to have the chromophore oriented parallel to the membrane plane. In contrast to "antenna"-type pigment molecules which are appropriately oriented in the polypeptide

structure and either re-emits photons or transfer excitons without
strong interaction with the protein, the chromophore in photosensory
and transducing receptors strongly interacts with the protein. The
initial photoexcitation has to induce a conformational change in the
protein. This, in the case of the sensory pigment, will create a
protein state which will store the information and allow its amplification by coupling to an enzymatic cascade. In the case of a
photoenergetic pigment (e.g. bacteriorhodopsin), the conformational
change itself generates the ion pumping activity. Therefore, in
both cases, the chromophore has to be totally caged within the protein structure. Considering the size of the retinal molecule and
its required orientation, 7 helices could simply be the minimum
number needed to encage it within the protein.

For the kinetics of the photocycles, there is a basic difference
of requirement between a photoenergetic receptor and a sensory receptor. In the former case, the efficiency is directly related to
the speed of the photocycle. Once the initial photoexcitation has
triggered the conformation change related to the vectorial transport
of protons or ions, the faster the relaxation, the sooner the regen-
erated pigment is able to activate a new transport and the higher
its efficiency. This is not true for a sensory receptor. There,
the goal is not to catch the maximum number of photons and convert
them in usable energy, but to store the information of a photon
capture and to generate (at the expense of energy) a maximally amplified signal with a minimum of dark noise. In the vertebrate visual
system, for example, the structural change induced by the photoisomerization of retinal converts rhodopsin into a long-lived active "R*"
state which is able to catalyse many biochemical cycles. The amplification, and therefore the sensitivity of the response, depends
directly on the lifetime of this active state. Indeed, the optimum
lifetime is the longest compatible with the time resolution needed
for the sensory response. For the visual signal in the retinal rod,
the time resolution is only of the order of a few tenths of a s. It
certainly does not need to be faster for the phototactic responses
of bacteria or other primitive organisms. If, for example, the time
resolution has to be of the order of one s, the same signal may be
generated by one photoreceptor molecule remaining active 1 s as well
as by 100 remaining active only 10 ms each. Therefore, one expects
such a sensory receptor to have a slow photocycle due to the existence
of a long-lived "photoactivated" protein state.

It is interesting in this context to realise now that the fast
photoenergetic ion pumping cycle of bacteriorhodopsin has no sensory
function, but that there exists also in halobacterium "slow" rhodopsins which may mediate, through a still unknown mechanism, the phototactic response (see the contribution by Hildebrand and by Spudich
in this volume). These "slow" pigments may have more analogies with
visual rhodopsin than the ion pumping one.

II. THE RETINAL ROD OUTER SEGMENT: A REMARKABLE "MODEL" SYSTEM.
     THE PHOTOINDUCED ENZYMATIC CASCADE CONTROLLING cGMP HYDROLYSIS

Presently, much is known about the biochemical cascade generated
in retinal rod outer segments by photoexcited rhodopsin (R^*), through
which one R^* molecule may control the hydrolysis of up to 10^5 molecules of guanine cyclic nucleotide (cGMP). It is a fast intracellular
process which precedes the electrophysiological response of the cell;
indeed, its connection with this response is not yet elucidated.
The mechanism of the amplifying cascade is identical to that used
by many membrane receptor molecules to mediate, in the cell, adenylate
cyclase activities responsible for a large variety of intracellular
responses. Some of the intermediate proteins are identical and inter-
changeable between the visual system and these other systems. This
cyclic nucleotide machinery certainly existed in simple organisms
prior to the evolution of the visual system. It is plausible that
the generation in the visual system of an electrophysiological and
neuronal response is but a recent adaptation in higher organisms of
an ancient photosensory signal already mediated by cyclic nucleotides
in more primitive organisms. But surprisingly, the molecular mechanism of this first stage of phototransduction might be much easier
to study in the more elaborate visual receptors of the vertebrates
than in primitive organisms.

In particular, the retinal rod cell presents many advantages
for molecular studies. The differentiation and morphological separation of the photosensitive machinery in a large outer segment
easily separated from the genetic and metabolic machinery in the
inner segment, added to the fact that rod outer segments (ROS) represent up to 95% of the photoreceptor mass in many vertebrate retinas,
makes ROS an ideal material for extraction and purification of proteins involved in phototransduction. Rhodopsin was long considered
the only protein significant for visual transduction. This was influenced by oversimplified ideas on the phototransduction mechanism.
It is now realized that rhodopsin is a photoactivable enzyme which
interacts with many other proteins on the disc membrane and initiates
an enzymatic amplification cascade. A recent major achievement has
been the detailed analysis of the mechanisms of this amplification.
If rhodopsin is by far not the only protein in the rod, it remains
true that the few identified proteins involved with rhodopsin in the
amplification and regulation of the photoinduced reactions constitute
the near totality of the rod protein content.

The compact arrangement of the internal membrane vesicles, the
discs, in the rods has three major implications: (1) The separation
of the discs, which contain the photopigment, from the rod cell
membrane in which the first electrophysiological signal is generated,
imposed the concept of an internal soluble transmitter. (2) The
highly oriented and quasi crystalline stacking of the discs over the

whole ROS volume gives the isolated organelle physical properties characteristic of the crystalline state; for example, a high diamagnetic anisotropy which allows perfect orientation of suspensions of ROS in magnetic fields. This has been capital for biophysical approaches such as diffraction, linear dichroism and even light scattering. (3) The compartmentation of the cytoplasm in extremely thin layers between the discs is probably essential for the speed and gain of enzymatic amplification. In this respect, the interdisc space could be compared to a synaptic cleft.

III.   FROM THE PHOTOISOMERIZATION OF RETINAL TO THE HYPERPOLARIZATION
       OF THE ROD: A MISSING LINK BETWEEN cGMP HYDROLYSIS AND CALCIUM
       RELEASE?

On the transduction mechanism, two major points were already accepted in the late sixties: (1) The initial photochemical event is the photoisomerization of the retinal chromophore tightly bound to rhodopsin, an intrinsic membrane protein embedded into the disc membrane. This very fast photoisomerization initiates a cascade of fast dark reactions in the protein, detected through characteristic spectral changes. The dogma of initial photoisomerization was later challenged for a while on kinetic grounds, after the first ps measurement (Peters et al., 1977). But new studies, mainly by Raman spectroscopy, have put the ancient dogma on much firmer ground (Eyring et al., 1980). (2) The electrophysiological response results from the reduction of an Na^+ inward current that flows in the dark through the ROS external membrane. The high Na^+ conductance in the dark explains the low polarization of the rod cell at rest. Calcium has soon been recognized as a potent effector of the light sensitive sodium current. This, however, was based on interpretations of extracellular effects of calcium ions. Then Hagins (1972) postulated the classical calcium hypothesis. The photoexcitation of rhodopsin led, by an undefined mechanism, to the release of calcium ions stored inside the discs. Calcium then diffuses in the cytoplasm toward the external membrane where it interacts with the sodium channels. Although this was not explicitly stated, the calcium hypothesis was often taken as implying that rhodopsin itself was the calcium channel. Numerous attempts at demonstrating such an ionic channel property for photoexcited rhodopsin have dominated the field for many years, with only very disappointing results.

The analysis of the kinetics of the electrophysiological response of the rod had, however, already indicated that many sequential steps of amplification should exist between the fast spectroscopic changes in photoexcited rhodopsin and the delayed and slow rising cellular response (Baylor et al., 1974). On the other hand, cyclic nucleotides made their entry in the early seventies with the observation that a modulation of cyclic nucleotide concentration in the rod was correlated with illumination, at least in vitro (Bitenski et al., 1971; Goridis and Virmaux, 1974). Cyclic nucleotides were usually denied a direct role in visual excitation because of the assumed slowness of their regulation mechanisms, from what was known of cyclic nucleotide dependent hormonal responses. But evidence accumulated to demonstrate a fast and intense increase in cyclic GMP hydrolysis in response to illumination (Yee and Liebman, 1978; Liebman and Pugh, 1979; Biernbaum and Bownds, 1979) and cGMP became a plausible competitor of calcium for the role of a soluble transmitter of visual excitation. All possible schemes have been proposed: the exclusive role of either of the candidates for the excitatory pathway; the other one being confined to the adaptative pathway; or a sequential role of cGMP and calcium. All recent developments seem to reinforce an apparent conflict. On one hand, the whole machinery around rhodopsin on the disc membrane seems exclusively geared toward the activation and regulation of a cGMP phosphodiesterase and is totally insensitive to calcium; there is evidence that manipulation of internal cGMP in rods can act on cell polarization, but there is no proof that it has a direct effect of cGMP on the sodium channels (Miller, 1982). On the other hand, at the level of the cell membrane, now the evidence is solid that a significant amount of calcium is released from the rod outer segment upon illumination (Gold and Korenbrot, 1980; Yoshikami et al., 1980). The kinetics of this release is compatible with a role of calcium in the control of the photosensitive sodium channels, and recent work strongly suggests that this calcium originates from the interior of the discs (Schroeder and Fain, 1984).

A link relating calcium release from the disc to the activation of the cGMP phosphodiesterase still seems to be missing. Eventual cGMP dependent calcium channels in the disc membrane or at the periphery of the discs will be extremely difficult to detect, as the discs are not accessible to microelectrodes or patch clamp techniques. It has been suggested that the protons released upon hydrolysis of cGMP would initiate a calcium-proton exchange through the disc membrane (Pugh and Muller, 1983; Goldberg et al., 1983). The signal would then be the increased rate of cGMP hydrolysis, not the decreased concentration of cGMP.

If an intermediate step remains obscure between cGMP hydrolysis and calcium release, considerable progress has been made, and substantial agreement has now been reached on the structure of rhodopsin, the mechanism of photoexcitation upon the isomerization of retinal (Miller, 1982) and the mechanism of amplified control of cGMP phosphodiesterase activity. At the other end of the transduction pathway, electrophysiologists mainly using the suction electrode technique (Baylor et al., 1979; Yau et al., 1981; Yau and Nakitani, 1984) and the patch clamp technique (Detwiler et al., 1982) are progressing regarding analysis of the photosensitive channels in the cell membrane. Many of these topics are reviewed in the excellent book edited by Miller (1981) and other recent papers (Birge, 1981; Dratz and Hargrave, 1983; Kühn, 1984; Chabre, 1983). Here, I will concentrate on the mechanisms of triggering, amplification and regulation of the light induced enzymatic cascade.

IV. RHODOPSIN STRUCTURE, THE OTHER PROTEINS INVOLVED IN THE CASCADE
     AND THE ORGANIZATION OF THE INTERDISC SPACE

     The rod disc membrane has often been taken as the archetype of
the fluid mosaic model, and it was the first natural membrane in
which lateral and rotational mobility of an intrinsic protein was
measured (Cone, 1972; Poo and Cone, 1974). This mobility remains
the highest ever observed for an intrinsic protein. This could suggest that mobility is important for the function of rhodopsin.

     Biophysical and biochemical studies have now converged toward
a well-defined model for the structure of vertebrate rhodopsin (see
Fig. 1). It consists of a bundle of 7 transmembrane α helices en-
caging the retinal in a hydrophobic core (Dratz and Hargrave, 1983;
Chabre, 1981, 1982). Two hydrophilic domains of equivalent size
($\sim$10 kD) protrude on both sides of the membrane.

     There are at least four proteins involved with photoexcited
rhodopsin (R^*) in the light-induced reaction cascade: (i) a guanine
nucleotide binding protein called transducin (T), which is the analog
of the "G" protein or hormonal Ns protein controlling the adenylate
cyclase system, (ii) an ATP dependent kinase (K), specific for photoexcited rhodopsin, (iii) a so-called "48 k" protein and (iv) cGMP
phosphodiesterase (PDE). All of these proteins are peripherally
bound to the disc membrane or soluble in the cytoplasm. The first
three interact directly at some stage with R^*.

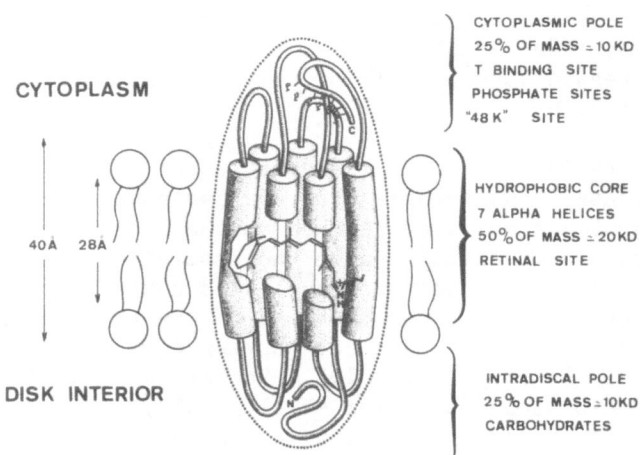

Fig. 1. Structure of visual rhodopsin from vertebrate retinal rod.
         (Adapted from a Figure kindly given by P. Hargrave.)

Table 1. The Major Proteins Identified in the Retinal Rod
          Outer Segment.

Proteins (Symbol)	Relation to Membrane	Molecular Weight (kDaltons)	Stoichiometry |Rhodopsin	Equivalent Molar Concentration in Cytoplasm
Rhodopsin (R)	Intrinsic	39 +2(glycocong.)	1	-
Transducin (T = Tα+Tβ+Tγ)	Peripheral or Soluble	≅80 (39+37+6)	≅10^{-1}	≅500 μM
"48 k"	Soluble	≅50	≅3.10^{-2}	≅150 μM
cGMP Phosphodi-esterase (PDE) + inhibitor (I)	Peripheral	≅180 (88+84+13)	≅10^{-2}	≅50 μM
ATP dependent R* Kinase (K)	Soluble	68	⩽10^{-3}	⩽5 μM
"Rim Protein" (P)	Intrinsic	240	≅3.10^{-3}	-

In Fig. 2, an attempt is made at sketching, on scale, the different compartments in the ROS. The stoichiometries and concentrations of the main identified proteins are listed in Table 1. The domain of free diffusion extends over a few square microns, limited either by the disc edge or by deep incisures in the large discs of the

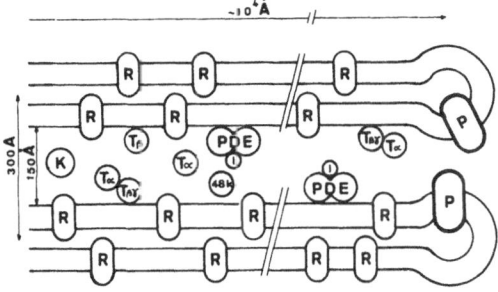

Fig. 2. The proteins involved in the photoinduced reaction cascade.
         The proteins are figured by spheres or cylinders approxi-
         mating their size and positions with respect to the disc
         membranes.

batracians ROS, for example. The cytoplasmic layer is less than
150 Å thick, extends over microns and is also limited by the thicker
rim region of the discs. This is a confined space in which local
concentrations of interacting proteins may reach high values. The
transducin and PDE molecules are peripherally bound and probably
mobile on the disc membrane; the kinase and the "48 k" protein are
soluble. The concept of molar concentration is not valid for proteins
which are not in solution. But, for example, as we shall see later,
when a subunit of transducin becomes soluble upon its interaction
with R*, its molar concentration in the cytoplasmic layer between the
discs may reach 500 M. This high concentration is probably critical
for the speed of the photoinduced reactions in the rod.

The high molecular weight intrinsic protein labelled "P" is
confined in the rim region (Papermaster et al., 1978) which is probably less fluid, and is the site of attachment of undefined cytoskeletal proteins which connect contiguous discs and possibly also connect
the discs to the cell membrane (Roof and Heuser, 1982). The protein
P has recently been shown to be phosphorylated upon illumination of
the rods (Szutz, 1983). This suggests that it has a role in the
visual process and is not a simple cytoskeletal component. It could
be a good candidate for a calcium channel through the disc. Illumination influences the phosphorylation of other low molecular weight
proteins present in minor quantities in the ROS also (Polans et al.,
1979); however, the location of these proteins is not known.

V.   THE LIGHT-INDUCED BIOCHEMICAL CASCASE (FIG. 3)

a) Triggering: Photoisomerization and Energy Uptake

It is now solidly established by Raman and other spectroscopic
techniques that the photoisomerization of retinal is the primary
event, occurring in ps and leading to the bathorhodopsin ground state.

At this stage, the all-trans chromophore is highly constrained
by the protein which has not yet changed its conformation. In the
point charge model of Honig et al. (1979), the chromophore is in
close proximity of charged groups in the protein, and its isomerization implies the separation of an ion pair. This is correlated with
a 35 kcal/mol energy uptake in the transition (Cooper, 1979), to be
compared with 0.5 kcal/mol for the cis-trans isomerization of free
retinal. This is more than 60% of the energy of the absorbed photon,
a very high efficiency for photoenergetic conversion. However, the
function of the visual pigment (by contrast with that of bacteriorhodopsin) is to memorize and transmit the information of the capture
of a photon, not to store its energy. This high energy barrier provides very efficient protection against spurious triggering of the
system by "thermal noise." In later dark reactions, this stored
energy will force conformational changes in the protein to relax the
strain on the chromophore.

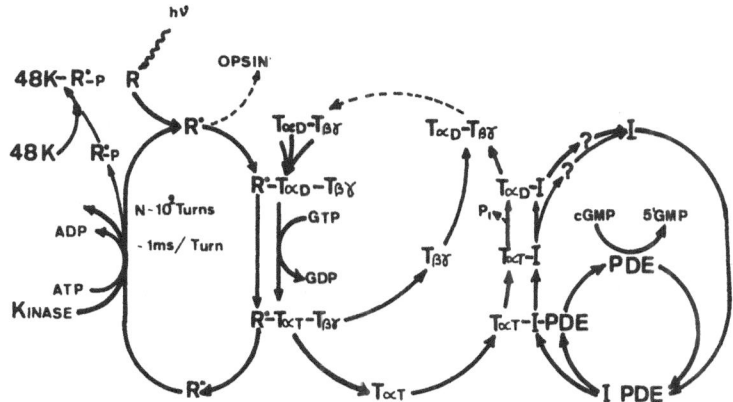

Fig. 3. The photoinduced amplifying reaction cascade.

b) The Bathorhodopsin ⟶ R* Transconformation: Flipping the Memory

The conformational changes of the protein lead to the active R* state through a cascade of successive steps characterized by their absorption spectra, which extends over about a ms. This seems long enough to allow for major rearrangements in the protein structure, but physical techniques have failed to demonstrate such large structural changes of the protein backbone (Chabre, 1982). The active state R*, definitively identified with the "meta II rhodopsin" spectral state (Bennett, 1983; Emeis et al., 1982), is characterized by its enzymatic properties. Two specific sites have developed on the cytoplasmic surface: a site of binding and catalytic activation for transducin (Kühn, 1980), and a site of multiple phosphorylation by an ATP dependent kinase near the C terminal end of rhodopsin (Wilden and Kühn, 1982). The kinase is ineffective in dark adapted rhodopsin.

c) First Step of Amplification: Catalysis of GDP/GTP Exchange by one R* on Many T

The binding of T onto R* perturbs the structure of T, having the results of loosening the nucleotide binding site on the T_α subunit. This catalyses the exchange of an originally bound GDP molecule for a high energy nucleotide triphosphate GTP (Fung et al., 1981). Upon completion of the exchange, the "activated" transducin dissociates from R*, which is, therefore, able to catalyse many exchanges in sequence. The kinetics of this reaction is controlled by the collision rate of R* and T, and, therefore, limited by the diffusion rates of these two proteins, respectively, in and on the disc membrane. It has recently been demonstrated by the infrared light scattering technique that in situ one R* could activate hundreds of T at the rate of about 1 per ms (Kühn et al., 1981; Vuong et al., 1984).

d) Second Amplification Stage: Transducin - cGMP Phosphodiesterase
   Coupling

      Upon activation by R* and GTP, the transducin molecules dissoci-
ate. The T GTP subunits diffuse in the cytoplasm and activate the
same number of PDE molecules by interacting with their inhibitory
subunit I. Each PDE molecule then becomes able to hydrolyse cGMP
at the rate of $3 \cdot 10^4$ molecules/s. Through this two step mechanism,
within the few hundred milliseconds delay which characterizes the
physiological response to a low intensity light flash, each R* formed
can induce the hydrolysis of more than 10^5 cGMP molecules (Yee and
Liebman, 1978; Liebman and Pugh, 1979). This is a signal sufficiently
fast and intense to be considered as significant for visual transduction.

e) First Regulation Mechanisms: R* Phosphorylation and Binding of
   "48 k" Protein

      The first regulation mechanism is on R*. The spontaneous decay
of meta II rhodopsin and the release of retinal from the protein part
of the pigment are much too slow to provide a significant turn off
mechanism; a faster "blocking" mechanism of R* is needed. It is
probably provided by the ATP dependent kinase and the subsequent
binding of the "48 K" protein to R*. The kinase and transducin compete for access to their binding sites, both located on the cytoplasmic surface of R*; steric hindrance prevents the two proteins to
bind simultaneously (Pfister et al., 1983). T is more abundant and
is initially predominant, but progressively the kinase will succeed
in phosphorylating R*. This phosphorylation does not in itself inhibit the catalytic R*-T interactions, but the "48 k" protein binds
preferentially and permanently to phosphorylated R* (Kühn et al.,
1985) and will soon block the access of T onto its site.

f) Second Regulation Mechanism on the PDE

      It is not sufficient to block R* if the already activated PDE
molecules are not rapidly deactivated. Such a switch off mechanism
is provided by the hydrolysis of the GTP molecule in T_α. $T_\alpha GDP$ loses
then its affinity for the PDE inhibitor and regains its affinity for
$T_{\beta\gamma}$. The PDE is reinhibited, and $T_\alpha GDP$ recombines with $T_{\beta\gamma}$. This
recombination is required to allow for the catalytic interaction with
a new R* in a subsequent cascade; only the holoenzyme $T_\alpha GDP-T_{\beta\gamma}$ is
able to bind to R* (Fung, 1983).

CONCLUSIONS

      Past the enzymatic amplification cascade, speculation is open
on a variety of alternative routes, which, at some yet unknown point,
would eventually connect with the calcium pathway. I will make only
a few concluding remarks.

(1) The activation of the cGMP cascade seems to be the main, and probably, the exclusive function of photoexcited rhodopsin. Rhodopsin is a monomeric protein of average size. Already, it seems very busy with its functions related to the triggering and control of the cGMP cascade. It is hard to conceive that it could simultaneously trigger an entirely different process to release calcium ions. Rhodopsin is definitely not a calcium channel. The analysis of the time course of the single photon response makes it very unlikely that rhodopsin would directly control the opening of channels in the disc membrane. There must be an amplification step before the eventual calcium release is triggered. It is probable that it is the R^*-T-PDE cascade which provides this amplification.

(2) It is not proven by far that the signal would be related to the decrease in cGMP concentration in the rod. It could be related to the hydrolysis of cGMP. The hydrolysis of cGMP is a highly exoenergetic reaction which might be directly or indirectly coupled to another process which remains to be found.

REFERENCES

Baylor, D. A., Hodgkin, A. L., and Lamb, T. D., 1974, J. Physiol. (London), 242:685.

Baylor, D. A., Lamb, T. D., and Yau, K. W., 1979, Physiology, 288:589.

Bennett, N., 1983, B.P.S.-ARVO meeting Bristol (abstract).

Biernbaum, M. S., and Bownds, M. D., 1979, J. Gen. Physiol., 74:649.

Birge, R. R., 1981, Annu. Rev. Biophys. Bioeng., 10:315.

Bitenski, M. W., Gorman, R. E., and Miller, W. H., 1971, Proc. Natl. Acad. Sci. USA, 68:561.

Chabre, M., 1981, in: "Membrane and Intercellular Communication," Balian et al., eds., North Holland.

Chabre, M., 1982, in: "Trends in Photobiology," C. Hélène, ed., Plenum Press, New York, p. 399.

Chabre, M., 1983, in: "Symposium of the Society for Experimental Biology, Vol. XXXVI, The Biology of Photoreception," Cambridge University Press, Cambridge, pp. 87-108.

Chabre, M., and Cavaggioni, A., 1975, Biochim. Biophys. Acta, 382:336.

Cone, R. A., 1972, Nature (New Biol.), 236:39.

Cooper, A., 1979, Nature, 282:531.

Detwiler, P. B., and Conner, J. D., and Bodia, R. D., 1982, Nature, 300:59.

Dratz, E. A., and Hargrave, P. A., 1983, Trends in Biochem. Sci., 8:128.

Emeis, D., Kühn, H., Reichert, J., Hofmann, K. P., 1982, FEBS Lett., 143:92.

Eyring, G., Curry, B., Mathies, R., Fransen, R., Paling, I., and Lugtenburg, G., 1980, Biochemistry, 19:2410.

Fung, B. K. K., 1983, J. Biol. Chem., 258:10495.

Fung, B. K., Hurley, J. B., and Stryer, L., 1981, Proc. Natl. Acad. Sci. USA, 78:152.

Gold, G. H., and Korenbrot, J. I., 1980, Proc. Natl. Acad. Sci. USA, 77:5577.

Goldberg, N. D., Ames, A., Gander, J. E., and Walseth, T. F., 1983, J. Biol. Chem., 258:9213.

Goridis, C., and Virmaux, N., 1974, Nature, 248:57.

Hagins, W. A., 1972, Annu. Rev. Biophys. Bioeng., 1:131.

Honig, B., Ebrey, T., Callender, R. H., Diner, V., and Ottolenghi, M., 1979, Proc. Natl. Acad. Sci. USA, 76:2503.

Kaupp, U. B., Schnetkamp, P. P. M., and Junge, W., 1981, Biochemistry, 20:5500.

Kühn, H., 1980, Nature, 283:587.

Kühn, H., 1984, in: "Progress in Retinal Research," Vol. 3, Osborne, and Chader, eds., Pergamon Press.

Kühn, H., Bennett, N., Michel-Villaz, M., and Chabre, M., 1981, Proc. Natl. Acad. Sci. USA, 18:6873.

Kühn, H., Hall, S. W., and Wilden, U., 1985, FEBS Lett., in press.

Liebman, P. A., and Pugh, Jr., E. N., 1979, Vis. Res., 19:375.

Miller, W. H. (Ed.), 1981, "Current Topics in Membrane and Transport Vol. 15, Molecular Mechanism of Photoreceptor Transduction," Academic Press, New York.

Miller, W. H., 1982, J. Gen. Physiol., 80:103.

Papermaster, D. S., Schneider, B. G., Zorn, M. A., and Kraehenbuhl, J. P., 1978, J. Cell Biol., 78:415.

Peters, K., Applebury, M. L., and Rentzepis, P. M., 1977, Proc. Natl. Acad. Sci. USA, 74:3119.

Pfister, C., Kühn, H., and Chabre, M., 1983, Eur. J. Biochem., 136: 489.

Polans, A. S., Hermolin, J., and Bownds, M. D., 1979, J. Gen. Physiol., 74:595.

Poo, M., and Cone, R. A., 1974, Nature, 247:438.

Pugh, E. N., and Muller, P., 1983, Proc. Natl. Acad. Sci. USA, 80:103.

Roof, D. J., and Heuser, J. E., 1982, J. Cell Biol., 95:487.

Saibil, H., Chabre, M., and Worcester, D., 1976, Nature, 262:266.

Schroeder, W. H., and Fain, G. L., 1984, Nature, 309:268.

Szutz, E. T., 1983, Biophys. J., 41:340a (Abstract).

Vuong, T. M., Chabre, M., and Stryer, L., 1984, Nature, in press.

Wilden, U. A., and Kühn, H., 1982, Biochemistry, 21:3014.

Yau, K. W., McNaughton, P. A., and Hodgkin, A. L., 1981, 292:502.

Yau, K. W., and Nakitani, K., 1984, Nature, 309:352.

Yee, R., and Liebman, P. A., 1978, J. Biol. Chem., 253:8902.

Yoshikami, S., George, J. S., and Hagins, W. A., 1980, Nature, 286: 395.

PARTICIPANTS

Members of the Advisory and Organizing Committees indicated by an asterisk.

ANDREOZZI, LAURA
Dip. Genetica e Biologia
Molecolare
Univ. "La Sapienza"
P. le A. Moro, 7
10185 Roma
Italy

BALDOCCHI, M. ANTONIA
CNR - Istituto di Biofisica
Via S. Lorenzo, 26
56100 Pisa
Italy

BARSANTI, LAURA
CNR - Istituto di Biofisica
Via S. Lorenzo, 26
56100 Pisa
Italy

BENEDETTI, PIER ALBERTO
CNR - Istituto di Biofisica
Via S. Lorenzo, 26
56100 Pisa
Italy

BERG, HOWARD C.
Division of Biology
Cal Tech
Pasadena, CA 91125
USA

BIGNETTI, ENRICO
Ist. Biologia Molecolare
Univ. Studi di Parma
Via del Taglio
Quart. Cornocchio
43100 Parma
Italy

BOON, JEAN PIERRE
Service de Chimie Physique II
Code Postal 231
Campus Plaine U.L.B.
Boulevard du Trionphe
1050 Bruxelles
Belgique

BORSELLINO, ANTONIO
Ist. Scienze Fisiche
Dip. Fisica
Via Dodecaneso, 33
16146 Genova
Italy

BRIGGS, WINSLOW R.
Carnegie Institution of
Washington
Department of Plant Biology
290 Panama Street
Stanford, CA 94305-1297
USA

CAUBERGS, ROLAND JULIEN
Rijksuniversitair Centrum
Universität Antwerpen
2020 Antwerpen
Groenenborgelaan 171
Belgium

CHABRE, MARC
Laboratoire Biologie
Moleculaire et Cellulaire
DRF Ceng 85X
38041 Grenoble
France

COHEN, ROBERT J.
Department of Biochemistry
and Molecular Biology
J. Hills Miller Health Center
Gainesville, FL 32610
USA

*COLOMBETTI, GIULIANO
CNR - Istituto di Biofisica
Via S. Lorenzo, 26
56100 Pisa
Italy

COLTELLI, PRIMO
CNR - Istituto di Biofisica
Via S. Lorenzo, 26
56100 Pisa
Italy

CONTI, FRANCO
Ist. Cibernetica e Biofisica
Corso Mazzini, 20
16032 Camogli (GE)
Italy

CREUTZ, CHARLES
Department of Biology
University of Toledo
Toledo, OH 43606
USA

CRIPPA, PIER RAIMONDO
Istituto Fisica
Via M. D'Azeglio, 85
43100 Parma
Italy

DIEHN, BODO
4100 E. Howe Road
Bath, MI 48808
USA

DI LENA, M. ROSARIA
CNR - Istituto di Biofisica
Via S. Lorenzo, 26
56100 Pisa
Italy

DUSCHL, ALBERT
Botanisches Institut I
Justus-Liebig-Universität
SenckenbergstraBe 17-21
D-6300 Giessen
W. Germany

ESEN, HAMZA
A.U. Tip Fakültesi
Fizyoloji Anabilim Dali
Shihhiye, Ankara
Turkey

EVANGELISTA, VALTERE
CNR - Istituto Biofisica
Via S. Lorenzo, 26
56100 Pisa
Italy

FEINLEIB, MARY ELLA
Biology Department
Tufts University
Medford, MA 02155
USA

FREDIANI, CARLO
CNR - Istituto di Biofisica
Via S. Lorenzo, 26
56100 Pisa
Italy

FREY, IRIS
1701 Andy Holt Avenue
POB 389
University of Tennessee
Knoxville, TN 37916
USA

GAMOW, IGOR
Department of Chemical
Engineering
University of Colorado
Boulder, CO 80309
USA

GHETTI, FRANCESCO
CNR - Istituto di Biofisica
Via S. Lorenzo, 26
56100 Pisa
Italy

GOMEZ-VILDA, PEDRO
Department de Electronica
Faculted de Informatica
Km 7 Carretera de Valenci
Madrid 31
Spain

GORDON, PATRICIA
Via La Tana
56042 Crespina
Pisa
Italy

GRIVSKY, EUGENE MICHAEL
4407 Eastwood Court
Fairfax, VA 22032
USA

GROLIG, FRANZ
Botanisches Inst. I
Justus-Liebig-Universität
Senckenbergerstraße 17-21
D-6300 Giessen
W. Germany

GUALTIERI, PAOLO
CNR - Istituto Biofisica
Via S. Lorenzo, 26
56100 Pisa
Italy

*HÄDER, DONAT-P.
Fachbereich Biologie Botanik
Marburg University
Lahnberge
D-Marburg/Lahn
W. Germany

HÄDER, MARIA
Fachbereich Biologie Botanik
Marburg University
Lahnberge
D-Marburg/Lahn
W. Germany

HAUPT, WOLFGANG
University of Erlangen-Nürnberg
Botanisches Institut
Schlossgarten 4
8250 Erlangen
W. Germany

HILDEBRAND, EILO
Inst. Neurobiologie der
Kernforschungsanlage
Postfach 1913
Jülich 1
W. Germany

KUNZELMANN, PETER
Inst. fur Biologie II Botanik
Albert Ludwigs Univ.
Schanzlestrasse 1
D-7800 Freiburg
W. Germany

*LENCI, FRANCESCO
CNR - Istituto Biofisica
Via S. Lorenzo, 26
56100 Pisa
Italy

LERCARI, BARTOLOMEO
Ist. Orticultura Floricultura
dell'Università
Viale delle Piagge, 23
56100 Pisa
Italy

LEVI, GIOVANNI
Department of Neurophysiology
Weizmann Inst. of Science
76100 Rehovot
Israel

LORK, WOLFRAM
Inst. fur Biologie II Botanik
Albert-Ludwigs-Univ.
Schanzlestrasse 1
D-7800 Freiburg
W. Germany

MACHEMER, HANS
Ruhr-Univ. Bochum
Arbeitsgruppe Rezeptoren
Gebäude ND6/25
4630 Bochum
W. Germany

MACNAB, ROBERT M.
Yale University
Department of Molecular
Biophysics
Box 6666
New Haven, CT 06511
USA

MEYER-WEGENER, JENS
P.O. Box 4037 Alexander Hall
North Carolina State University
Raleigh, NC 27607
USA

MIGLIORE, LUCIANA
Cattedra di Ecologia
Ist. Genetica, Fac. Scienze
Città Universitaria
00185 Roma
Italy

MOREL, NICOLE
Tufts University
Department of Biology
Medford, MA 02155
USA

NAITOH, YUTAKA
University of Tsukuba
Sakura-mura
Ibaraki 305
Japan

NEUBACHER, HARALD
Inst. für Biophysik
Justus-Liebig-Univ. Giessen
Leihgesterner Weg 217
D-6300 Giessen
W. Germany

NULTSCH, WILHELM
Fachbereich Biologie
Botanik Marburg Univ.
Lahnberge
D-3550 Marburg/Lahn
W. Germany

OMODEO, PIETRO
Univ. Padova
Istituto Biologia Animale
Via Loredan, 10
35100 Padova
Italy

PASSARELLI, VINCENZO
CNR - Istituto Biofisica
Via S. Lorenzo, 26
56100 Pisa
Italy

PELOSI, PAOLO
Ist. Industrie Agrarie
Via S. Michele degli Scalzi, 4
56100 Pisa
Italy

PHILIPPS, GLYNN O.
School of Natural Sciences
North E. Wales Inst.
Connah's Quay
Deeside, Clwyd CH5 4BR
UK

PICCOLINO, MARCO
CNR - Ist. Neurofisiol.
Via S. Zeno, 49/A
56100 Pisa
Italy

PICCINNI, ESTER
Ist. Biologia Animale
Via Loredan, 10
35100 Padova
Italy

PODBILEWICZ, BENJAMIN
Fte. del Emperador 25
Lomas de Tecamachalco
Neucalpan de Juarez
53950 Mexico

POFF, KENNETH L.
Michigan State University
MSU/DOE Plant Research Lab.
East Lansing, MI 48824
USA

PRUSTI, RABI K.
Texas Tech University
Box 4260
Lubbock, TX 79409
USA

RENARD, MICHEL
Inst. de Physique Bât. B5
Sart Tilman par Liege 1
B-4000 Liege
Belgium

RODELLAR-BIARGE VICTORIA
Department Electronica
Facultad de Informatica
Km 7 Carretera Valencia
Madrid 31
Spain

ROHNISCH-MACHEMER, SIEGRUN K.
Rühr Universitaet Bochum
Lehrstuhl für All Gemeine
Zoologie
Postfach 102148
4630 Bochum
W. Germany

RUSSO, VINCENZO E. A.
Max-Planck-Inst. für
Molekulare Genetik
Inhestrasse 63-73
1000 W. Berlin 33
W. Germany

SAHINOGLU, BABUR
A.U. Tip Fakültesi
Fizyoloji Anabilim
Dali
Sinhiye-Ankara
Turkey

SCEVOLI, PARTIZIA
CNR - Istituto Biofisica
Via S. Lorenzo, 26
56100 Pisa
Italy

SCHÄFER, EBERHARD
Biological Institute II
Univ. of Freiburg
Schanzlestrasse 1
D-7800 Freiburg
W. Germany

SHROPSHIRE, JR., WALTER
Smithsonian Institute
Radiation Biology Lab.
12441 Parklawn Drive
Rockville, MD 20852
USA

*SONG, PILL-SOON
Texas Tech University
Chemistry Department
Lubbock, TX 79409
USA

SPONGA, FEDERICA
Ist. Orticultura Floricultura
dell'Università
Viale delle Piagge, 23
56100 Pisa
Italy

SPUDICH, JOHN L.
A. Einstein College of Medicine
Department of Anatomy
1300 Morris Park Avenue
Bronx, NY 10461
USA

STEINHARDT, ALFRED R.
Inst. Biologie II Botanik
Albert-Ludwigs-Univ.
Schanzlestrasse 1
D-7800 Freiburg
W. Germany

SUNDBERG, STEVEN A.
Department of Anatomy
A. Einstein College of Medicine
1300 Morris Park Avenue
Building F, Room 6020
Bronx, NY 10461
USA

TIRINDELLI, ROBERTO
Fac. Medicina Veterinaria
Ist. Biologia Molecolare
Via del Taglio
Quart. Cornocchio
43100 Parma
Italy

URFUOGLU, NEVIN
Billur Sokak 38/4
Gaziosmanpasa-Ankara
Turkey

VIERSTRA, RICHARD D.
University of Wisconsin
Department of Botany
139 Birge Hall
Madison, WI 53706
USA

WAGNER, GOTTFRIED
Giessen University
Senckeberg Strasse 17
D-6300 Giessen
W. Germany

WALNE, PATRICIA L.
University of Tennessee
Department of Botany
Knoxville, TN 37996-1100
USA

INDEX (Key Words)

Abscisic acid, 1
Absorption spectra, 147, 251
Acetabularia mediterranea, 281
Action spectra, 75, 93, 147,
        231, 251
Actin, 281
Action potential, 179
Actomyosin, 281
Adaptation, 1, 31, 93, 299
Adenylate cyclase, 309
Algae, 231
Allo-phycocyanin, 147
Amanita phalloides, 281
Anabaena variabilis, 147
Amplification, 211, 309
Arabidopsis thaliana, 211
Attenuation, 211
Auxin, 1, 265
Avena, 211
Axoneme, 179

Bacteria, 19, 31, 93, 113, 147,
        299
Bacterial chemotaxis, 19
Bacterial taxis, 19, 31
Bacteriorhodopsin, 113, 309
Black box, 1
Blue light effects, 1, 211, 231
Brownian movement, 19
Bryopsis, 281

Ca^{2+} channels, 179
Ca^{2+} receptor current, 179
Calcium, 119, 165, 179, 281,
        299, 309
Calcium channel, 179, 281, 309
Calmodulin, 281

Carotene, 47, 147, 211
Carotenoids, 47, 147, 211, 231
Caulerpa prolifera, 281
cGMP, 309
Chara, 1, 281
Chemophobic responses, 165
Chemosensory responses, 31, 165
Chemotaxis, 19, 31
Chlamydomonas, 119, 251
Chlorophylls, 47, 147
Chloroplast development, 265
Chloroplast movements, 265, 281
Chloroplasts, 281
Chromophores, 47, 93, 113, 309
Ciliates, 299
Cilium, 179
Circadian rhythms, 231
Coleoptiles, 1, 211, 265
Color-sensing, 113
Color vision, 1
Computer simulation, 75
Conidia, 231
Cryptochrome, 231
Cyanobacteria, 147
Cylindrosperum licheniforme, 147
Cytoplasmic streaming, 281

Dichroism, 251
Dictyostelium discoideum, 75,
        251, 299
Dictyostelium mucoroides, 231
Dictyota dichotoma, 231
Differentiation, 231
Dynein, 179
Dynein ATPase, 281

Electric potential, 147, 179

327